Editorial Board Members' Collection Series: Improving Structural Integrity of Metals: From Bulk to Surface

Editorial Board Members' Collection Series: Improving Structural Integrity of Metals: From Bulk to Surface

Guest Editors

Francesca Borgioli
Denis Benasciutti
Umberto Prisco
Tomasz Tański

MDPI

Basel • Beijing • Wuhan • Barcelona • Belgrade • Novi Sad • Cluj • Manchester

Guest Editors
Francesca Borgioli
Dipartimento di Ingegneria Industriale (DIEF)
Università degli Studi di Firenze
Firenze
Italy

Denis Benasciutti
Dipartimento Politecnico di Ingegneria e Architettura (DPIA)
Università degli Studi di Udine
Udine
Italy

Umberto Prisco
Dipartimento di Ingegneria Chimica, dei Materiali e della Produzione Industriale
Università degli Studi di Napoli Federico II
Napoli
Italy

Tomasz Tański
Faculty of Mechanical Engineering
Silesian University of Technology
Gliwice
Poland

Editorial Office
MDPI AG
Grosspeteranlage 5
4052 Basel, Switzerland

This is a reprint of the Special Issue, published open access by the journal *Metals* (ISSN 2075-4701), freely accessible at: www.mdpi.com/journal/metals/special_issues/16D16B111T.

For citation purposes, cite each article independently as indicated on the article page online and using the guide below:

Lastname, A.A.; Lastname, B.B. Article Title. *Journal Name* **Year**, *Volume Number*, Page Range.

ISBN 978-3-7258-2906-4 (Hbk)
ISBN 978-3-7258-2905-7 (PDF)
https://doi.org/10.3390/books978-3-7258-2905-7

Contents

About the Editors

Francesca Borgioli

Francesca Borgioli is Associate Professor of Materials Science and Technology at the Department of Industrial Engineering (DIEF) of the Università di Firenze (University of Florence). She graduated in Physics in 1995 and earned a Ph.D. in Materials Science and Technology at the Università di Firenze in 1999. She has been an associate professor since 2006, and she has lectured on materials engineering courses at the School of Engineering and the School of Architecture. Since 2018, she has been a board member of the Electron Microscopy and Microanalysis Service Centre of the Università di Firenze. Since 1998, she has been a member of the Italian National Interuniversity Consortium for Materials Science and Technology (INSTM). Her research activity concerns the surface engineering of metallic materials for improving characteristics such as corrosion resistance, tribological properties, and biocompatibility. Her studies are mainly focused on low-temperature thermochemical treatments of stainless steels and on the so-called "expanded" phases obtained with these treatments, and she is the author of four reviews on these topics. Since 2019, she has been serving as an editorial board member of the journal *Metals*, and she is an editorial board member for the section "Structural Integrity of Metals".

Denis Benasciutti

Denis Benasciutti is a Full Professor of Machine Design at the University of Udine, Italy. In 2001 he graduated cum laude in Materials Engineering at the University of Ferrara, where he also received his Ph.D. in 2005. In 2006, he earned a master's in Welding Engineering and the diploma as European/International Welding Engineer (EWE/IWE). From 2006 to 2015, he was Assistant Professor at the University of Udine, Italy. From 2015 to 2024 he was appointed Associate Professor at the University of Ferrara, Italy. His research interests involve the structural integrity of mechanical components subjected to stochastic uniaxial and multiaxial loadings (with special focus on frequency-domain spectral methods) and the cyclic plasticity and low-cycle fatigue behavior of traditional and additive manufactured materials. He published over 95 scientific articles in peer-reviewed journals on the topics of metal fatigue and structural durability, finite element method, energy harvesting, and cyclic plasticity of metals. Since 2021, he has been an editorial board member and section board member (section: Structural Integrity of Metals) of the journal *Metals*.

Umberto Prisco

Umberto Prisco is Professor of Material Science and Engineering at the University of Napoli Federico II. Dr. Prisco graduated in Chemical Engineering in 1999. In 2003, he received his Ph.D. from the Department of Materials and Production Engineering at the University of Napoli Federico II. In 2003, Prisco was a research fellow at the Nanoscale Physics Research Laboratory, School of Physics and Astronomy, University of Birmingham, United Kingdom, where he also was assistant professor of the summer courses in experimental physics. He moved to the University of Napoli in 2005. His research interests and publications are in the area of the material science and engineering of structural materials, particularly metals and metal alloys; the structure and constitution of these materials and their properties; the processing and manufacturing of structural materials and components made of them; the development of physical models to simulate the effect of manufacturing processes on the structure and resultant properties of these materials; the validation of the proposed models through targeted experimentation based on novel techniques.

Tomasz Tański

Tomasz Tański has over 600 publications, including 28 items constituting monographs, didactic scripts, and chapters in books; 152 publications on WoS (times cited: 1,933; without self-citations: 1,580). He was a speaker, chairman, organizer, and a participant in 113 conferences, including 106 international ones. He has 160 article reviews published in the *JCR*, including 90 from the WoS list and 3 book reviews; he is an editor of 18 international book monographs, including 6 from the WoS list and 9 national patents. He has managed three national research projects and one national didactic project. He was a group leader of four projects at the Silesian University of Technology (financed by national and European funds).

He was a promoter in seven completed doctoral dissertations and eight open doctoral theses and a co-promoter in three doctoral dissertations in cooperation with the Universidade Federal De Pelotas, Brasil, and Jimma University, Ethiopia. He was also a supervisor of 17 international scientific projects carried out by Ph.D. students and adjuncts from foreign partner universities as part of European and non-European programs (Brazil, Ethiopia, Slovakia, and the Czech Republic); a reviewer of 4 doctoral dissertations and 1 habilitation thesis; and the supervisor/promoter of 150 engineering and master's theses.

Interests: nanotechnology; thin layers; material properties; structure; heat and surface treatment; plastic deformation.

metals MDPI

Editorial

Editorial Board Members' Collection Series: Improving Structural Integrity of Metals: From Bulk to Surface

Francesca Borgioli [1,*], Denis Benasciutti [2], Umberto Prisco [3] and Tomasz Tański [4]

1 Department of Industrial Engineering (DIEF), Università di Firenze, Via di S. Marta 3, 50139 Firenze, Italy
2 Department of Engineering, Università di Ferrara, Via Saragat 1, 44122 Ferrara, Italy; denis.benasciutti@unife.it
3 Department of Chemical, Materials and Production Engineering, Università di Napoli Federico II, Piazzale Tecchio, 80, 80125 Napoli, Italy; umberto.prisco@unina.it
4 Department of Engineering Materials and Biomaterials, Faculty of Mechanical Engineering, Silesian University of Technology, 44-100 Gliwice, Poland; tomasz.tanski@polsl.pl
* Correspondence: francesca.borgioli@unifi.it; Tel.: +39-055-275-8734

check for updates

Citation: Borgioli, F.; Benasciutti, D.; Prisco, U.; Tański, T. Editorial Board Members' Collection Series: Improving Structural Integrity of Metals: From Bulk to Surface. *Metals* **2024**, *14*, 1120. https://doi.org/10.3390/met14101120

Received: 6 September 2024
Accepted: 19 September 2024
Published: 1 October 2024

1. Introduction

Metals and alloys continue to play a crucial role in the design and construction of load-bearing structures and mechanical components. Ferrous and non-ferrous alloys find countless applications in various industrial sectors, such as that of automotives, aerospace, the marine sector, construction, and manufacturing.

When it comes to guaranteeing the structural integrity and safety of critical parts, a variety of protection and strengthening mechanisms may be used both at the bulk level and the surface level. At the bulk level, it is possible to take advantage of the correlation between the microstructure and mechanical properties of the material [1]. The chemical composition and content of alloy elements can be controlled to form new alloys or to improve existing ones. Manufacturing processes and heat treatments can be used to tailor the microstructure to possess the desired combination of mechanical properties (e.g., strength, ductility, hardness, and fracture toughness). For example, this can be done by controlling the grain size and/or phase type and distribution. Further, innovative manufacturing techniques, such as additive manufacturing, allow for the fabrication of components with complex geometries and specific mechanical behaviors [2].

Aside from material properties, the subject of structural integrity is also relevant to a variety of topics related to analysis methods [3]. For complex structures, advanced reliability methods and Monte Carlo simulations prove to be efficient tools to assess the probability of failure, and for improving the safety of structures in service [4]. When structures are assembled by welding, the structure integrity is controlled by the fatigue behavior of the welded joints, where high stress concentrations and material inhomogeneities make fatigue cracks more likely to nucleate and grow [5]. Analyses techniques, often supported by finite element calculations, can be carried out by considering the nominal, structural, and local stress or strain, as well as the notch stress intensity factor or fracture mechanics [6]. For structures subjected to tens of millions of fatigue cycles, the structure behavior in the Very High Cycle Fatigue (VHCF) regime is also of interest [7].

In many service conditions, the surface may play a key role in determining the performance and life of a component. Phenomena occurring on metal surfaces, such as wear, wet corrosion, and high-temperature oxidation, may act in synergy with external or even residual stresses, such that a premature failure occurs [8]. Causes of failure, such as corrosion fatigue, stress corrosion cracking, and hydrogen embrittlement, are of major concern in many applications, and many efforts have been made to study their mechanisms in different environments, as well as to model and predict the component behavior.

As a consequence, the modification of the topmost layers of metals and alloys, in order to change their characteristics, has become an important step in the manufacturing of industrial components, with the aim of extending their life. Surface engineering techniques allow us to improve surface hardness, hence wear and fatigue resistance, and corrosion resistance in many environments by means of coating processes [9], such as physical vapor deposition (PVD), chemical vapor deposition (CVD) and thermal spray, as well as diffusion processes [10], such as carburizing and nitriding. Surface engineering treatments, either as a single process or as a combination of different processes, can be tailored to achieve the most suitable characteristics for preserving the structural integrity of components.

The purpose of this Special Issue is to gather articles presenting up-to-date methods and approaches for analyzing, preserving, and improving the structural integrity of metallic components; we also pay attention herein to the phenomena occurring at the bulk and surface level, while also considering the role of the manufacturing process as it correlates to the material microstructure, as well as advanced methods for reliability analysis. The outcomes of experimental, numerical, and theoretical approaches are also considered.

2. Overview of the Published Articles

A total of eleven articles have been published on the structural integrity of metallic components, written from different perspectives, ranging from the bulk to the surface level. Various damage and strengthening phenomena are comprehensively analyzed by means of experimental findings, which are often correlated to the outcomes of numerical and theoretical models. In most articles, uniaxial tensile tests and hardness measurements were used to evaluate the mechanical properties of materials. The Digital Image Correlation (DIC) technique was employed for measuring full-field displacements and strains during testing, while microstructures were analyzed using optical or scanning electron microscopy (SEM). Regarding numerical analysis, several articles took advantage of the finite element method as an efficient tool for assessing the mechanical response and safety analysis of structures with complex geometries. Not only did these articles consider alloys obtained by conventional manufacturing processes, they also included metallic materials and even structures obtained by advanced techniques such as additive manufacturing.

An example of this is provided in the work of Distefano et al. (Contribution 1), who investigated the mechanical properties of Triply Periodic Minimal Surface (TPMS) lattice structures made of Ti6Al4V ELI titanium alloy, and produced via Direct Metal Laser Sintering (DMLS). The mechanical response, elastic modulus, and failure mechanisms of cylindrical lattice specimens with different cell sizes, and subjected to uniaxial compressive tests, were analyzed by means of the DIC technique. The experimental results were also compared with predictions from non-linear finite element models, showing a promising correlation. The study also emphasized how the mechanical response is almost independent of the cell size of the lattice structure.

Titanium alloys were also the focus of the study by Han et al. (Contribution 2), which aimed to improve the mechanical properties of a β-Ti alloy for biomedical applications—namely, a Ti-39Nb-6Zr alloy with different oxygen contents—by means of crystal grain refinement based on recrystallization promoted by specific mechanical and thermal processes. Various microstructural analysis techniques (optical microscopy, X-ray diffraction, field emission scanning electron microscopy (FE-SEM), and electron backscatter diffraction (EBSD)) were used to interpret the findings of the mechanical tests (Vickers hardness, tensile/compressive uniaxial tests). An improvement in mechanical strength in titanium alloys with added oxygen was in fact observed, thanks to the combined effect of solid-solution hardening caused by oxygen addition and crystal grain refinement from static recrystallisation, promoted by a combination of cold swaging followed by solution heat treatments at 900 °C and above.

Methods to inhibit the onset of inhomogeneous plastic deformation (Lüders deformation) in metal forming process are proposed by Qiu et al. (Contribution 3). Besides conventional countermeasures (e.g., interstitial-free steels, plastic pre-strain by skin-pass rolling, bimodal grain size distribution, dual-steels, sandwich-like microstructure, TRIP effect), a new method to suppress Lüders deformation in a medium-carbon tempered martensite steel is proposed. The method promotes simultaneous yielding everywhere, rather than at several local sites, so that the formation of local plastic bands, and thus Lüders phenomenon, are inhibited. The multiplication of yield sites in the steel is achieved by controlling the dislocation density through a suitable heat treatment. The effectiveness of the proposed method was verified on five types of martensite phases with the same morphology, but with different initial dislocation densities obtained by various tempering conditions. The use of DIC allowed monitoring the evolution of the stress–strain field and the elasto-plastic behavior of dog-bone specimens under monotonic tensile testing. The experimental outcomes confirmed the favorable effect of a high initial dislocation density and work-hardening in suppressing the Lüders deformation.

Another study dealing with the improvement in strength of surgical-grade 316L austenitic stainless steels, obtained by microstructure tailoring, is presented by Rybalchenko et al. (Contribution 4). While 316L steel has properties that make it suitable for internal fixation devices in orthopedic surgery, its mechanical properties (e.g., yield strength and fatigue strength) are in fact not as impressive as other alloys also employed in orthopedics. Since 316L steel is a stable material that does not undergo phase transformation during heat treatment and deformation, strategies to increase its mechanical strength could take advantage of grain refinement. Other than by severe plastic deformation, grain refinement could also be achieved by high-pressure torsion (HPT), which triggers a nanocrystalline state in 316L steel. The effect of HPT nanostructuring at two temperatures (20 °C, 400 °C) on a surgical-grade 316LVM austenitic stainless steel (LVM = low-carbon vacuum melted) was evaluated with reference to the microstructure, fatigue strength, corrosion resistance, and in vitro biocompatibility. While the results highlighted the effect of nanostructuring by HPT at 20 °C in enhancing the properties of 316LVM steel for medical use, they also emphasized how additional strengthening through HPT at 400 °C was not fully successful.

Some articles in this Special Issue are more focused on studying aspects of structural design, or their application to engineering case studies. For example, a reliability assessment of an orthotropic steel-deck bridge based on a fatigue crack growth model and finite element analysis, which considers the coupling effect of multiple cracks and the randomness of vehicle loading, is developed by Liu et al. (Contribution 5). In place of time-consuming Monte Carlo simulations, more efficient approaches (such as the iHL-RF and AK-MCS methods) are used to reduce the overall computational time of the reliability analysis. A long-span steel box girder with two U-ribs, subject to random traffic loading, is used as a case study. The stress intensity factor and crack depth of a collinear double crack was calculated using the ABAQUS-FRANC3D interaction technique. The results show that, compared with a single-crack model, the coupling effect of double cracks causes the fatigue reliability of steel decks to decrease significantly. The increase in annual traffic also has a significant effect on bridge reliability. Among the considered sampling techniques, the iHL-RF and AK-MCS methods allow for the computational efficiency to be improved while ensuring the required calculation accuracy.

An application of a Ni–Ti Shape-Memory Alloy (SMA) in the design of a one-touch threaded mechanical coupler, integrated with conventional steel bars in concrete structures, is studied by Song et al. (Contribution 6). The coupler allows for the localized application of SMA bars only in the plastic hinge areas of structural members, where significant displacements occur under seismic loadings. The coupler takes advantage of the recovery properties of SMAs in eliminating the residual deformation in plastic hinge areas. Uniaxial tensile tests on round specimens, supported by DIC, were conducted to analyze the performance of the innovative coupler and the mechanical response of the SMA–steel-connected bar. The experimental results indicate the effectiveness of the proposed mechanical coupler in the localized application of SMA bars within the plastic hinge areas of structural members.

Within the topic of metallic component structural integrity, it is worth mentioning for its relevance the fatigue behavior of welded joints in the VHCF regime (for cycles to failure > 10^7), which is the focus of the review by England et al. (Contribution 7). After describing the ultrasonic testing method, the characteristic course of S-N lines, and the typical crack initiation modes observed in VHCF, the review summarized the literature findings on the fatigue behavior of welded joints for various types of ferrous (low-carbon steel, alloyed steel, stainless steel) and non-ferrous (e.g., aluminum, magnesium, titanium) alloys. Also discussed therein are recommendations provided in design codes released by various organizations for standardization (BS, IIW, AWS, EN, DNV). This is done to emphasize analogies and differences, particularly, regarding the existence of a fatigue limit.

The structural integrity of metal-based components is influenced also by failure phenomena that start at the surface, such as those promoted by corrosion and wear. Therefore, it is very important to study how these phenomena occur, as well as the surface engineering techniques that allow the service life of components to be prolonged, taking advantage of the changes in surface characteristics that occur.

Aluminum alloys are used in many engineering applications because they have an ideal combination of being lightweight with medium strength and good corrosion resistance in many environments. However, when chloride ions are present, as in coastal environments, corrosion phenomena may become severe, and may play a crucial role in reducing the service life of the component. The fracture behavior of aluminum alloys under coastal environmental conditions is reviewed by Alqahtani et al. (Contribution 8). Changes in fracture toughness and fatigue-crack growth rates in coastal conditions are discussed, highlighting the influence of saltwater exposure, humidity, and temperature on fatigue and fracture behavior. Saltwater promotes corrosion, and changes in humidity and temperature affect corrosion behavior. Corrosion phenomena may cooperate in crack initiation and accelerate crack propagation, as in stress-corrosion cracking, corrosion fatigue, and hydrogen embrittlement. Modeling and predictive approaches for estimating the fracture toughness and fatigue crack growth of aluminum alloys, together with their applications and limitations, are also reviewed in their study.

Surface engineering techniques can be used to improve corrosion resistance, as well as surface hardness. Among these different strategies, low-temperature thermochemical treatments have attracted increasing interest in recent years for their surface modification of stainless steels. Traditional nitriding, carburizing, or nitrocarburizing processes, as usually applied to low-alloy steels or tool steels, cannot be transferred directly to stainless steels, since, at the used treatment temperatures, chromium (Cr) compounds can form, prejudicing the corrosion resistance of the alloys. However, by using treatment temperatures so low that nitrogen (N) and/or carbon (C) atoms can easily diffuse, while interstitial (in particular, Cr) atoms can be relatively immobile in the lattice, the formation of Cr compounds is inhibited. In these para-equilibrium conditions, N and C atoms are retained in solid solution in the iron lattices (austenite, ferrite, martensite) beyond the solubility limit, and supersaturated solid solutions, known as "expanded" phases, form. For austenitic stainless steels, low-temperature treatments are known to significantly improve surface hardness, as well as maintaining or even increasing corrosion resistance. The corrosion behavior in different environments of austenitic stainless steels subjected to low-temperature thermochemical treatments is reviewed by Borgioli (Contribution 9). The effects of the presence of N and C on the formation of the passive layers, and on the corrosion behavior of austenitic stainless steels, are considered, and the characteristics of the so-called expanded austenite, or S phase, produced by nitriding, carburizing, and nitrocarburizing treatments are discussed. An analysis of the international literature highlights that the formation of N- or C-rich expanded austenite causes the delay or even inhibition of the occurrence of localized corrosion phenomena (pitting, crevice) in solutions containing chloride-ions. In presence of sulfuric acid solutions, the corrosion behavior is influenced by the microstructure and phase composition of the modified surface layers. The formation of a modified surface layer consisting of only N- or C-rich expanded austenite allows for the maintenance or even increase in corrosion resistance. On the contrary, when N-induced h.c.p. martensite, nitrides, or carbides also form, a significant decrease in corrosion resistance is observed.

Low-temperature plasma carburizing and nitriding were performed by Adachi et al. on AISI 316L stainless-steel tungsten carbide composite layers, which were fabricated using laser metal deposition (Contribution 10). The authors studied the effects of two processes types, single carburizing and continuous nitriding after carburizing. Both treatments were able to produce a supersaturated solid solution of interstitial atoms (C, N) in the expanded and distorted austenite lattice, known as the S phase or expanded austenite. The formation mechanism of this phase was investigated by considering the influence of the carbon, which was present in solid solution in the as-deposited layers, and in eutectic carbides. They observed surface microhardness enhancement of up to about 1365 HV and an improvement in corrosion resistance in a 3.5 wt.% NaCl solution, with a significant reduction in pitting and crevice phenomena, for the treated samples.

Hard coatings are very important for increasing the performance of cutting tools. The synergic effects of nanosecond laser ablation, used as a precision machining ablation technique, and of an AlTiN coating deposited using a PVD technique, are studied by Fang et al. for cemented carbides consisting of WC carbides embedded in a Co matrix (Contribution 11). Also investigated in their study were both the surface integrity, assessed by changes in morphology, microstructure, and roughness, and mechanical integrity, evaluated on the basis of surface Vickers hardness measurements and scratch tests. In most of the sample types, nanosecond laser ablation had a beneficial effect on the mechanical integrity of the coated cemented carbides, as it enhanced surface hardness, and in the scratch tests it reduced the penetration depth and increased the critical load values for the emergence of specific damage/failure events.

Acknowledgments: The Guest Editors wish to thank all of the authors for their high-quality articles, as well as all of the Reviewers for their relevant comments. The Editors have appreciated the support of the *Metals* Editorial Office throughout the publication process.

Conflicts of Interest: The authors declare no conflicts of interest.

List of Contributions

1. Distefano, F.; Rizzo, D.; Briguglio, G.; Crupi, V.; Epasto, G. Full-Field Strain and Failure Analysis of Titanium Alloy Diamond Lattice. *Metals* **2024**, *14*, 775. https://doi.org/10.3390/met14070775
2. Han, C.-B.; Lee, D.-G. Effect of Oxygen on Static Recrystallization Behaviors of Biomedical Ti-Nb-Zr Alloys. *Metals* **2024**, *14*, 333. https://doi.org/10.3390/met14030333
3. Qiu, H.; Ueji, R.; Inoue, T. Suppression of Inhomogeneous Plastic Deformation in Medium-Carbon Tempered Martensite Steel. *Metals* **2024**, *14*, 306. https://doi.org/10.3390/met14030306
4. Rybalchenko, O.; Anisimova, N.; Martynenko, N.; Rybalchenko, G.; Tokar, A.; Lukyanova, E.; Prosvirnin, D.; Gorshenkov, M.; Kiselevskiy, M.; Dobatkin, S. Effect of Nanostructuring on Operational Properties of 316LVM Steel. *Metals* **2023**, *13*, 1951. https://doi.org/10.3390/met13121951
5. Liu, J.; Liu, Y.; Wang, G.; Lu, N.; Cui, J.; Wang, H. Fatigue Reliability Assessment for Orthotropic Steel Decks: Considering Multicrack Coupling Effects. *Metals* **2024**, *14*, 272. https://doi.org/10.3390/met14030272
6. Song, M.-K.; Choi, E.; Lee, J.-H. Development and Performance Evaluation of a Mechanical Connection for Steel and Shape Memory Alloy Bars. *Metals* **2024**, *14*, 300. https://doi.org/10.3390/met14030300
7. England, A.; Toumpis, A.; Gorash, Y. Very High Cycle Fatigue of Welds: A Review. *Metals* **2023**, *13*, 1860. https://doi.org/10.3390/met13111860
8. Alqahtani, I.; Starr, A.; Khan, M. Fracture Behaviour of Aluminium Alloys under Coastal Environmental Conditions: A Review. *Metals* **2024**, *14*, 336. https://doi.org/10.3390/met14030336
9. Borgioli, F. The Corrosion Behavior in Different Environments of Austenitic Stainless Steels Subjected to Thermochemical Surface Treatments at Low Temperatures: An Overview. *Metals* **2023**, *13*, 776. https://doi.org/10.3390/met13040776
10. Adachi, S.; Yamaguchi, T.; Tanaka, K.; Nishimura, T.; Ueda, N. Effects of Solid-Solution Carbon and Eutectic Carbides in AISI 316L Steel-Based Tungsten Carbide Composites on Plasma Carburizing and Nitriding. *Metals* **2023**, *13*, 1350. https://doi.org/10.3390/met13081350
11. Fang, S.; Llanes, L.; Guo, Y.B.; Bähre, D. Synergic Effects of Nanosecond Laser Ablation and PVD-Coating on Cemented Carbides: Assessment on Surface and Mechanical Integrity. *Metals* **2024**, *14*, 34. https://doi.org/10.3390/met14010034

References

1. Montanari, R.; Varone, A. (Eds.) *Processing-Structure-Property Relationships in Metals*; Multidisciplinary Digital Publishing Institute (MDPI): Basel, Switzerland, 2019. [CrossRef]
2. Gibson, I.; Rosen, D.; Stucker, B.; Khorasani, M. *Additive Manufacturing Technologies*; Springer: Cham, Switzerland, 2021.
3. Ferri Aliabadi, M.H.; Soboyejo, W.O. (Eds.) *Comprehensive Structural Integrity*, 2nd ed.; Elsevier Ltd.: Amsterdam, The Netherlands, 2023.
4. Kawai, R.; Song, C. Monte Carlo and variance reduction methods for structural reliability analysis: A comprehensive review. *Probab. Eng. Mech.* **2023**, *73*, 103479. [CrossRef]
5. Abdelbaset, H.; Zhu, Z. Behavior and Fatigue Life Assessment of Orthotropic Steel Decks: A State-of-the-art-Review. *Structures* **2024**, *60*, 105957. [CrossRef]
6. Radaj, D.; Sonsino, C.M.; Fricke, W. *Fatigue Assessment of Welded Joints by Local Approaches*, 2nd ed.; Woodhead Publishing Ltd.: Cambridge, UK, 2006.
7. Sakai, T. Historical Review and Future Prospect for Researches on Very High Cycle Fatigue of Metallic Materials. *Fatigue Fract. Eng. Mater Struct.* **2023**, *46*, 1217–1255. [CrossRef]
8. Miller, B.A.; Shipley, R.J.; Parrington, R.J.; Dennies, D.P. (Eds.) *Failure Analysis and Prevention—ASM Handbook*; ASM International: Materials Park, OH, USA, 2021; Volume 11. [CrossRef]
9. Ramezani, M.; Mohd Ripin, Z.; Pasang, T.; Jiang, C.-P. Surface Engineering of Metals: Techniques, Characterizations and Applications. *Metals* **2023**, *13*, 1299. [CrossRef]
10. Czerwinski, F. (Ed.) Thermochemical Treatment of Metals. In *Heat Treatment—Conventional and Novel Applications*; IntechOpen: Rijeka, Croatia, 2012; pp. 73–112. [CrossRef]

metals MDPI

Article

Full-Field Strain and Failure Analysis of Titanium Alloy Diamond Lattice

Fabio Distefano , Daniele Rizzo , Giovanni Briguglio , Vincenzo Crupi and Gabriella Epasto *

Department of Engineering, University of Messina, Contrada Di Dio—Vill. Sant' Agata, 98166 Messina, Italy; fabio.distefano1@unime.it (F.D.); daniele.rizzo@unime.it (D.R.); giovanni.briguglio1@studenti.unime.it (G.B.); crupi.vincenzo@unime.it (V.C.)

* Correspondence: gabriella.epasto@unime.it

Abstract: The advancement in additive manufacturing has significantly expanded the use of lattice structures in many engineering fields. Titanium diamond lattice structures, produced by a direct metal laser sintering process, were experimentally investigated. Two cell sizes were selected at five different relative densities. Morphological analysis was conducted by digital microscopy. The compressive tests and digital image correlation technique allowed the evaluation of elastic moduli to be used in the Gibson–Ashby model. Failure mechanisms of the structures have been analysed by digital image correlation, which represents a promising technique for strain evaluation of such structures. A non-linear finite element model of the lattice structures was developed and validated using the experimental data. The analysis of the results highlights the good mechanical properties of the Ti6Al4V alloy lattice structures.

Keywords: lattice structures; titanium alloy; additive manufacturing; mechanical properties; digital image correlation; finite element analysis

check for updates

Citation: Distefano, F.; Rizzo, D.; Briguglio, G.; Crupi, V.; Epasto, G. Full-Field Strain and Failure Analysis of Titanium Alloy Diamond Lattice. *Metals* **2024**, *14*, 775. https://doi.org/10.3390/met14070775

Academic Editors: Francesca Borgioli, Chonghe Li, Umberto Prisco, Tomasz Tański and Denis Benasciutti

Received: 31 March 2024
Revised: 24 June 2024
Accepted: 25 June 2024
Published: 30 June 2024

1. Introduction

The growing interest in advanced component manufacturing using 3D-printing technology opens new challenges and perspectives in the transportation industry, allowing the production of complex structures with high mechanical performance. Additive Manufactured Honeycomb Sandwich structures (AMHS) were recently proposed for marine structural applications [1]. Additive manufacturing (AM) lattice structures are well employed in ship, aircraft, automotive and biomedical devices [2–6]. Recently, additive manufacturing lattice structures have been employed inside structural ship's hulls for replacing inner planking layers. With the spread of the new unmanned aerial vehicles (UAVs), a lot of operations in the military and civilian sector have been simplified. These vehicles must be swift, full of agility and, at the same time, fuel efficient. These qualities could be possible to achieve firstly by using light-weight structures to reduce mass onboard, so the application of additive manufacturing technology may offer a good solution [7,8]. One of the most common techniques used in the manufacturing of Ti-6Al-4V lattice structures is Selective Laser Melting (SLM). Several studies on these structures have been conducted to assess crucial mechanical properties for various application fields, including medical [9,10]. Direct metal laser sintering (DMLS) is a further advanced 3D-printing process that uses laser technology to melt and layer metal powders, creating three-dimensional objects layer by layer. In this printing process, a substrate of metal powder is deposited and subsequently a precision laser selectively melts the powder according to 3D-model specifications. After, the lower substrate allows the next layer of powder to be added. This iteration continues until the entire object is created. DMLS printing is particularly advantageous for the manufacturing of complex and custom components in metal alloys, offering greater design and high mechanical performance compared with traditional metal fabrication methods.

The knowledge of the mechanical properties of these new AM structures is essential. The responses of bending-dominated and stretch-dominated lattice structures have been generally evaluated [11]. Stretch-dominated structures have higher stiffness rather than bending-dominated structures which are subjected to bending moments. Among the strut-based lattice structures that have been studied in the last years there are body-centred cubic (BCC) and face-centred-cubic (FCC) cells; in addition to these, there are other strut-based cells such as diamond and octet-truss cells. Triply periodic minimal surfaces (TPMS) structures are interesting for their energy adsorption applications. The challenge is to define their mechanical properties to choose the optimal design. Studies have pointed out the TPMS mechanical properties by means of finite element (FE) analyses and compressive tests for validating experimental data [12–15], highlighting the great advantages of employing TPMS structures not only for static purposes but also for dynamic applications with a high strain rate. TPMS show interesting fatigue behaviour especially for Gyroid cellular structures (GCS) [14]. The sheet-based Ti6Al4V alloy TPMS structures show great properties regarding the adsorption of energy, for example the TPMS-Diamond structure with a nominal thickness of 0.2 mm has a value of energy adsorbed of 37.9 MJ/m^3 with an adsorption efficiency of nearly 58% [15].

The methodology of this study is based on compressive tests supported by FE simulations and a digital image correlation (DIC) technique to assess the failure mechanisms of the lattice structures. Due to the high cost of the metallic powder materials [16] and of the AM processes [17,18], the implementation of an FE model alongside the experimental investigation is a common practice in the evaluation of the mechanical performance of the lattice structures [19–22]. A reliable FE model, in good agreement with the experimental results, can accurately predict the mechanical response of a lattice structure [23]. The combination of the DIC technique and FE analysis could unravel ambiguity in the mechanical properties of lattice structures [24–29]. Fila et al. [30] confirmed the importance of using the DIC technique for obtaining information about displacements, strain and velocities during quasi-static and dynamic tests of lattice structures. Recent cases reported by Drücker et al. [31] also support the hypothesis that DIC measurements could be useful for obtaining true stress–strain data for additively manufactured lattice structures. Köhnen et al. [32] investigated the plastic deformation behaviour of AISI 316L/1.44 lattice structures during tension, compression and fatigue testing using optical microscopy, SEM and DIC. Boniotti et al. [33] developed FE models for studying the effects of defects and geometrical irregularities of SLM AlSi7Mg lattice structures with the aid of DIC and micro-computed tomography. Neuhäuserová et al. [34] proposed an in-house algorithm for digital image correlation for evaluating the displacements of lattice structures subjected to quasi-static and dynamic loading conditions when analysing tetrakaidekahedral unit cells.

This study focuses on analysing the mechanical properties of lattice structures of Ti6Al4V ELI alloy, produced by the SLM process, through experimental compression tests, evaluating two different cell sizes. The use of the DIC technique enabled the mechanical characterization of the lattice materials and the construction of the Gibson–Ashby model. DIC was also employed for detecting failure mechanisms in lattice structures, allowing for further investigations about local strain approaches. The failure mechanisms on the tested specimens were observed by digital microscopy. In addition, FEM analyses have been performed and validated by experimental tests to evaluate their reliability in the design of such structures. The used approach provides insights about the performance of additive manufacturing lattice materials, offering potential advancements in material design and structural engineering.

2. Materials and Methods

2.1. Design and Manufactuting of the Specimens

TPMS diamond skeletal unit cell, described by Equation (1), is shown in Figure 1 [35].

$$\varphi_{\text{diamond}} \equiv \cos \times \cos y \cos z - \sin \times \sin y \sin z = C \quad (1)$$

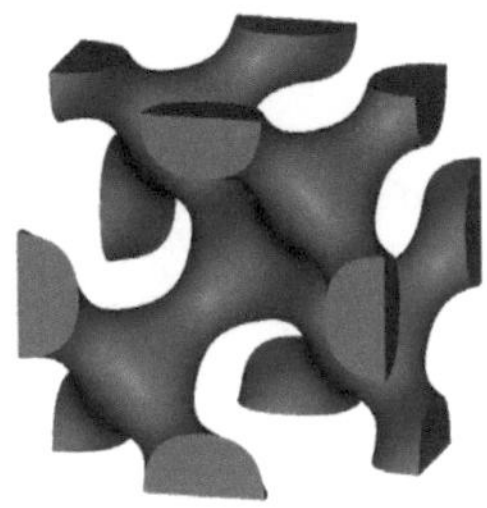

Figure 1. TPMS diamond skeletal unit cell.

TPMS diamond skeletal lattices were designed using the open-source ASLI software v0.1 [35]. Two cell dimensions were selected: one measuring 3 mm and the other 4 mm, with relative densities ranging from 10% to 50%. The analysis of the different cell sizes allows for the evaluation of its effect on the mechanical properties of the lattice material at various relative densities. Cylindrical specimens of 35 mm height and 30 mm diameter were fabricated in Ti6Al4V ELI (Grade 23) titanium alloy by a SLM 280 Twin 3D-printer (SLM Solutions Group AG, Lübeck, Germany). The used titanium alloy had a spherical shape and a particle size in the range 20–63 μm. The thickness of the powder layer was set to 30 μm. The additively manufactured specimens are shown in Figure 2.

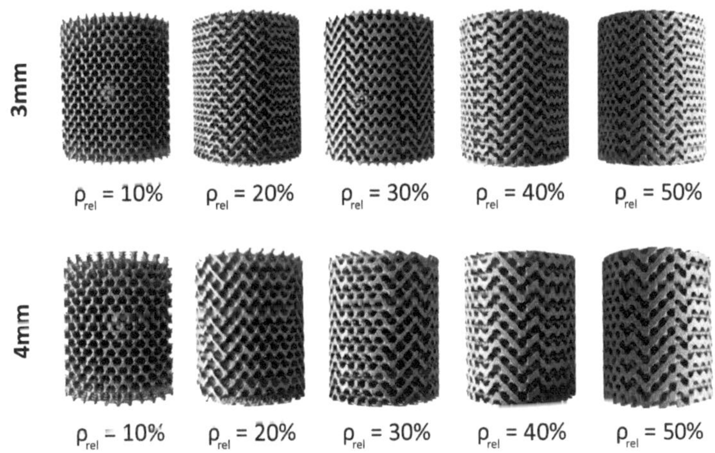

Figure 2. TPMS diamond skeletal specimens.

Digital microscopy was applied to evaluate the accuracy of the DMLS process before carrying out the compressive tests.

Uniaxial compressive tests were carried out on a universal testing machine (UTM), which was equipped with load cells of 25 kN to test specimen with relative density 10%, 250 kN to test specimens with relative densities 20% and 30%, and 600 kN to test specimens with relative densities 40% and 50%. The load cells were calibrated according to ISO 7500-1 [36], with those from 25 kN and 250 kN in Class 1 while those from 600 kN in Class 0.5. Compressive tests were carried out at a crosshead velocity of 2 mm/min.

DIC technique, with open source 2D DIC software Ncorr v1.2 [37], was used to calculate the elastic moduli of the diamond lattices and to evaluate their collapse mechanism. To this purpose, videos of the compressive tests were recorded with a super-macro-objective, at a resolution of 1280 × 720 pixels and an acquisition rate of 60 fps. The images were calibrated considering the known specimen's dimensions, obtaining a resolution of 0.08 mm/pixel. Due to the cylindrical specimen's shape, the ROI was analysed in the central portion of the images, with a dimension of 140 × 90 pixels and close to the distortion centre, where the distortion due to the geometry and the objective is minimal and can be considered negligible [38]. It was decided to not apply a speckle pattern on the specimen's surfaces, thus exploiting the inherent porosity of the structures. Gibson–Ashby model was applied to relate the mechanical properties of the lattices to their relative densities. According to the model, elastic moduli calculated from DIC and compressive strength obtained from UTM were plotted against the relative densities to obtain, respectively, the power Equations (2) and (3).

$$\frac{E^*}{E_s} = C_1\left(\frac{\rho^*}{\rho_s}\right)^{n_1} \quad (2)$$

$$\frac{\sigma^*}{\sigma_s} = C_2\left(\frac{\rho^*}{\rho_s}\right)^{n_2} \quad (3)$$

where E^* and σ^* are, respectively, the elastic modulus and the compressive strength of the lattice material, and E_s and σ_s are the elastic modulus and the compressive strength of the parent material. C_1, C_2, n_1, and n_2 are constant values, which are experimentally determined. The mechanical properties of the parent material, used to apply the Gibson–Ashby formulae, were obtained from the previous literature in which experimental tests were conducted on laser-based AM Ti-6Al-4V alloy specimens [39].

2.2. Finite Element Model

A non-linear FE model was developed by means of Altair HyperWorks 2022.3 (Altair Engineering, Troy, MI, USA) software package and Optistruct as implicit solver. The numerical analysis was conducted with the aim of evaluating the diamond lattice elastic modulus, making a comparison with the experimental results.

Numerical models were set up to exploit ASLI software v0.1 [35] to build a 3D mesh of first order tetrahedral elements with a dimension of 0.3 mm, following a mesh sensitivity study performed on the specimen with a relative density of 30%. It represents the worst case having the thinner strut among the specimens analysed in the FE model. The choice of the mesh size is based on the strut size, considering the software recommendations to have at least three elements on the smaller thickness. Three mesh sizes were analysed: 0.2 mm, 0.3 mm and 0.4 mm. Figure 3 reports the comparison of the analyses' results, in terms of von Mises stress.

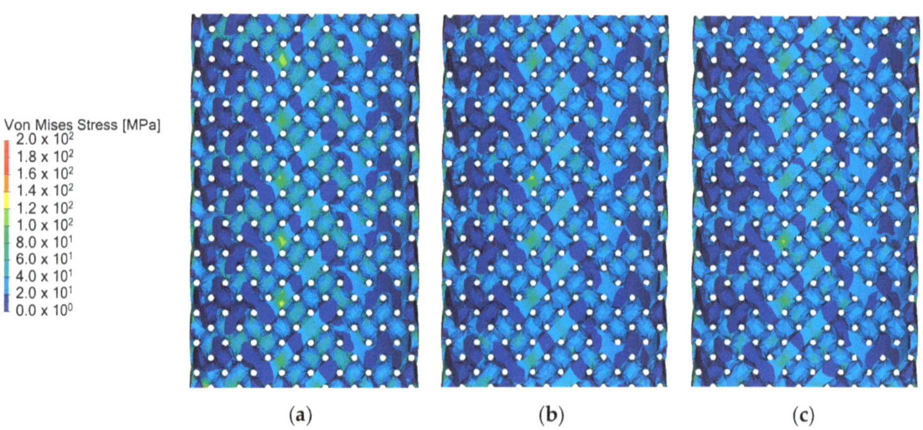

Figure 3. Von Mises stress results of the mesh sensitivity test for mesh sizes: (**a**) 0.2 mm; (**b**) 0.3 mm; (**c**) 0.4 mm.

Moreover, a comparison of the von Mises stress calculated in three different zones in the central section of the specimen was performed, as reported in Table 1 referring to Figure 4.

Table 1. Comparison of the von Mises stress evaluated in three different zone of the specimens for the mesh sensitivity test.

Mesh Size and Zone	Von Mises Stress [MPa]
0.2 mm zone 1	93
0.2 mm zone 2	96
0.2 mm zone 3	62
0.3 mm zone 1	82
0.3 mm zone 2	88
0.3 mm zone 3	58
0.4 mm zone 1	64
0.4 mm zone 2	69
0.4 mm zone 3	53

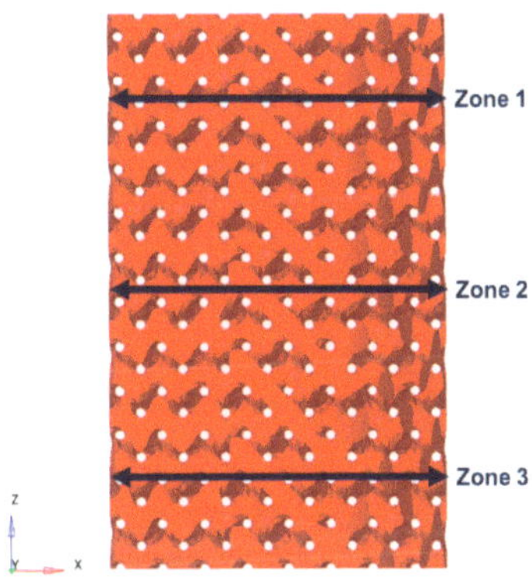

Figure 4. Zones of the specimen considered for the evaluation of the von Mises stress for the mesh sensitivity test.

The 0.4 mm mesh size was compared with the 0.3 mm mesh size and the von Mises stress percentage error, calculated in zones 1, 2 and 3, was, respectively: 22%, 22% and 8%. The mesh size 0.4 mm was therefore discarded since a percentage error higher than 20% was not considered suitable for the FE analysis. The comparison between mesh sizes 0.2 and 0.3 mm resulted in a von Mises stress percentage error in the central (zone 2) and lower (zone 3) parts of the specimens equal to 8% and 6%, respectively. These values are lower than 10% and therefore they are considered acceptable. A percentage error higher than 10% was found in the top part of the specimen and it is equal to 12%. However, the purpose of the proposed FE analysis is to calculate the lattice's elastic modulus, and the FE model does not aim to evaluate the diamond lattice's behaviour when subjected to high loads and deformation; therefore, the error was considered acceptable, and the mesh size of 0.3 mm was considered suitable for the FE analysis.

Due to computational time reason, smaller specimens were used compared with that of experimental tests. The use of smaller specimens does not affect the results, since the elastic modulus of the lattice structure does not depend on the specimen dimensions but only on the elastic modulus of the parent material and the lattice's relative density, as confirmed by the power relationship (2) of the Gibson–Ashby model. Cylindrical specimens of 16 mm diameter and 25 mm height were designed with diamond cells of 3 mm and 4 mm size and relative densities from 30% to 50%. The specimen dimension does not affect the evaluation of the lattice's elastic modulus. Two compression plates, modelled with quad elements of 0.3 mm size, were placed at the top and bottom surface of the specimens to apply boundary conditions. Lattice structures and compression plates were made of Ti6Al4V alloy and steel, respectively. Their mechanical properties, obtained from the literature [39], are reported in Table 2.

Table 2. Material properties used in the FE model.

Material	E [GPa]	ν	ρ [kg/m^3]
Ti6Al4V	110	0.34	4430
Steel	210	0.3	7850

A frictionless contact was applied at the interface between specimen and plates. A compressive load of 1 kN was applied on the top plate while the bottom plate was fixed, as depicted in Figure 5. The choice of 1 kN load is due to the fact that, for all the specimens experimentally tested, this value belongs to the elastic region in the load–displacement curve.

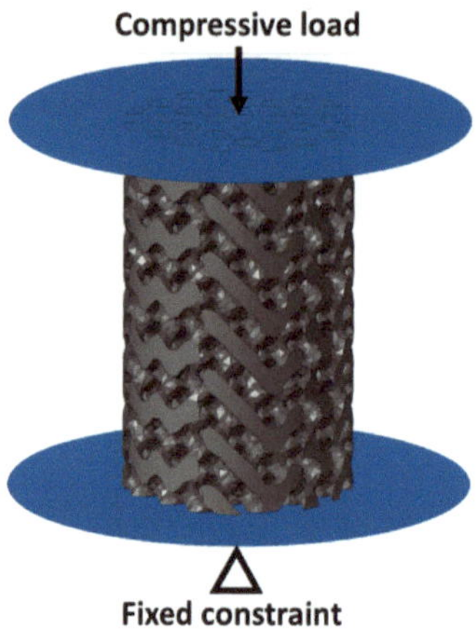

Figure 5. FE model boundary conditions.

3. Experimental Results and Discussion

3.1. Dimensional Check Results

The designed and the actual relative densities of the diamond lattices are reported in Table 3.

Table 3. Mass and relative density of the specimens.

Specimen	m [g]	ρ^*/ρ_s design [%]	ρ^*/ρ_s Actual [%]	ρ^*/ρ_s Error [%]
3 mm_10%	12.08	10	11.02	10.20
3 mm_20%	23.97	20	21.87	9.35
3 mm_30%	33.94	30	30.97	3.23
3 mm_40%	44.57	40	40.67	1.68
3 mm_50%	56.17	50	51.25	2.50
4 mm_10%	11.29	10	10.30	3.00
4 mm_20%	22.79	20	20.79	3.95
4 mm_30%	32.59	30	29.74	−0.87
4 mm_40%	43.35	40	39.55	−1.13
4 mm_50%	54.90	50	50.09	0.18

The actual relative densities of specimens with a cell size of 3 mm are higher than those with a cell size of 4 mm even if the designed relative density is the same (Table 3). Table 3 shows that, with equal cell size, the mismatch decreases at the increase in relative density. An increase in the cell size allows for a better quality of 3D printing, since with equal relative density specimens of a 4 mm cell size present a percentage mismatch lower than those with a 3 mm cell size.

Despite the overall specimen's dimensions being well respected, local defects were encountered during visual evaluation by digital microscopy, especially in specimens at lower densities with smaller strut diameters, which are more difficult to reproduce during the 3D-printing process, as shown in Figure 6.

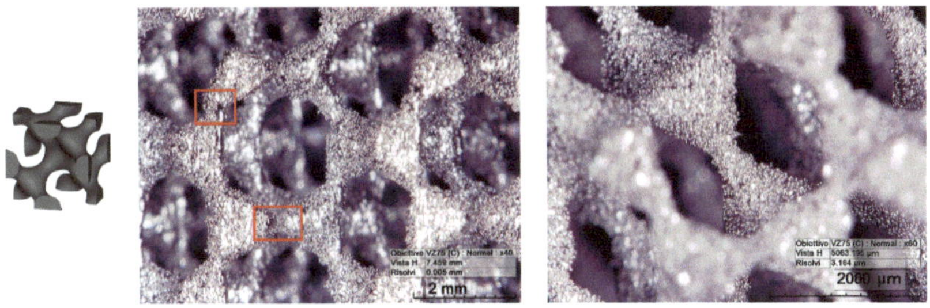

(**a**) (**b**) (**c**)

Figure 6. (**a**) diamond skeletal unit cell and digital microscopy of specimen 4 mm_10% with: (**b**) in the red boxes are highlighted inaccuracies in building the struts; (**c**) strut diameter correctly printed.

With reference to the diamond skeletal unit cell (Figure 6a), specimen 4 mm_10% presents defects in the central parts of the struts where the diameter is lower, as highlighted in Figure 6b. However, the microscopy observation allows for the assertion that, globally, the 3D-printing process well respects the strut's geometry, as visible in Figure 6c.

Specimens with a relative density of 50% and a higher strut diameter also showed bad reproducibility of the strut diameters in some parts of the geometry, leading to circularity defects in the specimen's pore shape, as observable in Figure 7.

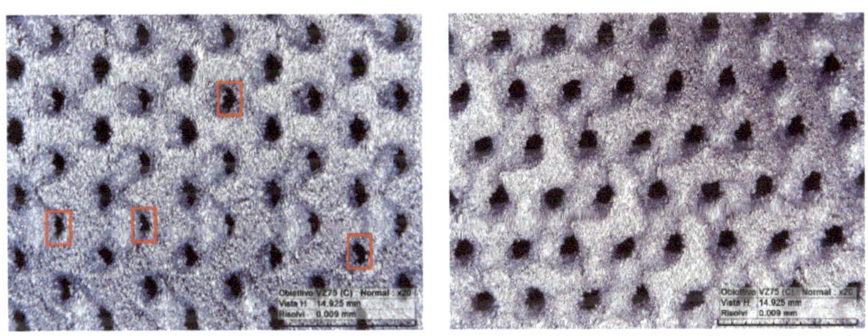

(**a**) (**b**)

Figure 7. Digital microscopy of (**a**) in the red boxes are highlighted discrepancies between designed and actual pore size geometry for specimen 3 mm_50%, (**b**) 3 mm_40%.

As highlighted in Figure 7a, the pore size geometry of the specimen 3 mm_50% is not always well reproduced; this problem is less observable in specimens with lower relative densities, as depicted in Figure 7b for the specimen 3 mm_40%.

The evaluation of the mismatch between designed and actual cell morphology is considered useful for the reliability of the FE analyses since those defects are generally not included in the simulations.

3.2. Compressive Tests

Figure 8 shows representative results of the compressive tests in terms of the stress–strain curves according to the various specimens' relative densities. According to the Gibson–Ashby theory [40], it is possible to highlight how the maximum force, and consequently the maximum stress, is similar for cells of 3 mm and 4 mm size. In the curves, the stress peak corresponds to failure of the single layer of the lattice structure. From the curves obtained from testing lattice structures with 10% relative density (Figure 8a), it is possible to highlight the stress trend during the entire test. It can be seen how the stress decreases with each layer failure until reaching a sort of plateau. The whole curve is reported as a representative trend of each specimen, which experienced a stress peak after a layer failure. For the other relative densities, the analysis was conducted focusing on the maximum peak, corresponding to the first layer failure. Also in this case, the curves measured for the specimens with 50% relative density are reported as examples (Figure 8b). The stress–strain curves at 10% relative density were plotted by measuring the strain via the UTM, and the other curves were plotted by measuring the strain using DIC equipment. From the analysis of the strains for each relative density, similar values were found in both cases of cells size analysed.

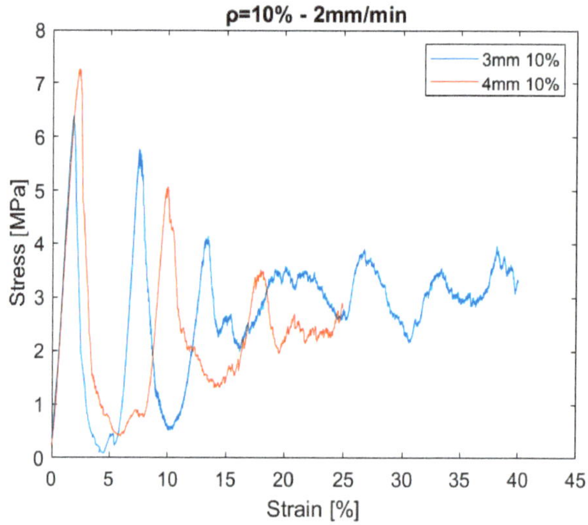

(a)

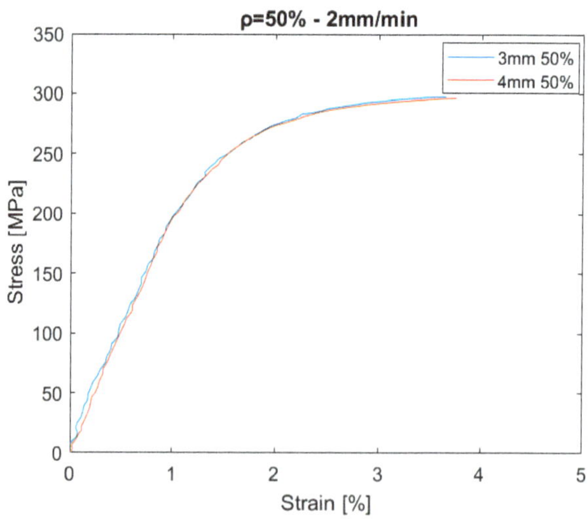

(b)

Figure 8. Stress—strain curves of the tested specimens: (**a**) whole curves obtained by UTM for 10% relative density; (**b**) curves at maximum stress obtained by DIC for 50% relative density.

The specific energy absorbed per unit mass (SEA_m) was calculated, using Equation (4), as the ratio of the stress–stain curve integral to the lattice structure density (ρ).

$$SEA_m = \frac{1}{\rho}\int_0^{\epsilon_{\sigma max}} \sigma(\epsilon)d\epsilon \quad (4)$$

The results, in terms of maximum force (F_{max}), maximum stress (σ_{max}) and elasticity modulus (E) and specific energy absorbed, are reported in Table 4. During the compressive test of the specimen 4 mm_20%, complete failure was not reached; therefore, it was not possible to evaluate the F_{max}, σ_{max} and SEA_m.

Table 4. Results of the compressive tests.

Specimen	F_{max} [kN]	σ_{max} [MPa]	E_{DIC} [MPa]	SEA_m [J/g]
3 mm_10%	4.51	6.38	548	0.09
3 mm_20%	47.85	67.69	3743	0.93
3 mm_30%	90.41	127.90	6983	1.67
3 mm_40%	141.93	200.79	13,175	3.06
3 mm_50%	210.42	297.69	20,437	3.65
4 mm_10%	5.14	7.26	738	0.13
4 mm_20%	/	/	3177	/
4 mm_30%	88.48	125.17	7484	2.74
4 mm_40%	139.81	197.79	12,955	3.00
4 mm_50%	209.61	296.54	20,567	3.83

The analysis of the results in terms of specific energy absorbed, shown in Figure 9, underlines how the values increase as the relative density increases. Furthermore, the specimens' comparison with the same relative density shows that cell size has a limited influence on the energy adsorption properties of the lattice structures under investigation.

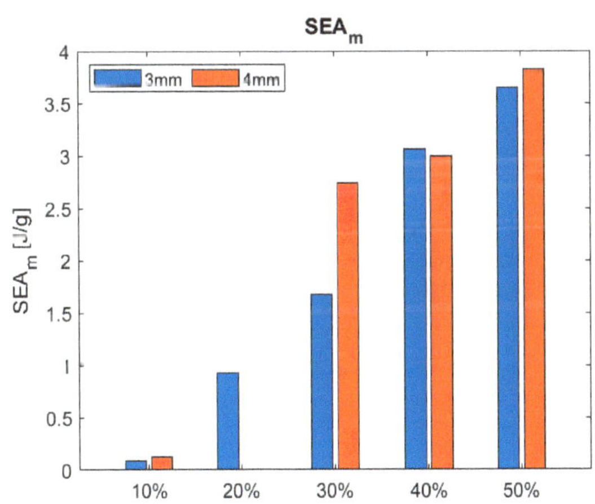

Figure 9. SEA comparison of the tested specimens.

There is a significant difference in terms of specific energy absorbed only for the specimens with a value of 30% of relative density. This could be due to the presence of defects inside the 3 mm_30% specimen.

The elastic modulus was calculated by DIC images, considering from each DIC frame the mean strain evaluated in the region of interest (ROI) of the specimen. The obtained strain values were then related to the stresses calculated from UTM and therefore the elastic modulus was calculated.

It was chosen to carry out tests at different density values to perform repetitions on specimens under the same conditions. The repeatability of the values of the compressive tests was validated by carrying out two repetitions only on specimens with 40% relative density. Figure 10 shows the elastic region, up to the compressive strength, of the specimens 3 mm_40% (Figure 10a) and 4 mm_40% (Figure 10b).

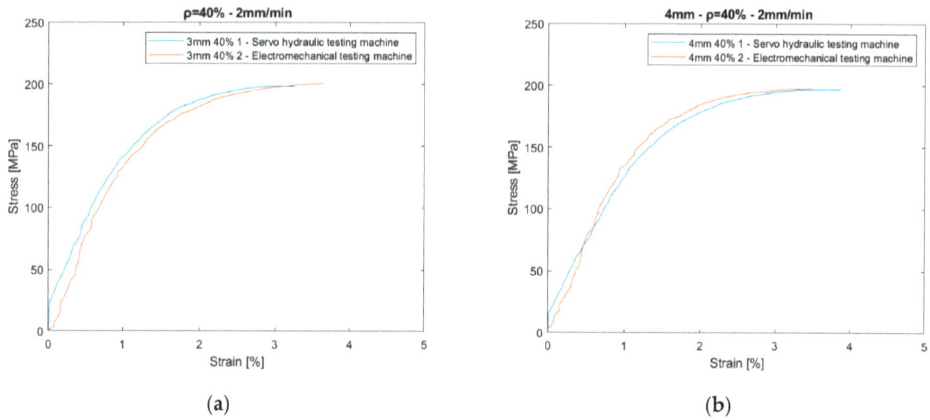

Figure 10. Stress—strain curves of the repetition tests of specimens: 3 mm_40% (**a**); 4 mm_40% (**b**).

For the specimens with 40% relative density, the two repetitions were carried out with a servo hydraulic testing machine and an electromechanical testing machine.

Table 5 reports the mechanical properties obtained by the repeated tests; the elastic moduli were calculated through DIC technique. The experimental tests confirm the repeatability of the results.

Table 5. Mechanical properties of the specimens 3 mm_40% and 4 mm_40%.

Specimen	E_{DIC} [MPa]	σ_{max} [MPa]
3 mm_40%_1	13,192	198
3 mm_40%_2	13,175	201
4 mm_40%_1	11,301	197
4 mm_40%_2	12,955	198

The mechanical properties obtained from compressive tests were related to the relative densities through the application of the Gibson–Ashby model. The results are reported in Figure 11 and summarised in Table 6, together with data from the literature for the same lattice topology.

Table 6. Gibson–Ashby model constants.

Specimen Type	C_1	n_1	R^2	C_2	n_2	R^2
3 mm [this work]	0.98	2.34	0.99	1.92	2.48	0.97
4 mm [this work]	0.84	2.12	0.99	1.76	2.40	0.97
Yan et al. adapted from [41]	0.17	1.64	0.99	1.39	1.95	0.99
Alabort et al. adapted from [42]	0.7	2.7	0.93	1.17	2.6	1

The results showed a limited effect of the cell size on the mechanical properties. The Gibson–Ashby model [43] predicts the range of the coefficient C_1 and C_2 to be, respectively, in the ranges of 0.1–4 and 0.1–1, while the exponent n_1 and n_2 are, respectively, 2 and 1.5. The calculated coefficient for the relative modulus falls within the predicted range, and Gibson–Ashby report that exponent values around 2 are considered acceptable. The coefficient and exponent of the relative strength fall out of the predicted range; however, data reported in literature [41,42] show discrepancies in the determination of these parameters. The coefficients C_2 evaluated in the present work are higher than those from the literature, while exponents n_2 are in accordance with the reported data.

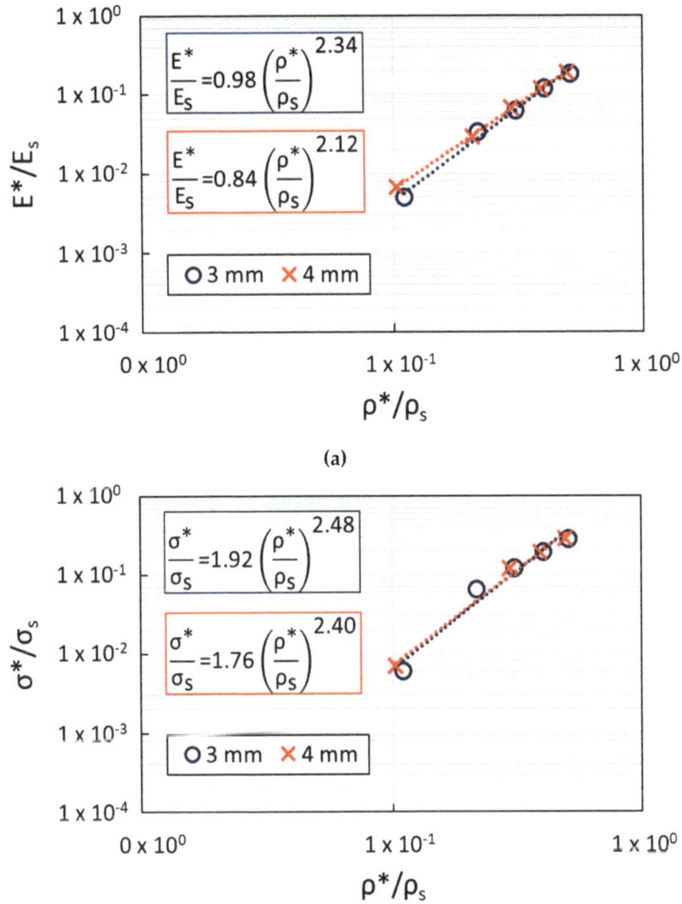

(a)

(b)

Figure 11. Gibson–Ashby curves. (**a**) relative modulus versus relative density; (**b**) relative strength versus relative density.

3.3. Collapse Mode

The tested lattice structures are shown in Figure 12. From visual inspection, it is possible to highlight various compressive failure modes. At 10% of relative density, for both cell sizes, it can be seen how the lattice structure collapses on itself in a brittle mode without the formation of macrocracks. As the relative density increases (from 20% to 50%), there is a layer's densification and therefore failure through 45° diagonal paths. It is possible to correlate the collapse mode with the stress–strain curves described in Figure 8. The number of stress peaks detected on the curves corresponds to the number of collapsed layers visible on the broken specimens. It is also possible to underline how, for the 3 mm cell size, the layer's densification at 30 and 40% densities allows the structure to strengthen by recovering up to 93% and 86% of the initial load. For the 4 mm cell size, the structure recovers up to 95% and 87% of the initial load. The failure via the diagonal path was also verified by the DIC investigation, as shown in Figure 13.

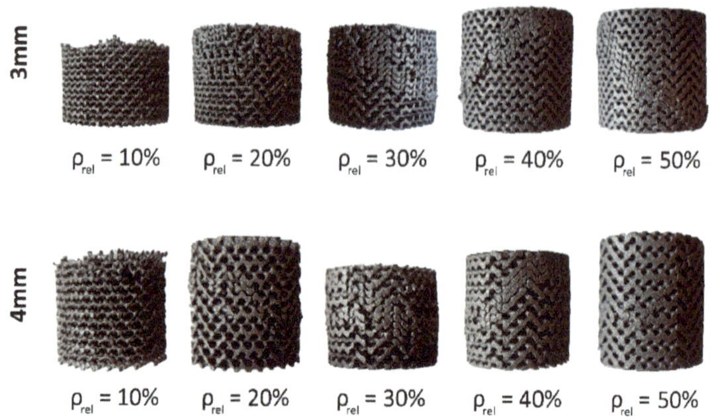

Figure 12. TPMS diamond skeletal tested specimens.

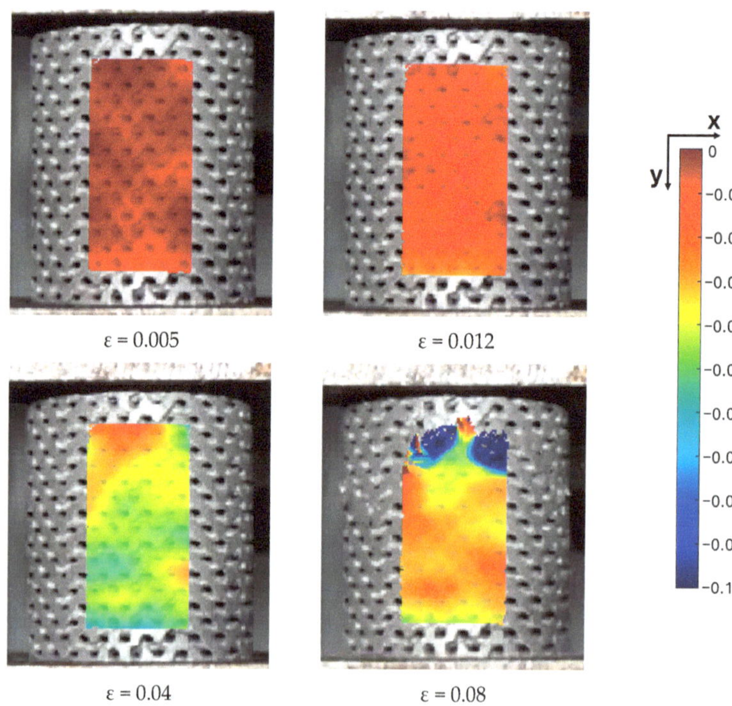

Figure 13. DIC results of specimen 4 mm_50%.

DIC images report the vertical strain in terms of Eulerian–Almansi strain [44], which is calculated with respect to the deformed state of the specimen. The strains reported in Figure 13 are evaluated as the mean strain of the DIC frame. The strains are almost constant in the elastic region ($\varepsilon = 0.005$), and they slightly increase in the bottom part after reached the elastic limit ($\varepsilon = 0.012$). From this point, the top region of the specimen starts to deform with a diagonal path (red part in Figure 13 at a strain of 0.04) while strains linearly increase in the central and bottom regions up to the maximum stress ($\varepsilon = 0.04$); however, no cracks formation and propagation were observed. After reaching the maximum stress, a sudden fracture appears in the top region with a bi-diagonal path, leading to a high strain increase ($\varepsilon = 0.08$). The observed behaviour is representative of the whole batch except for specimens with a relative density of 10%, which, as described in Figure 12, collapse on themselves layer-by-layer, as is also confirmed by the DIC vertical strain evolution presented in Figure 14.

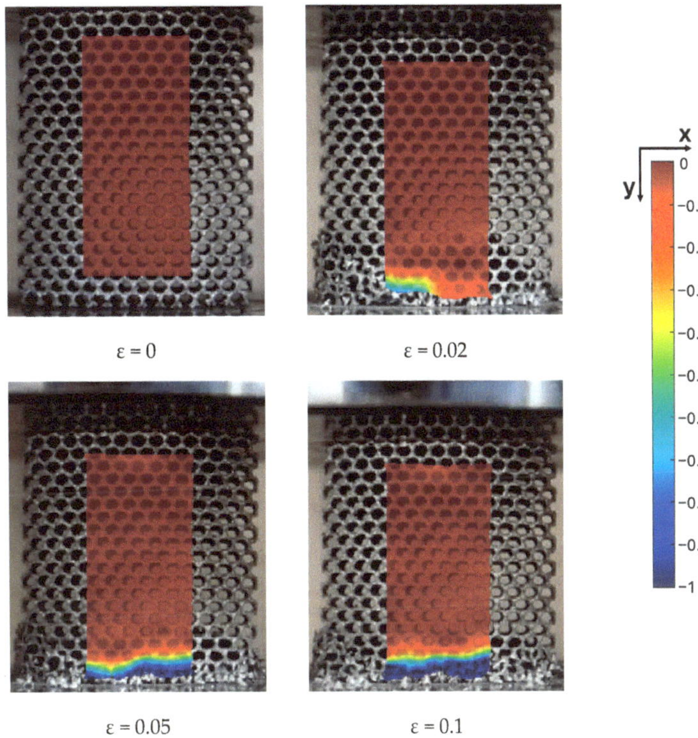

Figure 14. DIC results of specimen 3 mm_10.

Figure 14 shows the deformation process of the 3 mm_10% specimen during the compressive test up to a global strain of 10%. As visible, the deformation is concentrated in the lower layers while the upper layers remain undeformed. After the collapse of a specific layer, the adjacent one starts to deform, in a layer-by-layer mechanism of collapse.

3.4. Finite Element Model Results

Table 7 shows the results of the FE analysis, in terms of Young's Modulus, and the comparison with the experimental results obtained using the DIC technique. The elastic modulus was determined by evaluating the mean principal strain and plotting the result in a stress–strain curve. The elastic modulus was then evaluated as the slope of the stress–strain curve. The values of the elastic moduli, obtained by FE simulations, confirm the experimental results; the values increase as the relative density increases and are not influenced by the cell size. The values are very similar for specimens with the same density and different cell sizes; this difference is higher for the experimental values, considering that the FE models do not take into account the real defects [45] and the inaccuracies in the additive manufacturing production.

Table 7. FE results vs experimental data.

Specimen	E_{FEM} [MPa]	E_{DIC} [MPa]	E_{error} [%]
3 mm_30%	8548	6983	22.41
3 mm_40%	11,683	13,175	−11.32

Table 7. *Cont.*

Specimen	E_{FEM} [MPa]	E_{DIC} [MPa]	E_{error} [%]
3 mm_50%	15,317	20,437	−25.05
4 mm_30%	8490	7484	13.45
4 mm_40%	11,521	12,955	−11.07
4 mm_50%	15,066	20,567	−26.74

Table 7 shows that the 50% relative density presents the highest discrepancies between the experimental and FE results. The 30% relative density tends to overestimate the lattice's elastic moduli while the other relative densities tested underestimate this value. Overall, the presented results allow for the assertion that the proposed FE model can be applied as a good tool for a first approximation evaluation of the lattice's elastic modulus.

4. Conclusions

In this research, the 3D-printed titanium TPMS diamond lattice structures' characterization was carried out by compressive test and collapse mode analysis.

Before the mechanical characterization, a dimensional check was performed. Overall, specimens' dimensions were well reproduced during the DMLS printing process, since the calculated error between the designed and measured relative densities were lower than 10%. A slight size effect was encountered in the measurement of the relative density; indeed, specimens with 4 mm cell size showed lower percentage errors. Digital microscopy showed deformities in the strut diameter of the lowest (10%) and the highest (50%) relative densities analysed; however, these deformities were limited to a small number of struts.

From the stress–strain curves obtained by compressive tests, it was possible to observe the non-dependence of the cell size since, when comparing specimens with the same density, the load values were similar. DIC proved to be an effective and high-performance technique for evaluating the strains and elastic modulus of the material under examination and allowed for the building of the Gibson–Ashby predictive curves. The failure mode analysis highlights the various compressive failure modes: at a lower relative density, the lattice structure collapses on itself without a macrocrack formation, and for the other density, there is the formation of a 45° diagonal path. The densification phenomenon allows the structure to strengthen by recovering up to 95% in the 4 mm case with a 30% relative density.

The FE results showed that the FE model can be applied as a good tool for a first approximation evaluation of the lattice's elastic modulus. The proposed FE model turned out to be a fast test method since the mesh is automatically generated through the ASLI software v0.1 and the simulation time is reduced since load and deformation are applied in the specimen's elastic region.

Author Contributions: Conceptualization, G.E.; methodology F.D.; validation, V.C. and G.E.; formal analysis, F.D., D.R. and G.B.; investigation, F.D., D.R. and G.E.; resources, V.C. and G.E.; data curation, F.D., D.R. and G.B.; writing—original draft preparation, F.D., D.R. and G.B.; writing—review and editing, V.C. and G.E.; visualization, F.D., D.R. and G.B.; supervision, V.C. and G.E.; funding acquisition, V.C. All authors have read and agreed to the published version of the manuscript.

Funding: This research received no external funding.

Data Availability Statement: The raw data supporting the conclusions of this article will be made available by the authors on request.

Acknowledgments: This study shows the results of the research activities of the Research Project "EOLO" (Code: ARS01_01044), project funded by the PON (National Operative Programme) 2014–2020.

Conflicts of Interest: The authors declare no conflicts of interest.

References

1. Garbatov, Y.; Scattareggia Marchese, S.; Epasto, G.; Crupi, V. Flexural Response of Additive-Manufactured Honeycomb Sandwiches for Marine Structural Applications. *Ocean. Eng.* **2024**, *302*, 117732. [CrossRef]
2. Adelmann, B.; Hellmann, R. Mechanical Properties of LPBF-Built Titanium Lattice Structures—A Comparative Study of As-Built and Hot Isostatic Pressed Structures for Medical Implants. *Metals* **2022**, *12*, 2072. [CrossRef]
3. Jam, A.; du Plessis, A.; Lora, C.; Raghavendra, S.; Pellizzari, M.; Benedetti, M. Manufacturability of Lattice Structures Fabricated by Laser Powder Bed Fusion: A Novel Biomedical Application of the Beta Ti-21S Alloy. *Addit. Manuf.* **2022**, *50*, 102556. [CrossRef]
4. Boursier Niutta, C.; Ciardiello, R.; Tridello, A. Experimental and Numerical Investigation of a Lattice Structure for Energy Absorption: Application to the Design of an Automotive Crash Absorber. *Polymers* **2022**, *14*, 1116. [CrossRef] [PubMed]
5. Ferro, C.G.; Varetti, S.; Maggiore, P. Experimental Evaluation of Mechanical Compression of Lattice Trusses Made with Ti6Al4V for Aerospace Use. *Chin. J. Aeronaut.* **2024**, *37*, 520–532. [CrossRef]
6. Armanfar, A.; Alper Tasmektepligil, A.; Kilic, R.T.; Demir, S.; Cam, S.; Karafi, Y.; El Majd, B.A.; Gunpinar, E. Embedding Lattice Structures into Ship Hulls for Structural Optimization and Additive Manufacturing. *Ocean. Eng.* **2024**, *301*, 117601. [CrossRef]
7. Moon, S.K.; Tan, Y.E.; Hwang, J.; Yoon, Y.-J. Application of 3D Printing Technology for Designing Light-Weight Unmanned Aerial Vehicle Wing Structures. *Int. J. Precis. Eng. Manuf.-Green Technol.* **2014**, *1*, 223–228. [CrossRef]
8. Palomba, G.; Crupi, V.; Epasto, G. Additively Manufactured Lightweight Monitoring Drones: Design and Experimental Investigation. *Polymer* **2022**, *241*, 124557. [CrossRef]
9. Papazoglou, D.P.; Neidhard-Doll, A.T.; Pinnell, M.F.; Erdahl, D.S.; Osborn, T.H. Compression and Tensile Testing of L-PBF Ti-6Al-4V Lattice Structures with Biomimetic Porosities and Strut Geometries for Orthopedic Implants. *Metals* **2024**, *14*, 232. [CrossRef]
10. Turudija, R.; Stojković, M.; Stojković, J.R.; Aranđelović, J.; Marinković, D. Stiffness of Anatomically Shaped Lattice Scaffolds Made by Direct Metal Laser Sintering of Ti-6Al-4V Powder: A Comparison of Two Different Design Variants. *Metals* **2024**, *14*, 219. [CrossRef]
11. Maconachie, T.; Leary, M.; Lozanovski, B.; Zhang, X.; Qian, M.; Faruque, O.; Brandt, M. SLM Lattice Structures: Properties, Performance, Applications and Challenges. *Mater. Des.* **2019**, *183*, 108137. [CrossRef]
12. Zhang, L.; Feih, S.; Daynes, S.; Chang, S.; Wang, M.Y.; Wei, J.; Lu, W.F. Energy Absorption Characteristics of Metallic Triply Periodic Minimal Surface Sheet Structures under Compressive Loading. *Addit. Manuf.* **2018**, *23*, 505–515. [CrossRef]
13. Sun, Q.; Sun, J.; Guo, K.; Wang, L. Compressive Mechanical Properties and Energy Absorption Characteristics of SLM Fabricated Ti6Al4V Triply Periodic Minimal Surface Cellular Structures. *Mech. Mater.* **2022**, *166*, 104241. [CrossRef]
14. Yang, L.; Yan, C.; Cao, W.; Liu, Z.; Song, B.; Wen, S.; Zhang, C.; Shi, Y.; Yang, S. Compression–Compression Fatigue Behaviour of Gyroid-Type Triply Periodic Minimal Surface Porous Structures Fabricated by Selective Laser Melting. *Acta Mater.* **2019**, *181*, 49–66. [CrossRef]
15. AlMahri, S.; Santiago, R.; Lee, D.-W.; Ramos, H.; Alabdouli, H.; Alteneiji, M.; Guan, Z.; Cantwell, W.; Alves, M. Evaluation of the Dynamic Response of Triply Periodic Minimal Surfaces Subjected to High Strain-Rate Compression. *Addit. Manuf* **2021**, *46*, 102220. [CrossRef]
16. Flores, I.; Kretzschmar, N.; Azman, A.H.; Chekurov, S.; Pedersen, D.B.; Chaudhuri, A. Implications of Lattice Structures on Economics and Productivity of Metal Powder Bed Fusion. *Addit. Manuf.* **2020**, *31*, 100947. [CrossRef]
17. Richard, C.T.; Kwok, T.-H. Analysis and Design of Lattice Structures for Rapid-Investment Casting. *Materials* **2021**, *14*, 4867. [CrossRef] [PubMed]
18. Wei, K.; Yang, Q.; Ling, B.; Xie, H.; Qu, Z.; Fang, D. Mechanical Responses of Titanium 3D Kagome Lattice Structure Manufactured by Selective Laser Melting. *Extreme Mech. Lett.* **2018**, *23*, 41–48. [CrossRef]
19. Timercan, A.; Terriault, P.; Brailovski, V. Axial Tension/Compression and Torsional Loading of Diamond and Gyroid Lattice Structures for Biomedical Implants: Simulation and Experiment. *Mater. Des.* **2023**, *225*, 111585. [CrossRef]
20. Zhu, H.; Wang, P.; Wei, D.; Si, J.; Wu, Y. Energy Absorption of Diamond Lattice Cylindrical Shells under Axial Compression Loading. *Thin-Walled Struct.* **2022**, *181*, 110131. [CrossRef]
21. Guo, H.; Takezawa, A.; Honda, M.; Kawamura, C.; Kitamura, M. Finite Element Simulation of the Compressive Response of Additively Manufactured Lattice Structures with Large Diameters. *Comput. Mater. Sci.* **2020**, *175*, 109610. [CrossRef]
22. Shi, X.; Liao, W.; Li, P.; Zhang, C.; Liu, T.; Wang, C.; Wu, J. Comparison of Compression Performance and Energy Absorption of Lattice Structures Fabricated by Selective Laser Melting. *Adv. Eng. Mater.* **2020**, *22*, 2000453. [CrossRef]
23. Smith, M.; Guan, Z.; Cantwell, W.J. Finite Element Modelling of the Compressive Response of Lattice Structures Manufactured Using the Selective Laser Melting Technique. *Int. J. Mech. Sci.* **2013**, *67*, 28–41. [CrossRef]
24. Wu, C.; Luo, J.; Zhong, J.; Xu, Y.; Wan, B.; Huang, W.; Fang, J.; Steven, G.P.; Sun, G.; Li, Q. Topology Optimisation for Design and Additive Manufacturing of Functionally Graded Lattice Structures Using Derivative-Aware Machine Learning Algorithms. *Addit. Manuf.* **2023**, *78*, 103833. [CrossRef]
25. Wang, R.; Chen, Y.; Yan, X.; Cong, N.; Fang, D.; Zhang, P.; Liang, X.; Wu, W. Experimental Investigations on the Mechanical Performances of Auxetic Metal-Ceramic Hybrid Lattice under Quasi-Static Compression and Dynamic Ballistic Loading. *Appl. Sci.* **2023**, *13*, 7564. [CrossRef]
26. Somlo, K.; Chauhan, S.S.; Niordson, C.F.; Poulios, K. Uniaxial Tensile Behaviour of Additively Manufactured Elastically Isotropic Truss Lattices Made of 316L. *Int. J. Solids Struct.* **2022**, *246–247*, 111599. [CrossRef]

27. Hao, W.; Liu, Y.; Wang, T.; Guo, G.; Chen, H.; Fang, D. Failure Analysis of 3D Printed Glass Fiber/PA12 Composite Lattice Structures Using DIC. *Compos. Struct.* **2019**, *225*, 111192. [CrossRef]
28. Boniotti, L.; Foletti, S.; Beretta, S.; Patriarca, L. Analysis of Strain and Stress Concentrations in Micro-Lattice Structures Manufactured by SLM. *Rapid Prototyp. J.* **2019**, *26*, 370–380. [CrossRef]
29. Mauko, A.; Fíla, T.; Falta, J.; Koudelka, P.; Rada, V.; Neuhäuserová, M.; Zlámal, P.; Vesenjak, M.; Jiroušek, O.; Ren, Z. Dynamic Deformation Behaviour of Chiral Auxetic Lattices at Low and High Strain-Rates. *Metals* **2020**, *11*, 52. [CrossRef]
30. Fíla, T.; Koudelka, P.; Falta, J.; Zlámal, P.; Rada, V.; Adorna, M.; Bronder, S.; Jiroušek, O. Dynamic Impact Testing of Cellular Solids and Lattice Structures: Application of Two-Sided Direct Impact Hopkinson Bar. *Int. J. Impact Eng.* **2021**, *148*, 103767. [CrossRef]
31. Drücker, S.; Schulze, M.; Ipsen, H.; Bandegani, L.; Hoch, H.; Kluge, M.; Fiedler, B. Experimental and Numerical Mechanical Characterization of Additively Manufactured Ti6Al4V Lattice Structures Considering Progressive Damage. *Int. J. Mech. Sci.* **2021**, *189*, 105986. [CrossRef]
32. Köhnen, P.; Haase, C.; Bültmann, J.; Ziegler, S.; Schleifenbaum, J.H.; Bleck, W. Mechanical Properties and Deformation Behavior of Additively Manufactured Lattice Structures of Stainless Steel. *Mater. Des.* **2018**, *145*, 205–217. [CrossRef]
33. Boniotti, L.; Beretta, S.; Patriarca, L.; Rigoni, L.; Foletti, S. Experimental and Numerical Investigation on Compressive Fatigue Strength of Lattice Structures of AlSi7Mg Manufactured by SLM. *Int. J. Fatigue* **2019**, *128*, 105181. [CrossRef]
34. Neuhäuserová, M.; Fíla, T.; Koudelka, P.; Falta, J.; Rada, V.; Šleichrt, J.; Zlámal, P.; Jiroušek, O. Compressive Behaviour of Additively Manufactured Periodical Re-Entrant Tetrakaidecahedral Lattices at Low and High Strain-Rates. *Metals* **2021**, *11*, 1196. [CrossRef]
35. Perez-Boerema, F.; Barzegari, M.; Geris, L. A Flexible and Easy-to-Use Open-Source Tool for Designing Functionally Graded 3D Porous Structures. *Virtual Phys. Prototyp.* **2022**, *17*, 682–699. [CrossRef]
36. *ISO 7500-1*; Metallic Materials—Verification of Static Uniaxial Testing Machines—Part 1: Tension/Compression Testing Machines—Verification and Calibration of the Force-Measuring System. ISO: Geneva, Switzerland, 2004.
37. Blaber, J.; Adair, B.; Antoniou, A. Ncorr: Open-Source 2D Digital Image Correlation Matlab Software. *Exp. Mech.* **2015**, *55*, 1105–1122. [CrossRef]
38. Pan, B.; Yu, L.; Wu, D.; Tang, L. Systematic Errors in Two-Dimensional Digital Image Correlation Due to Lens Distortion. *Opt. Lasers Eng.* **2013**, *51*, 140–147. [CrossRef]
39. Beese, A.M.; Carroll, B.E. Review of Mechanical Properties of Ti-6Al-4V Made by Laser-Based Additive Manufacturing Using Powder Feedstock. *JOM* **2016**, *68*, 724–734. [CrossRef]
40. Gibson, L.J.; Ashby, M.F. *Cellular Solids—Structure and Properties*; Cambridge University Press: Cambridge, UK, 1997.
41. Yan, C.; Hao, L.; Hussein, A.; Young, P. Ti-6Al-4V Triply Periodic Minimal Surface Structures for Bone Implants Fabricated via Selective Laser Melting. *J. Mech. Behav. Biomed. Mater.* **2015**, *51*, 61–73. [CrossRef]
42. Alabort, E.; Barba, D.; Reed, R.C. Design of Metallic Bone by Additive Manufacturing. *Scr. Mater.* **2019**, *164*, 110–114. [CrossRef]
43. Ashby, M.F.; Evans, A.G.; Fleck, N.A.; Gibson, L.J.; Hutchinson, J.W.; Wadley, H.N.G. *Metal Foams: A Design Guide*; Library of Congress Cataloguing-in-Publication Data; Elsevier: Amsterdam, The Netherlands, 2000.
44. Attard, M.M. Finite Strain—Isotropic Hyperelasticity. *Int. J. Solids Struct.* **2003**, *40*, 4353–4378. [CrossRef]
45. Tamburrino, F.; Graziosi, S.; Bordegoni, M. The Design Process of Additively Manufactured Mesoscale Lattice Structures: A Review. *J. Comput. Inf. Sci. Eng.* **2018**, *18*, 040801. [CrossRef]

metals

MDPI

Article

Effect of Oxygen on Static Recrystallization Behaviors of Biomedical Ti-Nb-Zr Alloys

Chan-Byeol Han [1] and Dong-Geun Lee [2,*]

1 Advanced Metals Division, Korea Institute of Materials Science, Changwon 51508, Republic of Korea; byeol7763@naver.com

2 Department of Materials Science and Metallurgical Engineering, Sunchon National University, Suncheon 57922, Republic of Korea

* Correspondence: leechodg@scnu.ac.kr

Abstract: Titanium alloys that are used in biomedical applications must possess biocompatibility and a low elastic modulus so that they protect host bone tissue without causing stress shielding. As the elastic modulus of beta Ti alloys is close to that of bone (10–30 GPa), these alloys are considered potential orthopedic implant materials. The elastic modulus of the single β-phase Ti-39Nb-6Zr (TNZ40) alloy is approximately 40 GPa, whereas the strength is lower than that of other types of Ti alloys. Interstitial oxygen in a Ti matrix is well known to improve the matrix strength by solid-solution hardening. The desired mechanical properties can be optimized using a thermo-mechanical procedure to maintain a low elastic modulus. In order to enhance the strength, TNZ40 alloys were fabricated with different amounts of oxygen. The TNZ-0.16O and TNZ-0.26O alloys were cold swaged into 11 mm diameter bars, subjected to solution treatment at 900 °C and 950 °C for 2 h, and furnace-cooled to room temperature. As a result, recrystallized grains were clearly observed in the β matrix. The TNZ-0.26O alloy that was cold-worked by swaging followed by solution treatment at 900 °C exhibited the best mechanical properties (Vickers hardness: 247 HV, ultimate tensile strength: 777 MPa, elongation at rupture: 18.6%, and compressive strength: 1187 MPa). This study reports the effects of oxygen content on the recrystallization behavior and mechanical properties of these alloys.

Keywords: Ti-Nb-Zr alloy; static recrystallization; beta titanium alloy; solid solution hardening; cold swaging

Citation: Han, C.-B.; Lee, D.-G. Effect of Oxygen on Static Recrystallization Behaviors of Biomedical Ti-Nb-Zr Alloys. *Metals* **2024**, *14*, 333. https://doi.org/10.3390/met14030333

Academic Editor: Liang-Yu Chen

Received: 13 February 2024
Revised: 11 March 2024
Accepted: 13 March 2024
Published: 14 March 2024

1. Introduction

As human lifespans increase globally, the world is entering an era of aging in which the proportion of the elderly population increases. The demand for medical devices is expected to steadily rise, and interest in medical biomaterials is increasing [1–3]. When biomaterials are implanted in the human body, they must not cause stress shielding, in which the biomaterials suppress stress from being transmitted to the bone tissue. Studies are focusing on developing materials that include innocuous alloying elements that do not cause side effects such as allergic reactions and tissue inflammation [4,5].

The elastic modulus of a titanium alloy is not much different than that of bone tissue, and the alloy exhibits excellent bonding strength. Ti alloys are widely used as biomaterials owing to favorable properties, such as high corrosion resistance, low density, and elastic modulus [6–13]. Many studies have been focusing on the development of β-Ti alloys, which have better properties in terms of biomechanics and biochemical compatibility [14–19]. Poorly biocompatible materials can cause inflammatory reactions, corrode in the body, and leak metal ions, resulting in loss of osseointegration. In addition, if there is a large difference in elastic modulus between the bone and the implant material, most of the stress is transferred to the implant material, and the stress transferred to the bone is reduced, resulting in a stress-shielding effect. The bone is not stressed by the implant material, and bone loss occurs around the implant material. If this stress-shielding effect occurs for a long period of time, a large amount of bone loss causes many holes in the bone, resulting in low bone density and weak bone strength, leading to osteoporosis, a disease that increases the likelihood of fracture.

In order to improve the stress transfer between human bone tissue and implants, which is particularly important for biomaterials, it is important to develop materials that have an elastic modulus close to the elastic modulus of human bone tissue. Therefore, there is an increasing demand for the development of β-phase beta titanium alloys, which have a relatively low elastic modulus among titanium alloys. However, β-phase titanium alloys have a low modulus of elasticity, but there is a problem that their strength is lower than that of other alpha or alpha + beta Ti materials. Since the low modulus of elasticity is accompanied by a decrease in strength, it is necessary to study how to improve the compatibility of these opposing properties via microstructure control and mechanical processes. Typical β-Ti alloys have a lower elastic modulus than those of other biomaterials (e.g., stainless steel, cobalt-chrome, and polymer materials), and researchers are actively reporting on the development of β-Ti alloys such as Ti-Nb, Ti-Mo, Ti-Zr, Ti-Ta, and Ti-Nb-Zr that include biocompatible β-stabilizer elements such as Nb, Mo, Zr, and Ta [20–28].

The β-stabilizer element Nb is known to be effective in reducing the elastic modulus when added to titanium and is a biocompatible material that does not cause any harmful reactions in the human body [29]. In addition, as a β-isomorphous element, it stabilizes so that it exists in a β-alloy structure even at room temperature and does not cause phase decomposition during solution heat treatments. Zr is a biocompatible alloying element that has excellent corrosion resistance and neutral properties when used in Ti alloys. Furthermore, it has been reported that when Zr and Nb are added together, the β phase-stabilization effect is enhanced [30–32].

For this study, the Ti-39Nb-6Zr (TNZ40) alloy, which is a β-Ti alloy with a large amount of Nb and Zr content, was selected. This alloy has an elastic modulus of approximately 40 GPa and a single β phase [23,24]. The TNZ40 alloy has better cold formability than α + β Ti alloys, but it has lower strength, and research is needed to resolve this. To increase the strength of Ti alloys, interstitial elements such as O, N, and C are typically added, and strength improvements can be expected via the solid solution hardening mechanism [33]. However, studies are not actively being conducted on improving the mechanical properties of alloys by performing crystal grain refinement based on recrystallization via mechanical and thermal processes [34–40].

Therefore, in order to improve the strength of the TNZ40 alloy used in this study, TNZ40 alloys with different amounts of oxygen content were fabricated, and experiments were performed in which mechanical and thermal process conditions were designed to improve mechanical properties. The aim was to analyze changes in the microstructures and mechanical properties according to the oxygen content and process conditions.

2. Material and Methods

To manufacture the alloys used in this study, vacuum arc remelting was performed on Ti, Nb, and Zr raw materials to produce ingots containing Ti-39Nb-6Zr-0.16O (wt.%) and Ti-39Nb-6Zr-0.26O (wt.%) alloy compositions. These ingots had a diameter of Φ16. Cold swaging was performed to produce bars with a diameter of Φ11, which were then heat treated at 900 °C and 950 °C for 2 h. The bars were then furnace-cooled (FC) to produce the final specimens (referred to below as Ti-39Nb-6Zr-0.16O SW+ST900 °C (TNZ-0.16O SS900 °C), Ti-39Nb-6Zr-0.26O SW+ST900 °C (TNZ-0.26O SS900 °C), Ti-39Nb-6Zr-0.16O SW+ST950 °C (TNZ-0.16O SS950 °C), and Ti-39Nb-6Zr-0.26O SW+ST950 °C (TNZ-0.26O SS950 °C)). Four types of specimens were used in the experiment, and the microstructure and mechanical properties were analyzed comparatively to see how the microstructure and mechanical properties are affected by the difference in oxygen content as a function of the solution heat treatment temperature after cold swaging. The specimen fabrication process is shown in Figure 1.

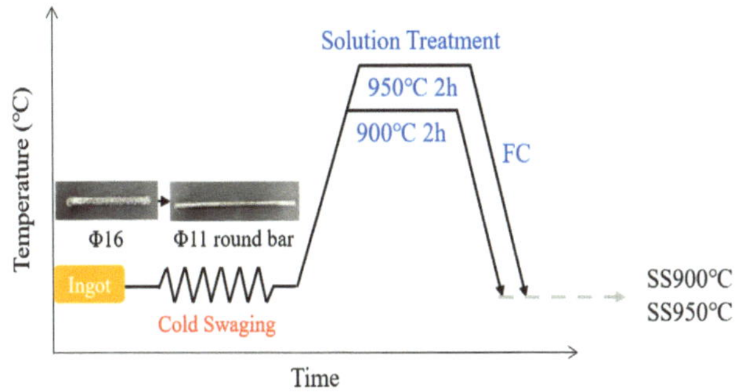

Figure 1. Experimental steps of the thermo-mechanical process for Ti-39Nb-6Zr-0.16O and Ti-39Nb-6Zr-0.26O specimens.

To observe the microstructure of each specimen, the specimens were micro-polished to 0.04 μm and then etched with an etching solution (60 mL H_2O_2, 30 mL H_2O, and 10 mL HF). To observe the microstructure, optical microscopy (OM) (BX53M, Olympus, Tokyo, Japan) was used. The samples were micro-polished to 0.04 μm to measure grain size, and grain orientation images were compared using field emission scanning electron microscopy (FE-SEM) and electron backscatter diffraction (EBSD) (JSM-7100F, JEOL, Tokyo, Japan) analysis. In general, SEM-EBSD analysis can be used to determine IPF, IQ, high-angle and low-angle grain boundaries, Kernel average misorientation (KAM), grain orientation spread (GOS), etc. The KAM value is the average value of the crystal rotation (crystal orientation difference) between the targeted measurement point and the surrounding measurement points, and the higher the value, the more strain is present in the material. GOS is used to determine the recrystallization region or to know the recrystallization fraction. The KAM and GOS values tend to increase as the dislocation density or internal strain energy accumulated in the specimen increases.

The presence of the phases was confirmed by X-ray diffraction (XRD) (XRD-7000, Brucker D8, Brucker, Ettlingen, German). To determine mechanical properties, a Vickers hardness tester (HM-200, Mitutoyo, Kawasaki, Japan) was used to measure hardness by pressing 12 points for 10 s with a load of 1 kgf. A dynamic universal materials testing machine (BESTUM-10MD, Ssaul Bestech, Seoul, Republic of Korea) was used for tensile and compression tests. The tensile samples had dimensions of 25 mm gauge length and Φ6 diameter according to ASTM E8 standard. The tensile tests were then performed at room temperature by applying tension at a rate of 10^{-3}/s until fracture occurred. The compression sample had a gauge length of 9 mm and a diameter of Φ6. The compression test was then performed at room temperature by applying compression at a strain rate of 10^{-3}/s until the strain reached 60%.

3. Results and Discussion

3.1. Analysis of Microstructure and Static Recrystallization

Images of the specimens that were observed with the optical microscope are shown in Figure 2. Sub-grains were observed within the large parent-phase grains of both specimens that experienced the SS900 °C conditions. For the SS950 °C conditions, it can be seen that the large parent-phase grain boundary disappeared, and the sub-grains were composed of main grains.

All specimens have a high internal deformation energy due to the accumulation of many internal dislocations due to cold swaging. This high internal deformation energy is directed to a lower energy state by recrystallization behavior due to the thermal energy supplied by successive solution treatment, i.e., the thermal energy and internal deformation energy of the solution treatment become the driving force for recrystallization nucleation and growth. Static recrystallization occurs in which new grains with lower energy are generated via dislocation annihilation by dislocation movement.

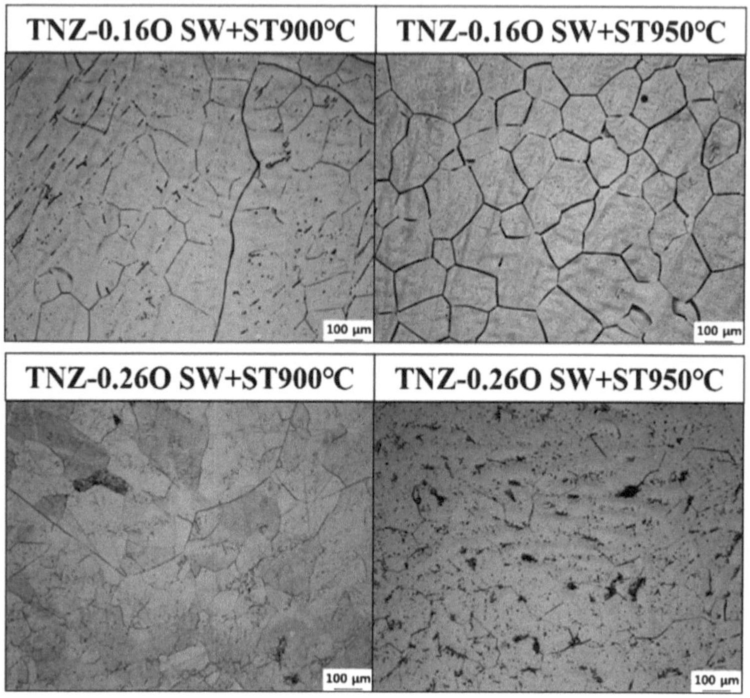

Figure 2. Optical microscope (OM) images of TNZ-0.16O SS900 °C, TNZ-0.16O SS950 °C, TNZ-0.26O SS900 °C, and TNZ-0.26O SS950 °C.

Figure 3 shows an image quality (IQ) map of TNZ-0.16O SS900 °C and TNZ-0.16O SS950 °C that was captured by EBSD, as well as an inverse pole figure (IPF) map that shows the high-angle grain boundary in black and the small-angle grain boundary in white. By analyzing the microstructure change process from TNZ-0.16O SS900 °C to TNZ-0.16O SS950 °C as shown in the IQ map, the microstructure of TNZ-0.16O SS900 °C simultaneously contains coarse-sized parent-phase grains and small-sized grains that have high-angle grain boundaries and low-angle grain boundaries within coarse-sized grains. It was determined that in the case of the fine-sized grains, recrystallization occurred due to the process that was performed. In TNZ-0.16O SS950 °C, the coarse-sized grain boundary disappeared, and it was composed of relatively equiaxed recrystallized grains with high-angle grain boundaries. As for the change in the average grain size, there was a reduction from 90 μm (min. 22 μm and max. 389 μm) in TNZ-0.16O SS900 °C to 80 μm (min. 13 μm and max. 365 μm) in TNZ-0.16O SS950 °C.

IQ and IPF maps of the microstructures and textural characteristics of TNZ-0.26O SS900 °C and TNZ-0.26O SS950 °C are shown in Figure 4. The microstructure of TNZ-0.26O SS900 °C consisted of coarse grains and small-sized recrystallized grains. In TNZ-0.26O SS950 °C, recrystallization occurred in most areas, and it consists of recrystallized grains and larger grains in which the size of the recrystallized grain has grown. It was observed that the areas where recrystallization occurred and their distribution were more even in TNZ-0.16O than in TNZ-0.26O, and the size and shape were uniform. The change in grain size was slight, with TNZ-0.26O SS900 °C having an average grain size of 84 μm (min. 16 μm and max. 460 μm) and TNZ-0.26O SS950 °C having an average grain size of 83 μm (min. 17 μm and max. 353 μm). From this, it was determined that the larger amount of oxygen that is dissolved in the crystal lattice is a factor that affects the occurrence of recrystallization and grain-growth behavior changes, and the change in the size of the average crystal grain was not large.

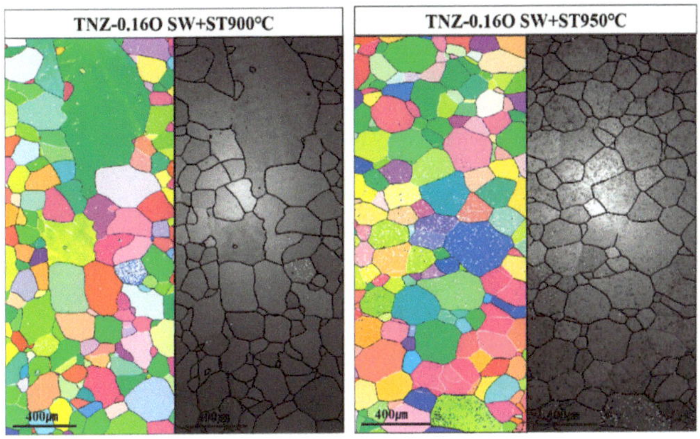

Figure 3. Inverse pore figure (IPF) map and image quality (IQ) map of TNZ-0.16O SS900 °C and TNZ-0.16O SS950 °C.

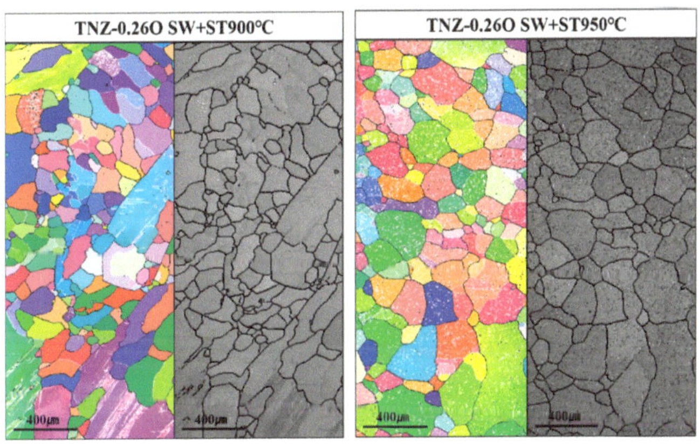

Figure 4. Inverse pore figure map and image quality map of TNZ-0.26O SS900 °C and TNZ-0.26O SS950 °C.

Overall, the microstructure of the TNZ-0.16O SW+ST950 °C specimen is composed of an equiaxed structure with a more uniform shape and size than the microstructure of the TNZ-0.26O SW+ST950 °C specimen and no coarse-sized crystal grains were observed. It was observed that the microstructure of the TNZ-0.26O SW+ST950 °C specimen was distributed with more uniformly shaped and sized equiaxed crystal grains than the TNZ-0.16O SW+ST950 °C specimen, and some regions had coarse-sized crystal grains. The oxygen atom, which has a very strong affinity for titanium, must be controlled very carefully in titanium materials. It was found that the static recrystallization phenomenon, depending on the oxygen content, resulted in inhomogeneous recrystallization grain size and distribution and that more oxygen employed in the crystal lattice delayed the recrystallization occurrence, resulting in a lower recrystallization fraction. It was found that controlling this inhomogeneous microstructure is necessary for the stability of the alloy's mechanical performance.

Figure 5 shows the IPF maps of TNZ-0.16O and TNZ-0.26O for each process, as well as the low-angle grain boundary fraction in which the misorientation between two grains is less than 15° and the high-angle grain boundary fraction in which the misorientation between two grain is 15° or more. There is also a grain orientation spread (GOS) map that shows the classified recrystallized regions. The GOS is the average value of orientation spread between all the points in a grain. The grains with low GOS values are considered recrystallized grains scanned with EBSD data [41]. In Figure 5, if the grain has a GOS value of 2 or less, it means that the grain is a recrystallized grain and is shown in blue. The regions with GOS values of 2 or greater are non-recrystallized grains, and they are shown in red. In TNZ-0.16O, the coarse-sized grains are non-recrystallized grains, and it can be seen that these regions decreased as the heat-treatment temperature increased. In TNZ-0.26O as well, the non-recrystallized grain regions decreased as the heat-treatment temperature increased, and the ratio of recrystallized regions was smaller than that of TNZ-0.16O.

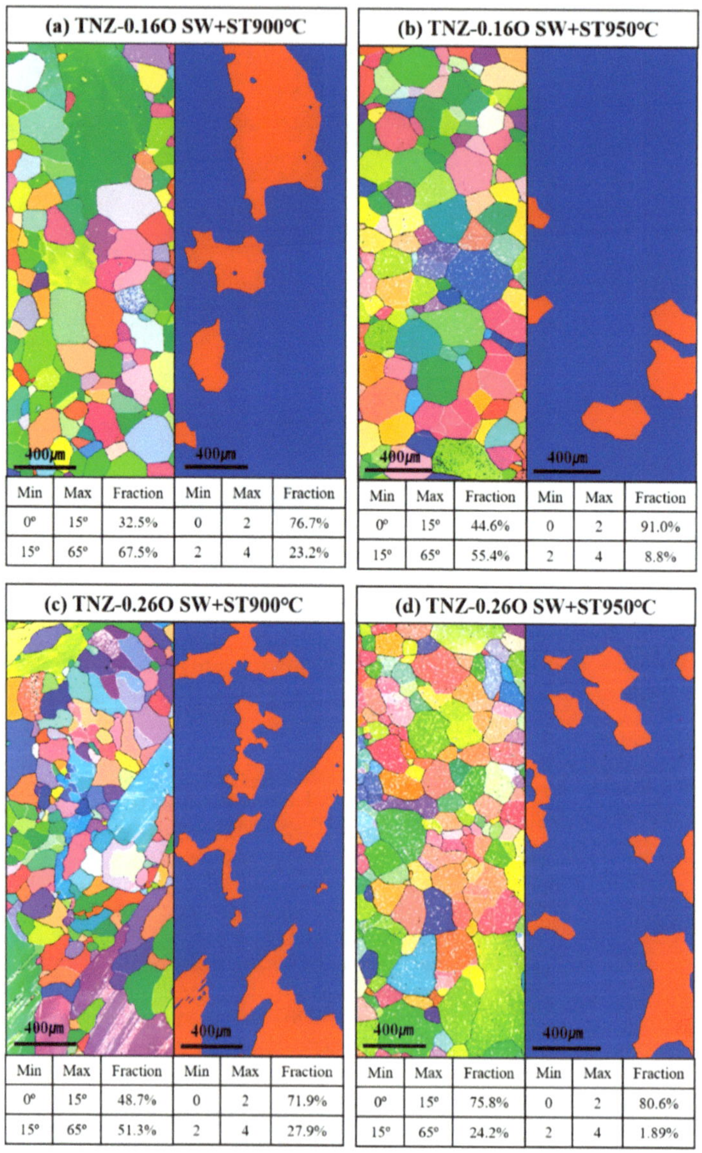

(a)

Min	Max	Fraction	Min	Max	Fraction
0°	15°	32.5%	0	2	76.7%
15°	65°	67.5%	2	4	23.2%

(b)

Min	Max	Fraction	Min	Max	Fraction
0°	15°	44.6%	0	2	91.0%
15°	65°	55.4%	2	4	8.8%

(c)

Min	Max	Fraction	Min	Max	Fraction
0°	15°	48.7%	0	2	71.9%
15°	65°	51.3%	2	4	27.9%

(d)

Min	Max	Fraction	Min	Max	Fraction
0°	15°	75.8%	0	2	80.6%
15°	65°	24.2%	2	4	1.89%

Figure 5. IPF maps indicating LAGB with white color, HAGB with black color and grain orientation spread (GOS) map indicating non-recrystallized grains with red region, recrystallized grains with blue region: (**a**) TNZ-0.16O SS900 °C, (**b**) TNZ-0.16O SS950 °C, (**c**) TNZ-0.26O SS900 °C, (**d**) TNZ-0.26O SS950 °C.

The shapes of the microstructures of each specimen observed in Figures 3 and 4 were different. It found that the difference in oxygen content is thought to affect recrystallization behavior. The material had high internal strain energy due to the cold swaging that was performed initially during processing. It can be seen that the heat energy that was obtained from the subsequent solution heat treatment was the driving force for recrystallized grain nucleation, and in order to move toward a stable energy state, recrystallization occurred, and the material changed to a low energy state. The nucleation and growth of new grains occurred during solution heat treatment and above after cold swaging. It can be concluded that static recrystallization occurred in this study's specimens due to the subsequent heat treatment. It can be inferred that recrystallization was delayed in the specimens with higher oxygen content because there were more obstacles to the recovery process, which relieved the accumulated strains.

Line graphs of all the specimen high-angle grain boundaries and recrystallized fractions, which are the areas where the GOS values were 2 or less, are shown in Figure 6. When the recrystallized fractions of the specimens are compared according to the process conditions, the recrystallized fraction was 76% in TNZ-0.16O SS900 °C, 71% in TNZ-0.26O SS900 °C, 91% in TNZ-0.16O SS950 °C, and 80% in TNZ-0.26O SS950 °C, indicating that TNZ-0.16O had a higher recrystallized fraction than TNZ-0.26O. In addition, the slope of the recrystallized fraction of TNZ-0.16O, according to the increase in the heat-treatment temperature, was steeper than that of TNZ-0.26O. Here, it can be concluded that the steep slope occurred because there was more active recrystallization due to the lower oxygen content.

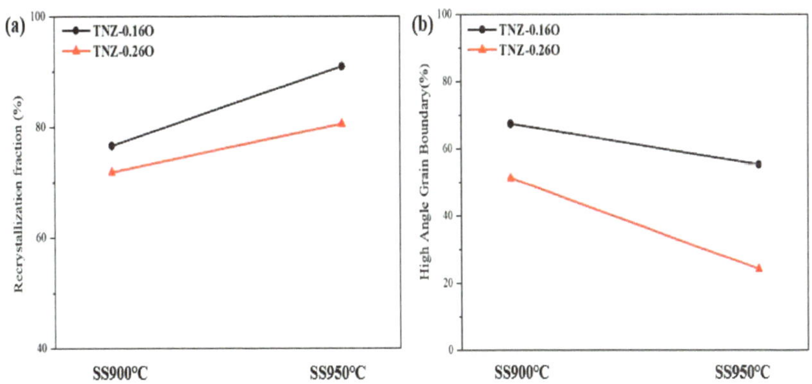

Figure 6. (**a**) Recrystallization fraction and (**b**) high angle grain boundary fraction of TNZ 0.16O SS900 °C, TNZ-0.16O SS950 °C, TNZ-0.26O SS900 °C and TNZ-0.26O SS950 °C.

Moreover, as the heat-treatment temperature increased, the recrystallized fraction increased, but the high-angle grain boundary fraction decreased. The change in the high-angle grain boundary fraction was 67% in TNZ-0.16O SS900 °C, 51% in TNZ-0.26O SS900 °C, 55% in TNZ-0.16O SS950 °C, and 24% in TNZ-0.26O SS950 °C. The reduction in the high-angle grain boundary fraction at the higher heat-treatment temperature occurred because recrystallization occurred in most areas of all specimens, and the recrystallized grains grew. Because of this, the total grain boundary area decreased.

The slope of the high-angle grain boundary fraction was steeper in TNZ-0.26O. It was determined that the reason for this is because TNZ-0.16O consisted of equiaxed grains, whereas TNZ-0.26O had a microstructure consisting of grains of non-uniform size and shape when the microstructures of the two specimens were compared; therefore, the range of change in the high-angle grain boundary fraction was larger. Also, as shown in Figure 5, in the case of TNZ-0.16O, static recrystallization occurred quickly throughout the entire area, and the recrystallization region rapidly grew as the temperature increased from 900 to 950 °C. The growth of the recrystallized grains was relatively slow, and the high-angle grain boundary fraction exhibited a gradual slope (Figure 6b). However, in the case of

TNZ-0.26O, which had a high oxygen content, the diffusion barrier caused by the oxygen atoms and the area where static recrystallization did not occur was larger than in the case of TNZ-0.16O. For this reason, a larger amount of heat energy from the solution heat treatment can be used to promote the growth of static recrystallized grains; therefore, as shown in Figure 6b, the high-angle grain boundary fraction decreased significantly as the temperature increased from 900 to 950 °C.

When the solution treatment is first performed at a temperature above the β transformation temperature, and then cold swaging is applied, the dislocation density increases and plastic deformation bands are formed in the region where the dislocation is integrated. As a result, brittleness increases, so high compressive strength can be obtained, but an appropriate elongation and yield strength value cannot be secured. Therefore, after cold swaging, a solution treatment was performed at a temperature above the β transformation temperature. A static recrystallization phenomenon occurred during the solution treatment process, and recrystallization initiation and growth were observed in the crystal grains.

The static recrystallization fraction increased as the heat treatment temperature increased, and the recrystallization fraction of the TNZ-0.16O specimen was higher than that of the TNZ-0.26O specimen. This means that the higher the oxygen content, the higher the dissolved oxygen in the β crystal lattice, which further hinders the movement of the dislocation. Therefore, less recrystallization was generated, and non-uniform recrystallization occurred.

The specimen XRD diffraction pattern results according to the oxygen content in each process are shown in Figure 7. It can be seen that only β-phase peaks were detected in both specimens for all processes, and no second phase other than the β phase was created. In addition, the β-phase peak intensity was slightly stronger in TNZ-0.16O than in TNZ-0.26O, and as the heat-treatment temperature increased, the peak intensity increased. It was determined that TNZ-0.16O, which includes a small amount of oxygen (an α-stabilizer element), exhibited strong peak intensity because it has a β phase in a stable energy state due to the larger amount of recrystallization.

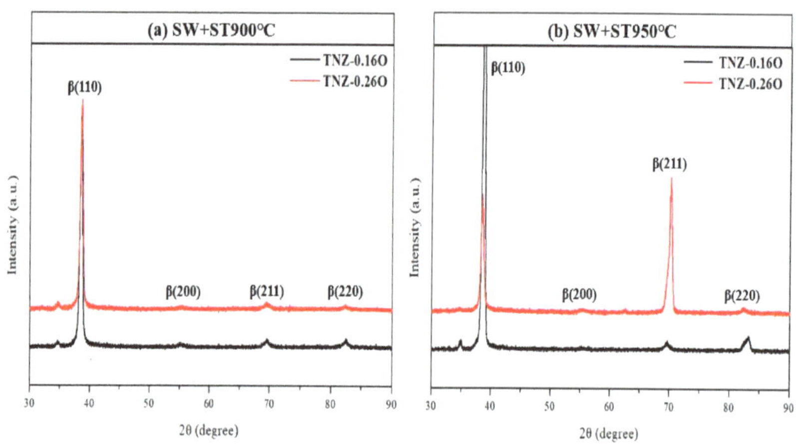

Figure 7. X-ray diffraction (XRD) pattern of (**a**) TNZ-0.16O SS900 °C, TNZ-0.26O SS900 °C and (**b**) TNZ-0.16O SS950 °C, TNZ-0.26O SS950 °C.

The recrystallization behavior was found to be different depending on the oxygen content, with the specimens with lower oxygen content showing a more homogeneous region of recrystallization. For the TZN-0.26O specimen, grains of inhomogeneous size grew as the heat treatment temperature increased. It is believed that oxygen is unevenly employed in the region and recrystallization occurs in localized areas, i.e., the parent grain disappears and the recrystallized grain becomes dominant, but uneven recrystallization and growth occurs.

3.2. Mechanical Properties

The measured Vickers hardness values of the specimens for all process conditions are shown in Figure 8. In all conditions, TNZ-0.26O had a higher hardness value than TNZ-0.16O due to the solid solution hardening effect. In a comparison of the hardness for each specimen, the hardness of TNZ-0.16O SS900 °C was 220 HV, and the hardness of TNZ-0.16O SS950 °C was 225 HV. When the solution heat-treatment temperature was high, the average grain size was reduced (90 μm → 80 μm) despite the increase in temperature, and the hardness increased slightly due to the changes in the microstructure, which consisted of equiaxed grains. The hardness of TNZ-0.26O SS900 °C was 247 HV, and the hardness of TNZ-0.26O SS950 °C was 242 HV. Unlike the trend of change in the hardness values of TNZ-0.16O, the hardness was lower at a higher solution heat-treatment temperature. Comparing the aforementioned microstructure shapes, it was determined that the reduction in the average grain size was small, and the hardness value was reduced slightly even as the temperature increased from 900 to 950 °C due to the non-uniform size and shape of the TNZ-0.26O grains.

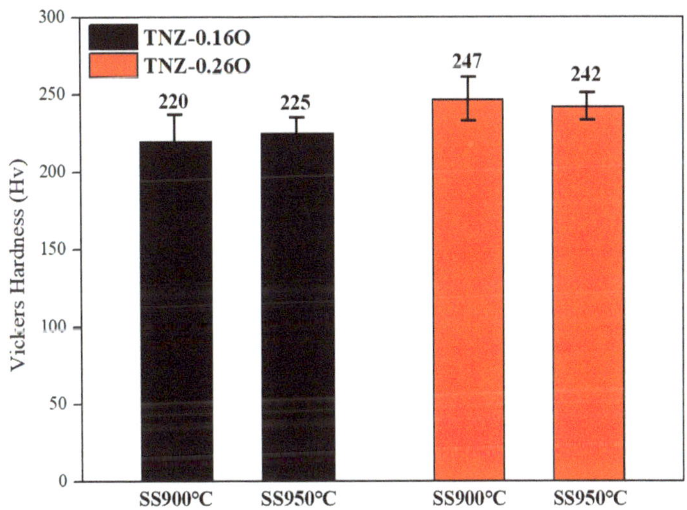

Figure 8. Vickers hardness of TNZ-0.16O SS900 °C, TNZ-0.16O SS950 °C, TNZ-0.26O SS900 °C and TNZ-0.26O SS950 °C.

Figures 9 and 10 and Tables 1 and 2 show the strain–stress curves and the numerical data obtained by performing room temperature tension and compression tests. The trend of results in the tensile and compressive strength is similar to that of the Vickers hardness. TNZ-0.260, which has a higher oxygen content than TNZ-0.16O, showed excellent overall tension and compression properties. The tension and compression properties of TNZ-0.16O improved at SS950 °C compared to SS900 °C, and this specimen exhibited an excellent elongation (15.8%). In TNZ-0.16O SS950 °C, it can be seen that the coarse parent-phase grains disappear, and that it consists of uniformly distributed equiaxed grains, giving it improved tensile and compressive strength.

In the TNZ-0.16O specimen, the average grain size decreased from 90 to 80 μm due to the increase in the solution heat-treatment temperature, and it showed a higher tensile strength, yield strength, and elongation at SS950 °C. However, the TNZ-0.26O specimen exhibited higher tension and compression properties at SS900 °C than at SS950 °C. More recrystallization occurred in TNZ-0.26O SS950 °C, but the grain size and shape were non-uniform, and the tensile strength (777 MPa → 755 MPa) and compressive strength (1187 MPa → 1155 MPa) were reduced.

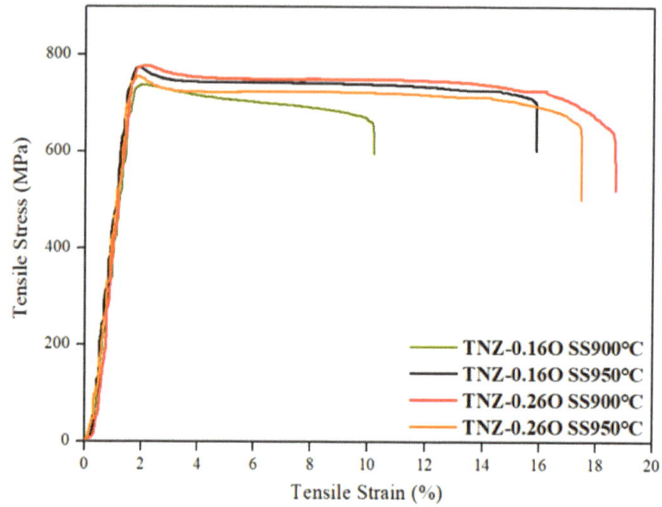

Figure 9. Tensile strain–stress curves of TNZ-0.16O SS900 °C, TNZ-0.16O SS950 °C, TNZ-0.26O SS900 °C and TNZ-0.26O SS950 °C.

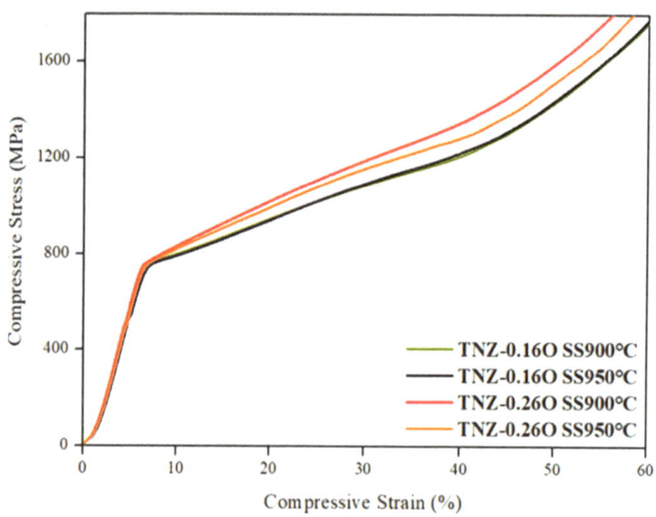

Figure 10. Compressive strain–stress curves of TNZ-0.16O SS900 °C, TNZ-0.16O SS950 °C, TNZ-0.26O SS900 °C, and TNZ-0.26O SS950 °C.

Table 1. Ultimate tensile strength, yield strength and elongation of Ti-39Nb-6Zr-0.16O SS900 °C, Ti-39Nb-6Zr-0.16O SS950 °C, Ti-39Nb-6Zr-0.26O SS900 °C, and Ti-39Nb-6Zr-0.26O SS950 °C.

Specimens	Ti-39Nb-6Zr-0.16O			Ti-39Nb-6Zr-0.26O		
	Ultimate Tensile Strength (MPa)	Yield Strength (MPa)	Elongation (%)	Ultimate Tensile Strength (MPa)	Yield Strength (MPa)	Elongation (%)
SW+ST900 °C	737 ± 5	734 ± 4	10.2 ± 2.2	777 ± 10	775 ± 3	18.6 ± 1.4
SW+ST950 °C	773 ± 3	764 ± 8	15.8 ± 1.5	755 ± 7	741 ± 6	17.4 ± 1.2

Table 2. Compressive yield strength and compressive strength at strain 30% of Ti-39Nb-6Zr-0.16O SS900 °C, Ti-39Nb-6Zr-0.16O SS950 °C, Ti-39Nb-6Zr-0.26O SS900 °C, and Ti-39Nb-6Zr-0.26O SS950 °C.

Specimens	Ti-39Nb-6Zr-0.16O		Ti-39Nb-6Zr-0.26O	
	Compressive Yield Strength (MPa)	**Compressive Strength (MPa) (at Strain 30%)**	**Compressive Yield Strength (MPa)**	**Compressive Strength (MPa) (at Strain 30%)**
SW+ST900 °C	774 ± 9	1086 ± 15	796 ± 13	1187 ± 21
SW+ST950 °C	777 ± 12	1092 ± 17	788 ± 14	1155 ± 15

As mentioned above, in order to control the microstructure of the Ti-39Nb-6Zr (TNZ40) β-Ti alloy and improve its mechanical properties, controlling the amount of oxygen, which is an α-stabilizer element, makes it possible to control the static recrystallization behavior, recrystallization fraction, and uniformity of grain-size distribution, thereby improving tension and compression properties at room temperature.

4. Conclusions

This study observed changes in the recrystallization behavior of Ti-39Nb-6Zr-0.16O (wt.%) and Ti-39Nb-6Zr-0.26O (wt.%) alloys, and the following conclusions were drawn.

(1) It was confirmed that static recrystallization, which creates new grains within coarse parent grains, occurred when solution heat treatments were performed at 900 °C and above after cold swaging a Ti-39Nb-6Zr alloy that had oxygen added. The microstructure of the Ti-39Nb-6Zr alloy with oxygen added at 0.16 wt.% consisted of equiaxed grains, and there were improvements in tensile strength (737 MPa → 773 MPa), elongation (10.2% → 15.8%) as the heat-treatment temperature increased, due to the grain refinement effect.

(2) It was confirmed that the Ti-39Nb-6Zr-0.26O alloy exhibited higher mechanical properties than Ti-39Nb-6Zr -0.16O due to the solid-solution hardening effect, but its recrystallized fraction was lower because the recrystallization behavior was delayed due to the greater amount of dissolved oxygen in the crystal lattice. Unlike the microstructure of Ti-39Nb-6Zr-0.16O, which consisted of equiaxed grains, the microstructure of Ti-39Nb-6Zr-0.26O exhibited a non-uniform grain size and distribution, and as the heat-treatment temperature increased, its mechanical properties decreased slightly.

(3) The solid-solution hardening caused by the addition of oxygen, which is an interstitial element, and the crystal grain refinement caused by static recrystallization had a combined effect, and it was found that the SW+ST900 °C conditions for the Ti-39Nb-6Zr-0.26O alloy were the optimal process conditions. These conditions resulted in excellent mechanical properties, including a tensile strength of 777 MPa, an elongation of 18.6%, and a compressive strength of 1187 MPa.

Author Contributions: Conceptualization, D.-G.L.; Methodology, C.-B.H.; Validation, C.-B.H.; Formal analysis, C.-B.H. and D.-G.L.; Investigation, D.-G.L.; Data curation, C.-B.H. and D.-G.L.; Writing–original draft, C.-B.H.; Writing–review & editing, D.-G.L.; Supervision, D.-G.L.; Funding acquisition, D.-G.L. All authors have read and agreed to the published version of the manuscript.

Funding: This work was supported by the Korean government MOTIE (the Ministry of Trade, Industry and Energy), the Korea Evaluation Institute of Industrial Technology (KEIT) (No. 20010047), and the Basic Science Research Program through the National Research Foundation of Korea (NRF) funded by the Ministry of Education (No. RS-2023-00244296).

Data Availability Statement: The original contributions presented in the study are included in the article, further inquiries can be directed to the corresponding author.

Conflicts of Interest: The authors declare no conflicts of interest.

References

1. Kim, M.S.; Kim, C.S. Design of Zr-7Si-xSn Alloys for Biomedical Implant Materials. *J. Korean Soc. Heat Treat.* **2022**, *35*, 8–19. [CrossRef]
2. Lee, K.M.; Park, S.W.; Lim, H.P.; Koh, J.T.; Kang, S.S.; Kim, H.S.; Park, K.B.; Ryoo, G.H.; Lee, K.K.; Lee, D.J. A Recent Research and Development Tendency of Dental Titanium Implant. *Trends Met. Mater. Eng.* **2009**, *22*, 33–40.
3. Geetha, M.; Singh, A.K.; Asokamani, R.; Gogia, A.K. Ti based biomaterials the ultimate choice for orthopedic implants—A review. *Prog. Mater. Sci.* **2009**, *54*, 397–425. [CrossRef]
4. Banerjee, D.; Williams, J.C. Perspectives on Titanium Science and Technology. *Acta Mater.* **2013**, *61*, 844–879. [CrossRef]
5. Lee, T.K.; Lee, S.W.; Kim, I.S.; Moon, Y.H.; Kim, H.S.; Park, C.H. Breaking the limit of Young's modulus in low-cost Ti-Nb-Zr alloy for biomedical implant applications. *J. Alloys Compd.* **2020**, *828*, 154401. [CrossRef]
6. Geetha, M.; Mudali, U.K.; Gogia, A.K.; Sokamani, R.A.; Raj, B. Influence of microstructure and alloying elements on corrosion behavior of Ti-13Nb-13Zr alloy. *Corros. Sci.* **2004**, *46*, 877–892. [CrossRef]
7. Shim, J.D.; Seok, H.K. A State of the Art on Research and Development of Biomedical Titanium Alloys. *Biomater. Res.* **2011**, *15*, 176–183.
8. Nunes, A.R.V.; Borborema, S.; Araújo, L.S.; de Almeida, L.H.; Kaufman, M.J. Production of a Novel Biomedical β-Type Titanium Alloy Ti-23.6Nb-5.1Mo-6.7Zr with Low Young's Modulus. *Metals* **2022**, *12*, 1588. [CrossRef]
9. Kaczmarek, M.; Jurczyk, K.; Koper, J.K.; Paszel-Jaworska, A.; Romaniuk, A.; Lipińska, N.; Żurawski, J.; Urbaniak, P.; Jakubowicz, J.; Jurczyk, M.U. In vitro biocompatibility of anodized titanium with deposited silver nanodendrites. *J. Mater. Sci.* **2016**, *51*, 5259–5270. [CrossRef]
10. Vazirian, S.; Farzadi, A. Dissimilar transient liquid phase bonding of Ti-6Al-4V and Co-Cr-Mo biomaterials using a Cu interlayer: Microstructure and mechanical properties. *J. Alloys Compd.* **2020**, *829*, 154510. [CrossRef]
11. Citlalli, G.T.; Manuel, J.M.J.; Demetrio, N.M.; Erick, M.B.; Maria, L.B.; Alejandro, L.M.M.; Hortensia, R.B.; Ángel, B.Z.M.; Laura, L.R.; Daimir, L.L.L.; et al. Corrosion Behavior of Titanium Alloys (Ti CP2, Ti-6Al-2Sn-4Zr-2Mo, Ti-6Al-4V and Ti Beta-C) with Anodized and Exposed in NaCl and H_2SO_4 Solutions. *Metals* **2024**, *14*, 160. [CrossRef]
12. Amirnejad, M.; Rajabi, M.; Jamaati, R. Importance of Individual Evaluation of Crystallographic Texture and Microstructure Effects on Biocompatibility and Corrosion Performance of Ti6Al4V Alloy. *Met. Mater. Int.* **2023**, *29*, 343–356. [CrossRef]
13. Luo, S.; Jiang, Y.; Thuillier, S.; Castany, P.; Zeng, L. Numerical Analysis on the Spatiotemporal Characteristics of the Portevin–Le Chatelier Effect in Ti-12Mo Alloy. *Met. Mater. Int.* **2023**, *29*, 269–279. [CrossRef]
14. Ozan, S.; Lin, J.; Li, Y.; Ipek, R.; Wen, C. Development of Ti-Nb-Zr alloys with high elastic admissible strain for temporary orthopedic devices. *Acta Biomater.* **2015**, *20*, 176–187. [CrossRef] [PubMed]
15. Miyazaki, S.; Imai, T.; Igo, Y.; Otsuka, K. Effect of Cyclic Deformation on the pseudoelasticity Characteristiecs of Ti-Ni Alloys. *Metall. Trans. A* **1986**, *17*, 115–120. [CrossRef]
16. Chaves, J.M.; Florencio, O.; Silva, P.S.; Marques, P.; Schneider, S. Anelastic relaxation associated to phase transformations and interstitial atoms in the Ti-35Nb-7Zr alloy. *J. Alloys Compd.* **2014**, *616*, 420–425. [CrossRef]
17. Park, H.B. A study on microstructure and corrosion resistance of Ti-Nb alloys by hot rolling. *J. Korean Acad. Dent. Technol.* **2002**, *23*, 223–230.
18. Xiong, C.; Xue, P.; Sun, B.; Li, Y. Effect of annealing temperature on the microstructure and superelasticity of Ti-19Zr-10Nb-1Fe alloy. *Mater. Sci. Eng. A* **2017**, *688*, 464–469. [CrossRef]
19. Yua, S.; Yuc, Z.; Guoa, D.; Zhua, H.; Zhang, M.; Han, J.; Yu, Z.; Cao, Y.; Wang, G. Enhanced bioactivity and interfacial bonding strength of Ti3Zr2Sn3Mo25Nb alloy through graded porosity and surface bioactivation. *J. Mater. Sci.* **2022**, *100*, 137–149. [CrossRef]
20. Kim, H.Y.; Miyazaki, S. Martensitic Transformation and Superelastic Properties of Ti-Nb Base Alloys. *Mater. Trans.* **2015**, *56*, 625–634. [CrossRef]
21. Lee, D.J.; Lee, K.K.; Cho, K.Z.; Yoon, T.R.; Park, H.B. Microstructure and Corrosion Resistance of Ti-15Sn-4Nb Alloy with Hf Adding Element. *J. Korean Acad. Dent. Technol.* **2001**, *23*, 55–64.
22. Han, C.B. Effect of Oxygen Content on Microstructure and Mechanical Properties of Ti-39Nb-6Zr Alloy. Master's Thesis, Sunchon National University, Suncheon, Republic of Korea, 2022.
23. Jang, J.H.; Lee, D.G. Effects of Silicon and Heat-Treatment on Microstructure and Mechanical Properties of Biomedical Ti-39Nb-6Zr Alloy. *Metals* **2021**, *11*, 168. [CrossRef]
24. Hwang, Y.J.; Choi, Y.S.; Hwang, Y.H.; Cho, H.W.; Lee, D.G. Biocompatibility and Biological Corrosion Resistance of Ti–39Nb–6Zr+0.45Al Implant Alloy. *Metals* **2021**, *11*, 2. [CrossRef] [PubMed]
25. Niemeyer, T.C.; Grandini, C.R.; Pinto, L.M.C.; Angelo, A.C.D.; Schneider, S.G. Corrosion behavior of Ti-13Nb-13Zr alloy used as a biomaterial. *J. Alloys Compd.* **2009**, *476*, 172–175. [CrossRef]
26. Sulkowski, B.; Panigrahi, A.; Ozaltin, K.; Lewandowska, M.; Mikułowski, B.; Zehetbauer, M. Evolution of strength and structure during SPD processing of Ti–45Nb alloys: Experiments and simulations. *J. Mater. Sci.* **2014**, *49*, 6648–6655. [CrossRef]
27. Jeong, H.J.; Luu, V.T.; Jeong, Y.H.; Hong, S.T.; Han, H.N. A Study on Uniaxial Tensile Deformation Behavior of Superelastic Titanium Alloy. *Korean J. Met. Mater.* **2020**, *58*, 162–168. [CrossRef]
28. Choi, I.H.; Oh, J.M.; Kim, D.Y.; Kim, J.W.; Kwon, H.J. Comparison of Phase Evolution and Sintering Properties of Ti-10Mo Alloys Prepared using Bulk Scrap and Blended Elemental (BE) Powder. *Korean J. Met. Mater.* **2022**, *60*, 180–187. [CrossRef]

29. Vicente, F.B.; Correa, D.R.N.; Donato, T.A.G.; Chavez, V.E.A.; Buzalaf, M.A.R.; Grandini, C.R. The Influence of Small Quantities of Oxygen in the Structure. Microstructure. Hardness. Elasticity Modulus and Cytocompatibility of Ti-Zr Alloys for Dental Applications. *Materials* **2014**, *7*, 542–553. [CrossRef] [PubMed]
30. Karre, R.; Niranjan, M.K.; Dey, S.R. First principles theoretical investigations of low Young's modulus beta Ti-Nb and Ti-Nb-Zr alloys compositions for biomedical applications. *Mater. Sci. Eng. C* **2015**, *50*, 52–58. [CrossRef]
31. Nie, L.; Zhan, Y.; Hu, T.; Chen, X.; Wang, C. β-type Zr-Nb-Ti biomedical materials with high plasticity and low modulus for hard tissue replacements. *J. Mech. Behav. Biomed. Mater.* **2014**, *29*, 1–6. [CrossRef]
32. Zhao, D.; Chang, K.; Ebel, T.; Nie, H.; Willumeit, R.; Pyczak, F. Sintering behavior and mechanical properties of a metal injection molded Ti–Nb binary alloy as biomaterial. *J. Alloys Compd.* **2015**, *640*, 393–400. [CrossRef]
33. Silva, M.R.D.; Gargarella, P.; Plaine, A.H.; Pauly, S.; Bolfarini, C. Influence of oxygen and plastic deformation on the microstructure and the hardness of a Ti–Nb–Ta–Zr–O Gum Metal. *Mater. Sci. Eng. A* **2021**, *828*, 142122. [CrossRef]
34. Ji, X.; Gutierrez-Urrutia, I.; Emura, S.; Tsuchiya, K. Deformation mechanisms and effect of oxygen addition on mechanical properties of Ti-7.5Mo alloy with α″ martensite. *MATEC Web Conf.* **2020**, *321*, 11059. [CrossRef]
35. Yokota, K.; Bahador, A.; Shitara, K.; Umeda, J.; Kondoh, K. Mechanisms of tensile strengthening and oxygen solid solution in single β-phase Ti-35 at.%Ta+O alloys. *Mater. Sci. Eng. A* **2021**, *802*, 140677. [CrossRef]
36. Ishiguro, Y.; Tsukada, Y.; Koyama, T. Phase-field study of the spinodal decomposition rate of β phase in oxygenadded Ti–Nb alloys. *Comput. Mater. Sci.* **2020**, *174*, 109471. [CrossRef]
37. Lee, Y.S.; Niinomi, M.; Nakai, M.; Narita, K.; Cho, K.; Liu, H. Wear transition of solid-solution-strengthened Ti-29Nb-13Ta-4.6 Zr alloys by interstitial oxygen for biomedical applications. *J. Mech. Behav. Biomed. Mater.* **2015**, *51*, 398–408. [CrossRef] [PubMed]
38. Li, S.; Nam, T. Superelasticity and tensile strength of Ti-Zr-Nb-Sn alloys with high Zr content for biomedical applications. *Intermetallics* **2019**, *112*, 106545. [CrossRef]
39. Tahara, M.; Kim, H.Y.; Inamura, T.; Hosoda, H.; Miyazaki, S. Lattice modulation and superelasticity in oxygen-added β-Ti alloys. *Acta Mater.* **2011**, *59*, 6208–6218. [CrossRef]
40. Hwang, Y.J.; Park, Y.K.; Kim, C.L.; Kim, J.Y.; Lee, D.G. Mechanical Properties Variation of Ti-6Al-4V Alloy by Microstructural Control. *J. Korean Soc. Heat Treat.* **2016**, *29*, 220–226. [CrossRef]
41. Kang, J.Y.; Park, S.; Park, J.Y.; Park, S.J.; Song, Y.H.; Park, S.T.; Kim, G.L.; Oh, K.W. Annealing Textures and Grain Size of Tantalum Sheet. *Korean Soc. Technol. Plast.* **2019**, *28*, 247–256. [CrossRef]

metals

MDPI

Article

Suppression of Inhomogeneous Plastic Deformation in Medium-Carbon Tempered Martensite Steel

Hai Qiu *, Rintaro Ueji and Tadanobu Inoue

Research Center for Structural Materials, National Institute for Materials Science, 1-2-1 Sengen, Tsukuba 305-0047, Ibaraki, Japan
* Correspondence: qiu.hai@nims.go.jp

Abstract: The Lüders phenomenon is one type of inhomogeneous plastic deformation occurring in the elastic-to-plastic transition region, and it is an undesirable plastic deformation behavior. Although conventional measures based on the chemical composition design, plasticity processing principle, or utilization of composited microstructures are used to suppress this phenomenon in engineering, demerits are present, such as high cost and low fracture behavior. The Lüders phenomenon begins with the formation of plastic bands (inhomogeneous yielding) at one or several local sites. If yielding simultaneously occurs everywhere rather than at several local sites, the formation of local plastic bands will be inhibited; as a result, the Lüders deformation will be suppressed. Based on this idea, a new approach was proposed in which the number of local yield sites was increased by heat treatments. A medium-carbon tempered martensite steel (Fe-0.3C-1.5Mn, in wt%) was used to verify the validity of the new approach, and the optimum heat-treatment conditions for the balance of mechanical property and deformation behavior were determined.

Keywords: Lüders deformation; inhomogeneous plastic deformation; medium-carbon steel; tempering; digital image correlation

Citation: Qiu, H.; Ueji, R.; Inoue, T. Suppression of Inhomogeneous Plastic Deformation in Medium-Carbon Tempered Martensite Steel. *Metals* **2024**, *14*, 306. https://doi.org/10.3390/met14030306

Academic Editors: Umberto Prisco, Tomasz Tański, Francesca Borgioli and Denis Benasciutti

Received: 15 February 2024
Revised: 28 February 2024
Accepted: 1 March 2024
Published: 4 March 2024

1. Introduction

Inhomogeneous plastic deformation occurs in some crystalline materials in the form of single or multiple plastic bands. Piobert [1] and Lüders [2] first reported this phenomenon that takes place in the elastic-to-plastic transition region in mild steel. This plastic instability that occurs in the elastic-to-plastic transition region is referred to as Lüders deformation (or Lüders phenomenon). It has a typical characteristic on the tensile curve, a yield plateau [3,4], as shown in Figure 1a. This yield plateau involves the whole Lüders deformation process, which is schematically illustrated in Figure 1b. The whole process is roughly divided into three stages: (Stage I) yielding occurs at one or several sites; (Stage II) one or several plastic bands form at those sites; (Stage III) these plastic bands propagate across the whole specimen. Lüders deformation in a material is dependent on its microstructural factors and working environments. The former involves the microstructure type, microstructural morphology and grain size [5–8], and the latter includes applied stress [9,10], strain rate [11–13], specimen size [13,14], and temperature [15].

The mechanism of the Lüders phenomenon has been widely studied. Cottrell [16,17] first proposed a dislocation model that assumes that the Lüders phenomenon is related to interactions between solute interstitials, such as C and N, and mobile dislocations. The dislocations are initially locked by solute interstitials, which tend to form Cottrell atmospheres around them. The pinning of dislocations is associated with an increase in yield strength (i.e., hardening). When the stress threshold for unlocking or multiplying these dislocations is exceeded, the dislocations are unpinned, and a rapid multiplication of mobile dislocations occurs; as a result, yield strength decreases (i.e., softening). The pinning and unpinning of dislocations result in strain aging [18]. It is generally believed that the Lüders phenomenon is caused by static strain aging (SSA) [19,20]. Onodera [21] and Hahn [22] modified dislocation models in which the dominant mechanism of Lüders band formation is attributed to rapid dislocation multiplication.

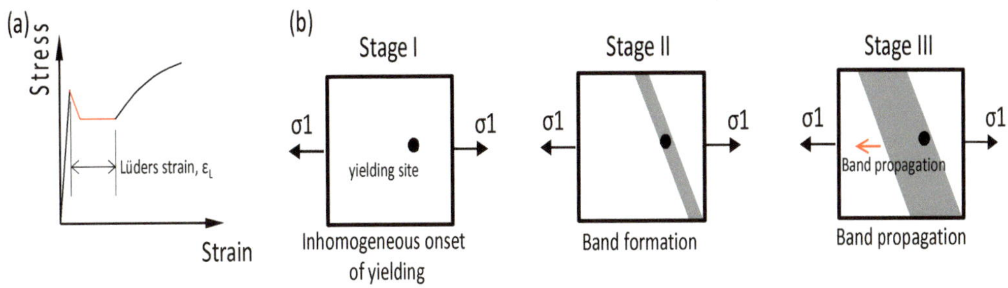

Figure 1. Exhibition of Lüders deformation on the stress–strain curve (**a**), and schematic illustration of the Lüders deformation process (**b**).

In the metal forming process, if the used plates have the Lüders phenomenon, stretcher–strain marks (i.e., plastic bands) will be formed on the surface of the cold-formed products. If the surface damage is severe, it will induce the failure of the forming product. Therefore, the Lüders phenomenon is an undesirable plastic deformation, and many measures have been applied to avoid it. Figure 2 illustrates the principles of two conventional methods widely used in engineering. Method 1 (chemical composition design) is based on Cottrell's dislocation model for plastic band formation. As aforementioned, plastic band formation is related to the interaction between interstitial atoms and dislocations. If the number of interstitial atoms is decreased, the possibility of forming a plastic band will be lowered. Under this guideline, IF (interstitial-free) steels have been developed in which interstitial elements, such as C and N, were controlled to be as low as possible [23]. Although decreasing the amount of C and N can lower the possibility of forming plastic bands, the Lüders phenomenon is still present. Method 2 (plasticity processing principle) originates from the characteristics of Lüders deformation on the stress–strain curve. As shown in Figure 1a, when the material deforms from the elastic state (elastic tension region) to the fully plastic state (work-hardening region), there is a stage in which the Lüders phenomenon occurs with a length of ε_L. If a material is pre-loaded and the plastic pre-strain is large enough, the plastically processed material will directly cross the inhomogeneous plastic region (yield plateau) during the tension process. Plastic pre-strain was introduced over the whole bulk material by using a skin-pass mill, as shown in Figure 2. For a steel with a chemical composition of 0.023C-0.018Si-0.175Mn (in mass%), when the pre-strain exceeds 2.23%, the skin-pass-processed steel has a round stress–strain curve, and the Lüders phenomenon is completely suppressed [24]. However, this method has its own demerit; for example, large plastic pre-strain decreases the resistance to fracture. Moreover, this cold rolling increases the production cost. It is better to avoid this method as far as possible. In the most recent decade, studies were performed on suppressing the Lüders deformation by utilizing composited microstructures [24–30]. Different phases have different mechanical properties. Grain size and microstructural morphology also affect mechanical properties. It is possible to eliminate plastic instability by appropriately combing the mechanical property of each microstructural factor. Those studies are summarized as follows:

(1) Inhomogeneous grain size distribution

A bimodal grain size distribution in an Fe-7Mn-0.05C (wt %) was produced to effectively inhibit the Lüders deformation [25].

(2) Multiple phases

Dual-phase (ferrite/martensite) steel was developed [26].

(3) Macroscopic inhomogeneous structure

A sandwich-like structure in which each layer has different mechanical properties was used [27].

(4) TRIP effect

In a multiphase (metastable austenite/ferrite/martensite) steel, each phase and the transformation-induced plasticity (TRIP) play a specific role in the plastic deformation process [28,29]. The TRIP effect was used to suppress the Lüders deformation [30–35].

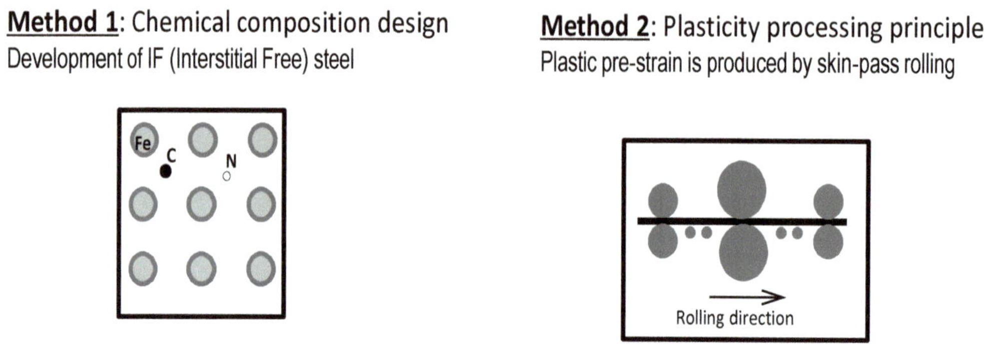

Figure 2. Two conventional engineering methods for suppressing the Lüders phenomenon.

Medium-carbon tempered martensite steel is a common structural steel. Figure 3 shows the stress–strain curve of a medium-carbon tempered martensite steel [36]. It has a good balance of strength and ductility. However, the Lüders phenomenon is present in the steel, and this inhomogeneous plastic deformation limits its application in engineering. This steel is composed of only one phase (martensite) and it has medium carbon content; thus, Method 1 in Figure 2 and the aforementioned composited microstructure method are not suitable for this steel. Because the strength is high, it is difficult to introduce plastic pre-strain by cold skin-rolling (Method 2 in Figure 2). To suppress the Lüders phenomenon in this steel, a new approach is needed.

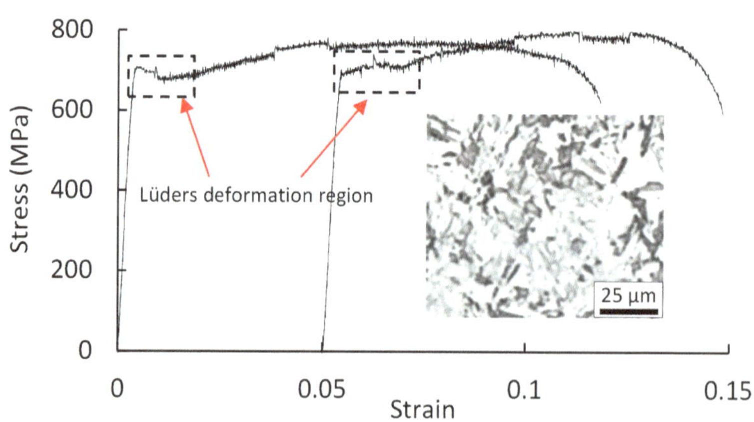

Figure 3. Stress–strain curves of two tension tests of a medium-carbon tempered martensite steel (Fe-0.3C-1.5Mn) (adapted from Ref. [36]).

The steel in Figure 3 was selected as the target material. The objectives of this work are as follows: to suppress the Lüders deformation in the target steel by using a new approach, as shown in Figure 4, and to determine the optimum conditions of heat treatment to balance the mechanical property and deformation behavior.

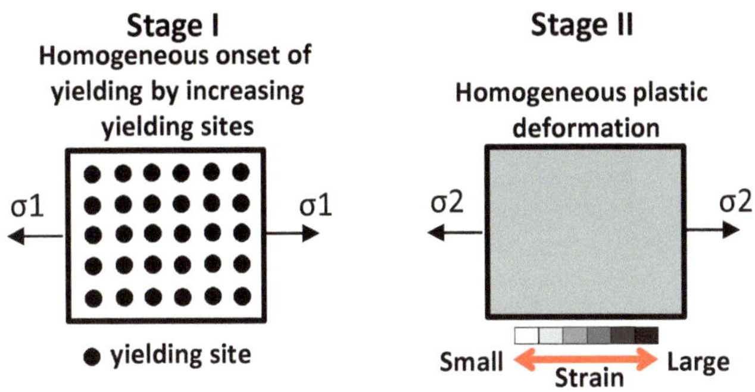

Figure 4. Concept of suppressing inhomogeneous plastic deformation.

2. Experimental Methods

2.1. Approach for Suppressing the Lüders Phenomenon

In the present study, we proposed a new approach for suppressing inhomogeneous plastic deformation. Figure 1b shows that Lüders deformation begins in one or several local yield sites. As illustrated in Figure 4, if sufficient yield sites are provided, local yielding will take place everywhere simultaneously instead of at several local sites; as a result, the formation of local plastic bands will be suppressed, and the material will plastically deform without the Lüders phenomenon. The multiplication of yield sites is suggested to occur by controlling the dislocation density through heat treatment. In this approach, heat treatment has two main roles: ① controlling the initial dislocation density to eliminate the Lüders phenomenon; ② adjusting the mechanical properties and fracture behavior.

2.2. Sample Preparation and Tensile Test

A hot-rolled medium-carbon steel (0.3C-1.5Mn-Fe balance, in mass%) was used as the as-received material. Heat treatments were applied to produce five types of samples (S0~S4), as shown in Figure 5. The as-received steel plates, which were 120 mm long × 40 mm wide × 1.6 mm thick, were heated at 850 °C for 5 min, followed by water quenching. The obtained samples (S0 sample) were used as the start steel for the heat treatments that followed. The S0 samples were reheated at 500 °C and kept for different times, and then water-cooled. The samples, corresponding to keeping times of 30 s, 2 min, and 10 min, were named S1, S2 and S3 samples, respectively. The difference in the heat treatment between the S3 and S4 samples was their cooling: water cooling for S3 and air cooling for S4 samples. A reference heat-treatment process used in Figure 4 is also given in Figure 5 [23]. The microstructures of S0~S4 were examined by electron back-scattered diffraction (EBSD) with a step of 0.1 μm.

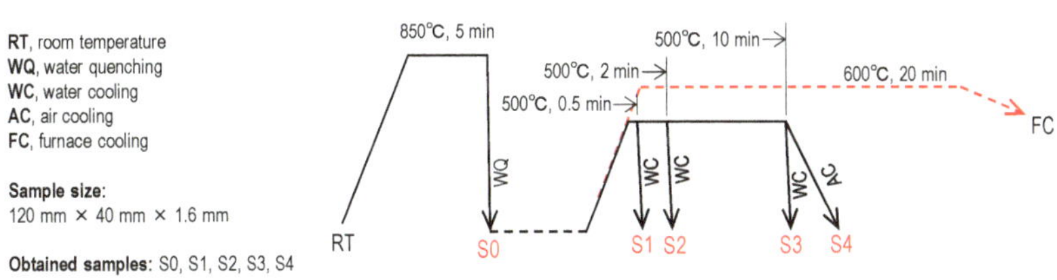

Figure 5. Heat-treatment processes (black lines) and obtained samples (S0~S4). A reference heat treatment (dotted red lines) is given (adapted from Ref. [36]).

Dog-bone-type tensile specimens shown in Figure 6 were machined from the steel plates of S0~S4. The front surfaces of the specimens were sprayed with white and black paint to make speckles for the digital image correlation (DIC) analysis before tensile testing. Tension tests were performed on those specimens at room temperature and at a crosshead speed of 0.01 mm/s. In the tension processes, the deformation processes on the front surfaces of the tensile specimens were recorded successively using a camera at time intervals of 0.5 s. The observed area (30 mm × 8 mm) is shown in Figure 6. Two methods were used to measure the strain during the tension process: (Method 1) An extensometer with a gauge length of 30 mm (equal to the length of the parallel part of the specimen) was attached to the back surface of the tensile specimen. The obtained strain by the extensometer was an average value over the gauge length of 30 mm. (Method 2) The strain and strain-rate field on the front surface of the tensile specimen were measured with the digital image correlation (DIC) technique. The digital images (area: 30 mm × 8 mm) obtained were processed using VIC-2D software (Correlated Solutions, Inc., USA) with a subset size of 9 pixels × 9 pixels (246 μm × 246 μm) and a step of 5 pixels (137 μm) to produce the displacement field, strain field, and strain-rate field. In the DIC operation, the displacement uncertainty is 0.02 pixels.

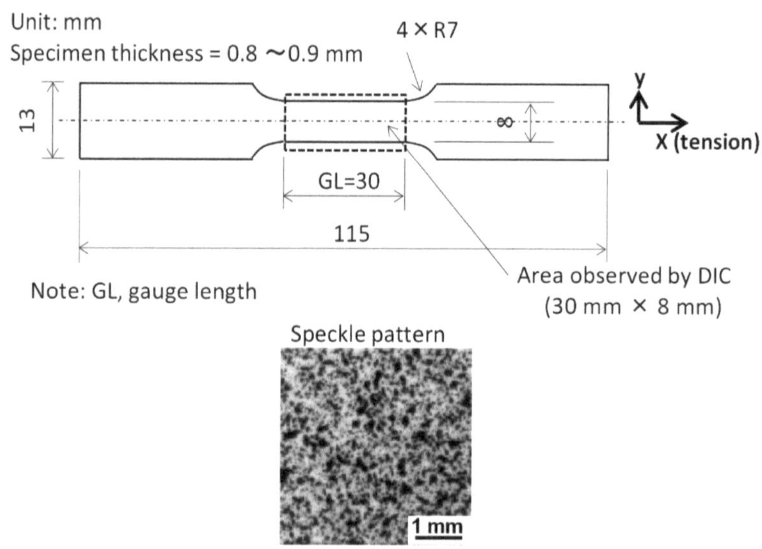

Figure 6. The size of the dog-bone-type specimen and the spackle patterns on the front surface of the specimen. x, tension direction.

Two results are expected from tension tests: (1) stress–strain curves, and (2) the evolution of strain and strain-rate field. The former result indicates the global image of macroscopic deformation behavior, and the latter shows the local deformation from which inhomogeneous plastic deformation, for example, plastic bands, can be identified.

3. Results and Discussion

3.1. Microstructures

The microstructures of the S0~S4 samples were examined by EBSD. Figure 7 shows the image quality (IQ) map and the inverse pole figure (IPF) of the S0 sample. The phase type and preferential phase orientation can be identified from the IQ map and the IPF, respectively. It can be seen that the S0 sample is fully composed of martensite. Phase orientation was measured from the rolling direction (RD) and transverse direction (TD) in Figure 7b, respectively. The color distribution in Figure 7b represents the orientation distribution. Various colors indicate the presence of random orientations. It is clear that there is almost no preferential orientation.

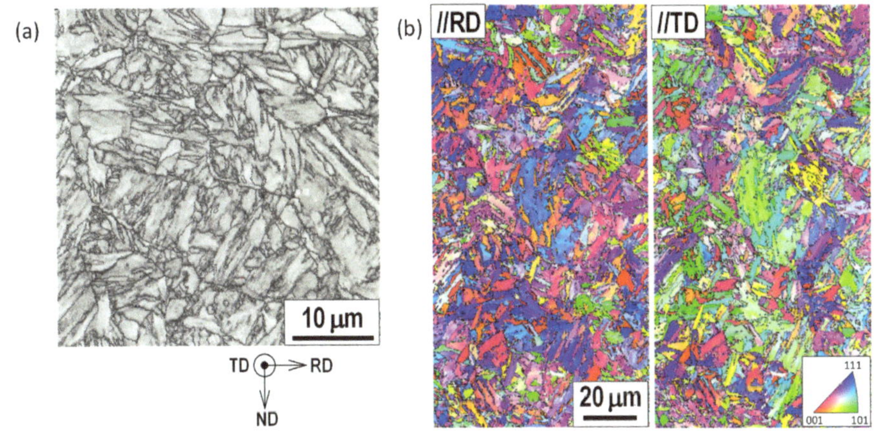

Figure 7. Image quality (IQ) map (**a**) and inverse pole figure (IPF) of the S0 sample (**b**).

S1~S4 samples were obtained by tempering S0 samples at 500 °C. The tempering temperature is much lower than the phase transformation temperature, and the tempering times were short. Therefore, the microstructures of the S0~S4 samples have the same morphology, and precipitates are hardly present.

The Kernel average misorientation (KAM) is often used to qualitatively evaluate the density of the geometrically necessary dislocations (GNDs) [37–39]. The greater the KAM value, the higher the GND density. The KAM map of martensite was measured and is shown in Figure 8. The color bar (0° to 2°) shows the distribution of the KAM values. The average KAM values in S0~S4 samples are plotted in Figure 9. Figure 9 shows that the KAM value (i.e., the dislocation density) decreases from S0 to S4.

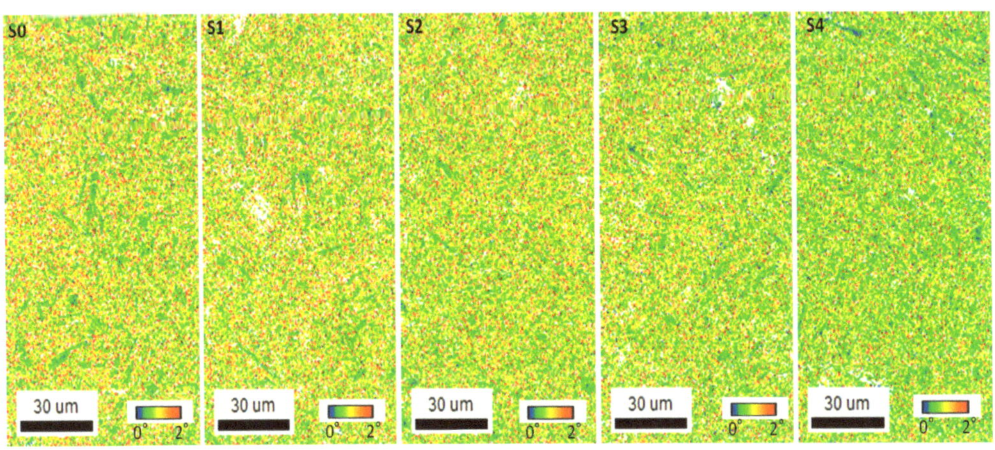

Figure 8. KAM maps of S0, S1, S2, S3, and S4 samples.

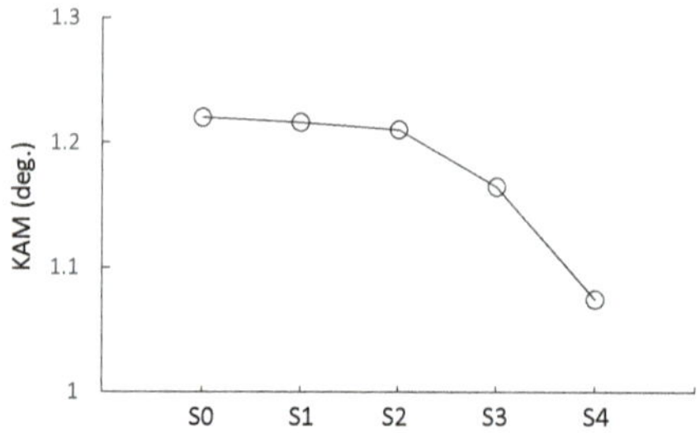

Figure 9. The KAM average values for a 1st neighbor rank estimated by EBSD analysis for various samples (S0~S4).

3.2. Stress–Strain Curves

The response of global deformation to applied force is expressed in terms of the stress–strain curve. The stress–strain curves of the S0~S4 samples are shown in Figure 10a. The strain in Figure 10a was measured with the extensometer attached to the back surface of the tensile specimen, and it is an average value over the length of 30 mm. This average strain reflects the global deformation behavior. By comparing the stress–strain curves of the five samples, the strength was found to decrease, while the fracture strain increased from the S0 to S4 samples. According to the Bailey–Hirsch equation, the strength linearly varies with $\sqrt{\rho}$ (ρ, the dislocation density) [40]. An increase in the dislocation density increases the strength. The trend of strength among the S0~S4 samples agrees with the amount of the initial dislocation density in those samples. The shapes of the five stress–strain curves beyond the elastic deformation region also differ. This indicates that the five samples have different work-hardening abilities. S0 and S1 samples have higher work-hardening abilities while the work-hardening ability is low in the S2~S4 samples, which is attributed to the difference in the dislocation density. To clearly show the region of transition from the elastic state to the plastic state, the stress–strain curves are enlarged around the point of 0.2% proof stress in Figure 10b–f. A clear yield plateau is present in Figure 10f. This indicates that the Lüders deformation occurred in the S4 sample. However, in addition to that, detailed information on the deformation, for example, plastic band formation and propagation, is unavailable from the stress–strain curve. The local deformation behavior in the S0~S4 samples will be investigated by DIC analysis in the following section.

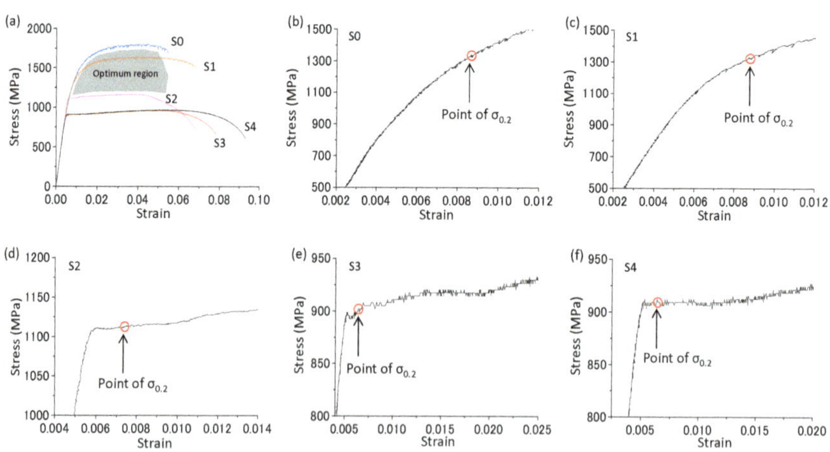

Figure 10. Stress–strain curves of S0~S4 samples (**a**) and the enlargement of those curves around the point of 0.2% proof stress ($\sigma_{0.2}$) for S0 (**b**), S1 (**c**), S2 (**d**), S3 (**e**), and S4 (**f**) samples.

3.3. Local Plastic Deformation

Plastic strain localization is a feature of inhomogeneous plastic deformation. The plastic strain localization in the Lüders deformation is in the form of one plastic band or several plastic bands. The occurrence of the Lüders deformation phenomenon is judged by whether there is a plastic band. Previous studies [3,4,36] showed that the strain-rate field can effectively identify the moving plastic bands. In the present study, DIC analysis was performed to determine the strain-rate fields of the S0~S4 samples.

The inhomogeneous plastic deformation process can be expressed by the evolution of the strain-rate field. In Figure 11, the S4 sample is selected to show the whole process of the Lüders deformation. Nineteen points (①~⑲) that cross the yield plateau were selected. Point ① and point ⑲ are the starting point and end point of the yield plateau, respectively. The strain (ε_x) along the x-direction (tension direction) and its strain rate ($\dot{\varepsilon}_x$) were determined. The strain-rate fields ($\dot{\varepsilon}_x$) over the parallel part of the tensile specimen (cf. Figure 6) at those points are shown in Figure 11. At point ①, the B1 band is formed. The B1 band continues to grow at point ②. At point ③, a branch band (B2) crossing the B1 band forms along a different direction. The B1 band gradually weakens from point ③, while the B2 band gradually grows from point ③ to point ⑤ and then weakens. At point ⑭, both B1 and B2 completely disappear. The positions of B1 and B2 hardly changed. This means that neither propagated.

B3 and B4 bands have formed at point ④. The B3 band grows from points ④~⑥, and then propagates to the left and right (⑦~⑪). The B4 band grows from points ④~⑦, and then degenerates from points ⑧~⑩. At point ⑫, the B5 band forms along another direction at the initial positions of the B3 and B4 bands. At point ⑬, a new band (B6 band) forms at the left side of the B5 band. The B7 band develops at point ⑮, and then propagates to the left side, and finally disappears at point ⑲.

The strain-rate fields of other samples (S0, S1, S2 and S3) were also measured using the DIC technique. In a stress–strain curve, the point of 0.2% proof stress ($\sigma_{0.2}$) is often used to evaluate the yielding ability. The $\sigma_{0.2}$ of the S0~S4 samples was taken as a reference point, and the strain-rate fields of the five samples at $\sigma_{0.2}$ are compared in Figure 12. It can be seen that plastic bands are present in the S2, S3, and S4 samples, while no plastic bands are found in the S0 and S1 samples. The average strain rate over the whole area shown in Figure 12 at $\sigma_{0.2}$ was determined for the five samples and is given in Table 1. The maximum strain rate within the plastic bands is also measured. The maximum value in the plastic bands is one order larger than the average value over the whole area. It is noted that in addition to the point of $\sigma_{0.2}$, plastic bands were not present at any point of the stress–strain curve in the S0 and S1 samples before the onset of specimen necking. Microstructure analysis shows that the S0~S4 samples have the same morphology but different dislocation densities. Figure 9 indicates that the S0 and S1 samples have higher initial dislocation densities. In Figure 8, red points show the sites of high local misorientation. From the S4 sample to S0 sample, the density of red points gradually increases, and their distribution is almost uniform. Those massive red points can easily provide sufficient micro-yielding sites; as a result, plastic bands are suppressed. As shown in Figure 10b–f, the shapes of the stress–strain curves in the elastic-to-plastic transition region are different; the S0 and S1 samples have round curves and the S2~S4 samples have flat curves. This shape variation is dependent on the occurrence of the Lüders phenomenon.

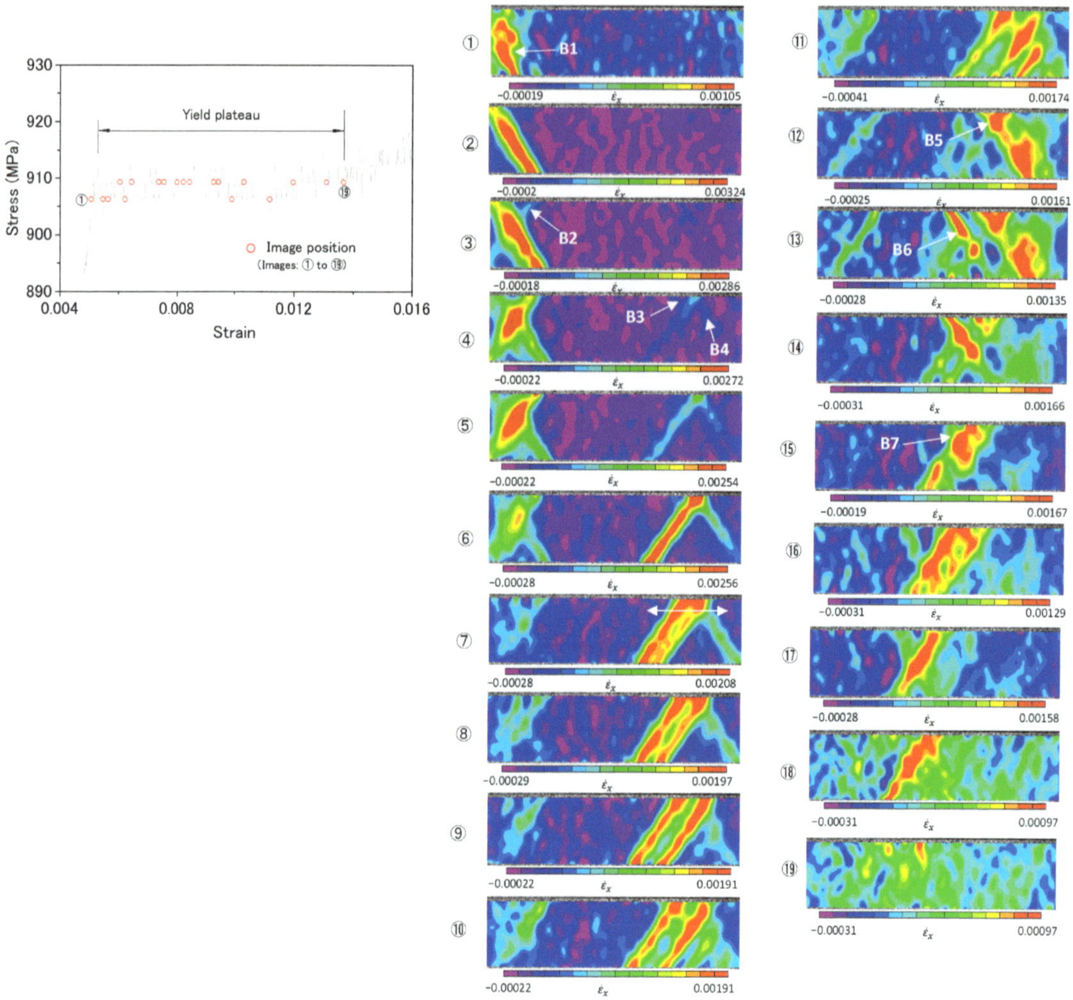

Figure 11. Evolution of plastic bands in the S4 sample from point ① to point ⑲ in terms of the strain-rate field.

Table 1. Strain rates obtained from Figure 12.

Sample No.	Average Strain Rate (s^{-1})	Max. Strain Rate in the Band (s^{-1})
S0	1.02×10^{-4}	—
S1	1.83×10^{-4}	—
S2	2.61×10^{-4}	1.09×10^{-3}
S3	3.21×10^{-4}	2.37×10^{-3}
S4	4.62×10^{-4}	3.33×10^{-3}

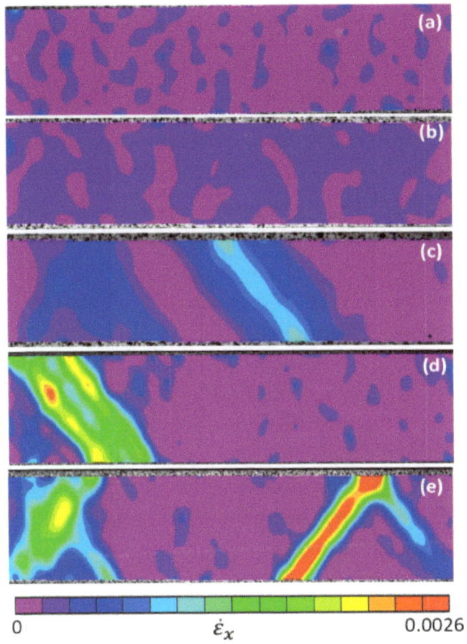

Figure 12. Strain-rate field over the whole area observed by DIC, as shown in Figure 6 at the point of 0.2% proof stress. (**a**) S0 sample, (**b**) S1 sample, (**c**) S2 sample, (**d**) S3 sample, (**e**) S4 sample.

The optimal material behavior for the steel used was judged by a standard: a good balance of strength and ductility as well as no Lüders phenomenon. The shaded region in Figure 10a meets the standard. Apparently, the S1 sample has the best material behavior among the five samples. The heat-treatment conditions corresponding to the shaded region are the optimal ones.

As shown in Figure 1a, the Lüders strain (ε_L) represents the length of the Lüders deformation process. When the Lüders strain is equal to zero, the Lüders deformation phenomenon completely disappears. Tsuchida et al. carried out an extensive study on the Lüders strain under different test conditions [6,11]. These experimental data were obtained from different steels (i.e., different chemical compositions and grain sizes) and under different tensile conditions, such as test temperature and strain rate. Those steels are mainly composed of ferrite. They proposed a model in which the Luders strain, ε_L, is a function of a parameter, the work-hardening rate at the lower yield stress (LYS), $(d\sigma/d\varepsilon)_{LYS}$ (where σ is the true stress, and ε is the true strain).

$$\varepsilon_L\ (\text{in }\%) = \frac{k}{(d\sigma/d\varepsilon)_{LYS}} \tag{1}$$

where constant k is 8000 MPa. Equation (1) is a single parameter model. By using it, the ε_L can be conveniently determined as long as the work-hardening rate is known, irrespective of test condition and microstructure (chemical composition, grain size). In fact, this model delicately integrates the effects of grain size, strain rate, temperature, and carbon content on the Lüders strain. An increase in the work-hardening rate decreases the Lüders strain. A similar conclusion was obtained by Cai et al., that high work-hardening ability favors the inhibition of the Lüders band [30]. When the work-hardening rate is big enough, the Lüders strain approaches zero, i.e., the Lüders deformation phenomenon will disappear. The work-hardening rate is the gradient of a true stress–true strain curve. Although the curves in Figure 10 are nominal stress–strain curves, not true stress–strain curves, their gradients can be used to compare the trend of the work-hardening rate among the five samples. Figure 10b–f can qualitatively examine this inverse relation between the Lüders strain and the work-hardening rate: the work-hardening rates of the S0 and S1 samples in which Lüders deformation phenomenon is not present are much higher than those of the S2~S4 samples.

4. Conclusions

Five types of martensite phases with the same morphology but different initial dislocation densities were produced by tempering. Their plastic deformation behaviors were investigated using the DIC technique. The results obtained regarding the inhomogeneous plastic deformation (i.e., the Lüders deformation) of tempered martensite were as follows:

(1) The amount of micro-yielding sites can be increased by increasing the initial dislocation density. When the initial dislocation density is high enough, the Lüders deformation can be suppressed.
(2) Among the five samples, the S1 sample has the best behavior; it has a good balance of strength and ductility and no Lüders deformation phenomenon. The heat-treatment conditions between S0 and S2 are the optimal region for balancing the mechanical property and deformation behavior.
(3) The Lüders deformation phenomenon is related to the work-hardening rate. Increasing the work-hardening rate is good for suppressing the Lüders deformation.

Author Contributions: Conceptualization, H.Q., R.U. and T.I.; methodology, H.Q.; software, H.Q.; validation, H.Q., R.U. and T.I.; formal analysis, H.Q.; investigation, H.Q.; resources, H.Q. and R.U.; data curation, H.Q.; writing—original draft preparation, H.Q.; writing—review and editing, H.Q., R.U. and T.I.; visualization, H.Q.; supervision, H.Q. All authors have read and agreed to the published version of the manuscript.

Funding: This research was partly funded by the Technical Research Aid of the JFE 21st Century Foundation.

Data Availability Statement: The data presented in this study are available on request from the corresponding author. The data are not publicly available due to legal reasons.

Conflicts of Interest: The authors declare no conflicts of interest.

References

1. Piobert, A. Expériences sur la pénétration des projectiles dans le fer forgé. *Mémoire L'artillerie* **1842**, *5*, 502.
2. Lüders, W. Uber die äeuerung der elasticität an stahlartigen eisenstäben und stahlstäben, und über eine beim biegen solcher stäbe beobachtete molecularbewegung. *Dingler's Polytech. J.* **1860**, *155*, 18–22.
3. Qiu, H.; Inoue, T.; Ueji, R. Experimental measurement of the variables of Lüders deformation in hot-rolled steel via digital image correlation. *Mater. Sci. Eng. A* **2020**, *790*, 139756. [CrossRef]
4. Qiu, H.; Inoue, T.; Ueji, R. In-situ observation of Lüders band formation in hot-rolled steel via digital image correlation. *Metals* **2020**, *10*, 530. [CrossRef]
5. Van Rooyen, G.T. Basic factors which influence the Lüders strain during discontinuous yielding. *Mater. Sci. Eng.* **1971**, *7*, 37–48. [CrossRef]
6. Tsuchida, N.; Masuda, H.; Harada, Y.; Fukaura, K.; Tomota, Y.; Nagai, K. Effect of ferrite grain size on tensile deformation behavior of a ferrite-cementite low carbon steel. *Mater. Sci. Eng. A* **2008**, *488*, 446–452. [CrossRef]
7. Zhang, Y.; Ding, H.; Zhu, H.K.; Zhou, X.D.; Zhang, Y.T.; Jia, N.; Zhang, N.X. Influence of microstructural morphology on the continuous/discontinuous yielding behavior in a medium manganese steel. *Mater. Sci. Eng. A* **2021**, *824*, 141746. [CrossRef]
8. Hu, B.; Ding, F.L.; Tu, X.; Guo, Q.Y.; Ding, C.C.; Wang, Y.; Luo, H.W. Influence of lamellar and equiaxed microstructural morphologies on yielding behavior of a medium Mn steel. *Materialia* **2021**, *20*, 101252. [CrossRef]
9. Lloyd, D.J.; Morris, L.R. Lüders band deformation in a fine grained aluminium alloy. *Acta Metall.* **1977**, *25*, 857–861. [CrossRef]
10. Delwiche, D.E.; Moon, D.W. Strain profile of a propagating Lüders front. *Mater. Sci. Eng.* **1972**, *9*, 347–354. [CrossRef]
11. Tsuchida, N.; Tomota, Y.; Nagai, K.; Fukaura, K. A simple relationship between Lüders elongation and work-hardening rate at lower yield stress. *Scr. Mater.* **2006**, *54*, 57–60. [CrossRef]
12. El-Magd, E.; Scholles, H.; Weisshaupt, H. Influence of strain rate on the stress–strain curve in the range of Lüders strain. *Steel Res.* **1996**, *67*, 495–500. [CrossRef]
13. VanRooyen, G.T. The stress and strain distribution in a propagation Lüders front accompanying the yield-point phenomenon in iron. *Mater. Sci. Eng.* **1968**, *3*, 105–117. [CrossRef]
14. Cai, Y.L.; Yang, S.L.; Fu, S.H.; Zhang, Q.C. The influence of specimen thickness on the Lüders effect of a 5456 Al-based alloy: Experimental observations. *Metals* **2016**, *6*, 120. [CrossRef]

15. Zhang, M.H.; Li, L.F.; Ding, J.; Wu, Q.B.; Wang, Y.D.; Almer, J.; Guo, F.M.; Ren, Y. Temperature-dependent micromechanical behavior of medium-Mn transformation-induced-plasticity steel studied by in situ synchrotron X-ray diffraction. *Acta Mater.* **2017**, *141*, 294–303. [CrossRef]
16. Cottrell, A.H.; Bilby, B.A. Dislocation theory of yielding and strain ageing of iron. *Phys. Soc.* **1949**, *62*, 49–62. [CrossRef]
17. Cottrell, A.H. A note on the Portevin-Le Chatelier effect. *Lond. Edinb. Dublin Philos. Mag. J. Sci.* **1953**, *44*, 829–832. [CrossRef]
18. Marais, A.; Mazière, M.; Forest, S.; Parrot, A.; Le Delliou, P. Identification of a strain-aging model accounting for Lüders behavior in a C-Mn steel. *Philos. Mag.* **2012**, *92*, 3589–3617. [CrossRef]
19. Wang, X.G.; Wang, L.; Huang, M.X. Kinematic and thermal characteristics of Lüders and Portevin-Le Chatelier bands in a medium Mn transformation-induced plasticity steel. *Acta Mater.* **2017**, *124*, 17–29. [CrossRef]
20. Tsukahara, H.; Iung, T. Finite element simulation of the Piobert–Lüders behavior in an uniaxial tensile test. *Mater. Sci. Eng. A* **1998**, *248*, 304–308. [CrossRef]
21. Onodera, R.; Nonomura, M.; Aramaki, M. Stress drop, Lüders strain and strain rate during serrated flow. *J. Jpn. Inst. Met.* **2000**, *64*, 1162–1171. [CrossRef]
22. Hahn, G.T. A model for yielding with special reference to the yield-point phenomena of iron and related BCC metals. *Acta Metall.* **1962**, *10*, 727–738. [CrossRef]
23. Ikeda, S.; Makii, K.; Shindo, Y.; Miura, M.; Hashimoto, S. The effect of precipitates on the plastic deformation behavior in high tensile strength IF sheet steels. *Res. Dev. Kobe Steel Eng. Rep.* **2002**, *52*, 39–42.
24. Grassino, J.; Vedani, M.; Vimercati, G.; Zanella, G. Effects of skin pass rolling parameters on mechanical properties of steels. *Int. J. Prec. Eng. Manufac.* **2012**, *13*, 2017–2026. [CrossRef]
25. Han, J.; Kang, S.-H.; Lee, S.-J.; Lee, Y.-K. Fabrication of bimodal-grained Al-free medium Mn steel by double intercritical annealing and its tensile properties. *J. Alloys Compd.* **2016**, *681*, 580–588. [CrossRef]
26. Mohsenzadeh, M.S.; Mazinani, M. On the yield point phenomenon in low-carbon steels with ferrite-cementite microstructure. *Mater. Sci. Eng. A* **2016**, *673*, 193–203. [CrossRef]
27. Liang, Z.Y.; Cao, Z.H.; Lu, J.; Huang, M.X.; Tasan, C.C. Influence of co-existing medium Mn and dual phase steel microstructures on ductility and Lüders band formation. *Acta Mater.* **2021**, *221*, 117418. [CrossRef]
28. Mao, W.Q.; Gao, S.; Gong, W.; Harjo, S.; Kawasaki, T.; Tsuji, N. Quantitatively evaluating the huge Lüders band deformation in an ultrafine grain stainless steel by combining in situ neutron diffraction and digital image correlation analysis. *Scripta Mater.* **2023**, *235*, 115642. [CrossRef]
29. Yin, W.J.; Briffod, F.; Hu, H.Y.; Yamazaki, K.; Shiraiwa, T.; Enoki, M. Quantitative investigation of strain partitioning and failure mechanism in ultrafine grained medium Mn steel through high resolution digital image correlation. *Scripta Mater.* **2023**, *229*, 115386. [CrossRef]
30. Cai, Z.H.; Jing, S.Y.; Li, H.Y.; Zhang, K.M.; Misra, R.D.K.; Ding, H.; Tang, Z.Y. The influence of microstructural characteristics on yield point elongation phenomenon in Fe-0.2C-11Mn-2Al steel. *Mater. Sci. Eng. A* **2019**, *739*, 17–25. [CrossRef]
31. Zhan, Z.D.; Shi, Z.G.; Wang, Z.M.; Lu, W.J.; Chen, Z.N.; Zhang, D.; Chai, F.; Luo, X.B. Effect of manganese on the strength–toughness relationship of low-carbon copper and nickel-containing hull steel. *Materials* **2024**, *17*, 1012. [CrossRef]
32. Sugimoto, K.; Shioiri, S.; Kobayashi, J. Effects of mean normal stress on strain-hardening, strain-induced martensite transformation, and void-formation behaviors in high-strength TRIP-aided steels. *Metals* **2024**, *14*, 61. [CrossRef]
33. Benzing, J.T.; Luecke, W.E.; Mates, S.P.; Ponge, D.; Raabe, D.; Wittig, J.E. Intercritical annealing to achieve a positive strain-rate sensitivity of mechanical properties and suppression of macroscopic plastic instabilities in multi-phase medium-Mn steels. *Mater. Sci. Eng. A* **2021**, *803*, 140469. [CrossRef]
34. Liu, R.D.; Hu, Z.P.; Lin, C.Q.; Yang, D.P.; Gu, X.L.; Xu, X.; Guo, J.Y. A novel design to eliminate Lüders band in medium-Mn steel and its microstructure-property relationship. *Crystals* **2023**, *13*, 936. [CrossRef]
35. Ma, J.W.; Lu, Q.; Sun, L.; Shen, Y. Two-step intercritical annealing to eliminate Lüders band in a strong and ductile medium Mn Steel. *Metall. Mater. Trans. A* **2018**, *49*, 4404–4408. [CrossRef]
36. Qiu, H.; Ueji, R.; Inoue, T.; Kimura, Y. Plastic instability in medium-carbon tempered martensite steel. *Materials* **2021**, *14*, 4609. [CrossRef]
37. Zhang, J.; Huang, M.X.; Sun, B.H.; Zhang, B.N.; Ding, R.; Luo, C.; Zeng, W.; Zhang, C.; Yang, Z.G.; van der Zwaag, S.; et al. Critical role of Lüders banding in hydrogen embrittlement susceptibility of medium Mn steels. *Scr. Mater.* **2021**, *190*, 32–37. [CrossRef]
38. Ozerov, M.; Sokolovsky, V.; Yurchenko, N.; Astakhov, I.; Povolyaeva, E.; Plekhov, O.; Tagirov, D.; Stepanov, N.; Zherebtsov, S. Effect of cold rolling on microstructure and mechanical properties of a cast TiNbZr-based composite reinforced with borides. *Metals* **2024**, *14*, 104. [CrossRef]
39. Yi, Y.X.; Yin, F.; Zhai, J.J.; Liu, Y.X. Microstructure evolution and numerical modeling of TC4 titanium alloy during ultrasonic shot peening process. *Metals* **2024**, *14*, 275. [CrossRef]
40. Kato, M. *Introduction to the Theory of Dislocations*; Shokabo: Tokyo, Japan, 2003; p. 120.

metals

MDPI

Article

Effect of Nanostructuring on Operational Properties of 316LVM Steel

Olga Rybalchenko [1], Natalia Anisimova [1,2,3], Natalia Martynenko [1], Georgy Rybalchenko [4], Alexey Tokar [1,3], Elena Lukyanova [1], Dmitry Prosvirnin [1], Mikhail Gorshenkov [3], Mikhail Kiselevskiy [2,3] and Sergey Dobatkin [1,3,*]

1 A.A. Baikov Institute of Metallurgy and Materials Science, Russian Academy of Sciences, 119334 Moscow, Russia; orybalchenko@imet.ac.ru (O.R.); n.u.anisimova@gmail.com (N.A.); nmartynenko@imet.ac.ru (N.M.); tokar.aa@misis.ru (A.T.); elukyanova@imet.ac.ru (E.L.)

2 N.N. Blokhin National Medical Research Center of Oncology (N.N. Blokhin NMRCO), Ministry of Health of the Russian Federation, 115478 Moscow, Russia

3 Institute of Advanced Materials and Nanotechnologies, National University of Science and Technology MISIS, 119049 Moscow, Russia; mvg@misis.ru

4 P.N. Lebedev Physical Institute, Russian Academy of Sciences, 119991 Moscow, Russia; rybalchenkogv@lebedev.ru

* Correspondence: sdobatkin@imet.ac.ru

Abstract: In this study, high-pressure torsion (HPT) was used to process austenitic 316LVM stainless steel at 20 °C and 400 °C. The effects of HPT on the microstructure, mechanical, and functional properties of the steel were investigated. By applying both HPT modes on the 316LVM steel, a nanocrystalline state with an average size of the structural elements of ~46–50 nm was achieved. The density of the dislocations and twins present in the austenite phase varied depending on the specific HPT conditions. Despite achieving a similar structural state after HPT, the deformation temperatures used has different effects on the mechanical and functional properties of the steel. After HPT at 20 °C, the yield strength of the 316L steel increased by more than nine times up to 1890 MPa, and the fatigue limit by more than two times up to 550 MPa, when compared to its coarse-grained counter-parts. After HPT at 20 °C, the 316LVM steel exhibited better ductility, higher low-cycle fatigue resistance, greater resistance to corrosion, and improved in vitro biocompatibility compared to processing at 400 °C. The reasons for the deterioration of the properties after HPT at 400 °C are discussed in the article.

Keywords: stainless steel; high-pressure torsion; ultrafine-grained microstructure; twinning; mechanical properties; corrosion resistance; biocompatibility in vitro

Citation: Rybalchenko, O.; Anisimova, N.; Martynenko, N.; Rybalchenko, G.; Tokar, A.; Lukyanova, E.; Prosvirnin, D.; Gorshenkov, M.; Kiselevskiy, M.; Dobatkin, S. Effect of Nanostructuring on Operational Properties of 316LVM Steel. *Metals* **2023**, *13*, 1951. https://doi.org/10.3390/met13121951

Academic Editors: Francesca Borgioli, Umberto Prisco, Tomasz Tański and Denis Benasciutti

Received: 1 November 2023
Revised: 22 November 2023
Accepted: 22 November 2023
Published: 28 November 2023

1. Introduction

Austenitic stainless steels (SSs) are widely used in various industrial applications and in the medical field due to their excellent combination of functional properties, including high strength, ductility, corrosion resistance, biocompatibility, and good formability [1]. These materials offer a versatile solution for a range of applications, providing both mechanical and chemical durability. At the same time, they meet the requirements for medical devices and implants [2]. Due to its low carbon content and molybdenum alloying, austenitic 316L stainless steel (SS) exhibits high resistance to various forms of corrosion. Additionally, its high nickel content stabilizes the austenitic phase, resulting in non-magnetic behavior. These properties, combined with its high processability, biocompatibility, and cost effectiveness, make 316L SS an ideal material for the production of internal fixation devices used in orthopedic surgery, as well as for the production of stents used in cardiovascular procedures [3]. The versatile nature of 316L SS allows for its wide application range in both industrial and medical fields. However, SS materials do have some limitations. For example, they have a low yield strength, which significantly restricts their application range. Additionally, in the field of biomedical usage, one of the lowest fatigue strengths of 316L SS among the alloys used in orthopedics is a disadvantage, since fatigue damage is a common cause of implant failure [4].

It should be noted that stainless steel is predominantly used for manufacturing implantable devices for osteosynthesis. Some of them have increased strength requirements. Thus, for skeletal pins made of deformable stainless steel with dimensions from 1 to 2.8 mm, the tensile strength should be at least 1240 MPa with a minimum elongation of about 3% [5]. Since surgical grade 316L SS is a stable material that does not undergo phase transformations during heat treatment and deformation, grain refinement is one of the few strategies used to increase the strength [6]. Among the effective methods of grain refinement is severe plastic deformation (SPD) [7,8]. Studies [9–13] have demonstrated improvements in the mechanical and operational properties, including fatigue strength, by obtaining an ultrafine-grained (UFG) structure in austenitic corrosion-resistant steels through equal channel angular pressing (ECAP) method. Another promising method that allows the ultrafine structure refinement without steel sample destruction is high-pressure torsion (HPT) [8,14–18]. This method has already been tested on austenitic stainless steels, including 316L SS [19–31].

All studies confirmed the achievement of a nanocrystalline state in 316L SS after HPT. Thus, in [21], a grain refinement up to 50 nm was noted during high-pressure torsion at 10 revolutions. Researchers [21] have also highlighted the occurrence of a martensitic $\gamma \rightarrow \varepsilon \rightarrow \alpha'$ transformation in the steel structure during HPT, which has resulted in a significant increase in hardness up to 6000 MPa. Furthermore, it has been noted that hardness is more sensitive to changes in grain size rather than changes in phase composition. On the contrary, in the study of the deformation mechanisms through HPT in 316L steel over a wide temperature range from −196 °C up to 720 °C, Scheriau et al. [22] determined that only at deformation temperatures below room temperature does the deformation-induced γ (fcc) $\rightarrow \varepsilon$ (hcp) transformation occur. At deformation temperatures above 450 °C, the dominant mechanism was dislocation slip, while in the range below 450 °C to room temperature, deformation twinning in the austenite phase was observed. At the same time, the authors in [22] noted a refinement of the average grain size to 100 nm after HPT at room temperature. Moreover, in many studies, authors have noted the increase in hardness both after the post-deformation annealing of 316L SS [23–26] and during deformation through HPT at elevated temperatures [26].

Wang et al. [23] investigated nanocrystalline plates of 316L stainless steel in the annealed austenitic state obtained through HPT at 500 °C for one hour using TEM analysis and specific electrical resistivity measurements. Thus, the study [23] concluded that the strengthening of 316L SS occurred through a combination of nanostructuring during HPT and the precipitation of segregations during annealing. The authors in their work made a comparative analysis of the mechanical characteristics of steel 316L. At that time, their results were the most impressive. The tensile strength of the samples subjected to HPT was 1790 ± 50 MPa and further increased with increasing annealing temperature to 500 °C, reaching a maximum value of 2390 ± 50 MPa.

In further research, Renk et al. [24,25] confirmed the precipitation of segregations in samples of 316L SS with a diameter of 35 mm and a height of 11 mm. The samples were nanostructured using the quasi-constrained high-pressure torsion method under a pressure of 3.60 GPa for 15 rotations. In the study [23], the authors demonstrated a 20% increase in the hardness of nanostructured 316L SS after annealing at 823 K for 30 min. After comparing the results of mechanical testing and atom probe tomography (APT) data, it was concluded that the increase in the hardness of 316L SS was not caused by segregations of dissolved elements. However, the particles of dissolved elements or second phases stabilize the nanocrystalline structure. Therefore, they allow for annealing processes such as annihilation and relaxation, which are necessary for strengthening phenomena during annealing. These particles play a role in maintaining the stability and enhancing the mechanical properties of the nanocrystalline material. In a subsequent study [25], the same authors attempted to improve the fatigue limit of 316L SS by annealing it at 823 K for 90 min, since this treatment resulted in the greatest increase in strengthening. They noted that the strain strength of 316L SS after 15 turns of quasi-constrained high-pressure torsion was 1865 MPa, which increased to 2226 MPa after annealing. During the experiment, the fatigue limit increased up to 1 GPa. The authors concluded that the reason for the increase was due to defect relaxation, which shifted the onset of plasticity to higher stress levels, combined with the stabilization of nanograin boundaries through segregations.

In the study [26], 316 steel was nanostructured using the HPT method at room temperature and at 400 °C. As a result, after applying HPT to the austenitic steel at different temperatures, an ultrafine-grained structure was obtained. This was characterized by different dislocation and twinning densities, different grain sizes of 40 and 90 nm at room temperature and at 400 °C, respectively, while demonstrating a similar level of increased strength of approximately 1700 MPa. Furthermore, the use of atom probe tomography (APT) revealed that HPT at elevated temperatures resulted in the formation of Mo–Cr–Si segregations in 316 stainless steel. Taking into account the various contributions to the material strengthening, the authors demonstrated that segregations could lead to a significant increase in strength. By impeding the dislocation glide, segregations also enhance the yield strength. Since the initiation of fatigue cracks is a result of dislocation glide, HPT at 400 °C is expected to result in a significant increase in the fatigue limit.

The aim of this study was to investigate the effect of HPT nanostructuring at two temperatures, room temperature and 400 °C, on the operational characteristics of surgical grade 316LVM stainless steel, including fatigue strength, corrosion resistance, and in vitro biocompatibility. It is expected that the high specific strength of 316LVM stainless steel after HPT will make it possible to decrease the size of implants and, as a result, reduce the toxic effect of the implanted device on the human body The promising results in [26] raised hopes for obtaining not only a high specific strength of surgical grade steel after HPT at 400 °C but also fatigue strength while maintaining or even improving corrosion resistance and in vitro biocompatibility.

2. Materials and Methods

The study utilized an austenitic 316LVM stainless steel with low carbon content, which is commonly used for implant manufacturing. The nominal composition of the 316LVM steel is presented in Table 1. The material was initially in the as-received state and was solution treated at 1050 °C for 1 h, followed by being quenched in water (solution heat treatment). High-pressure torsion (HPT) deformation was conducted using a Bridgman-type setup (JEOl, Tokyo, Japan) on samples with a diameter of 20 mm and a thickness of 1.5 mm, applying a pressure of 6 GPa. HPT was performed using two isothermal deformation modes: at temperatures of 20 °C and 400 °C. For all deformation modes, the samples were subjected to 10 rotations, which corresponds to a true strain of approximately 5.7 at the center of the sample radius.

Table 1. Chemical composition of the 316LVM stainless steel.

Elements	Amount (wt.%)								
	C	Cr	Ni	Cu	Si	Mn	Mo	N	S, P	Fe
316LVM	0.012	17.3	14.11	0.07	0.24	1.77	2.75	0.07	0.001/0.024	balance

TEM analysis of the structure was conducted using a JEM-200CX microscope (JEOl, Tokyo, Japan), while the X-ray diffraction (XRD) analysis was performed using a DRON 4.07 diffractometer in the 2θ angle range from 10 to 130° at a scanning step value 0.1° with CoKα radiation (wave length: λ = 1.7902 Å) at rated operation X-ray sources of 60 kV and focus size of 1.0 mm × 10 mm.

For the foils, samples were taken from the middle of the sample radius. Microhardness measurements were conducted using a 402 MVD Wolpert Wilson instrument with a 1 N load. Mechanical properties were determined using an "INSTRON 3382" machine (INSTRON, Norwood, MA, USA) on flat samples that were 1.0 mm thick, 1.0 mm wide, and had a working length of 2.0 mm (Figure 1), in accordance with [32]. High-cycle fatigue tests were performed under repeated tension conditions on an INSTRON 8801 servo-hydraulic machine with a maximum load capacity of 100 kN, a testing frequency of 40 Hz, and a stress ratio of R = 0.1 on flat samples 1.0 mm thick, 1.0 mm wide, and with a length of the working part of 1.6 mm.

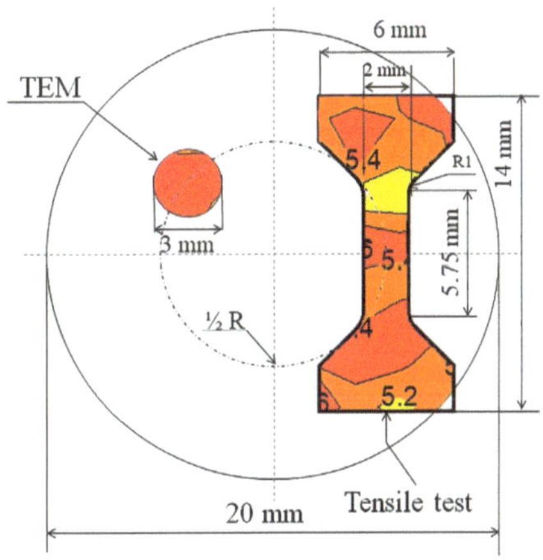

Figure 1. Schematic representation showing the locations of all investigation and test samples on the HPT disk. Red, orange, and yellow colors indicate different levels of microhardness. Thus, it is noted that we used the distribution of microhardness values over the surface for choosing the place to cut the samples for testing. The sample was cut from the area with uniform deformation.

Corrosion potential, corrosion current density, and corrosion rate were evaluated from the results of electrochemical tests by the potentiodynamic polarization (PDP) method using an SP-300 potentiostat (Bio-Logic SAS; Seyssinet-Pariset, France). They were obtained using EC-Lab software (BioLogic, Seyssinet-Pariset, France), in accordance with [33]. Corrosion studies were carried out in physiological solution at room temperature. A flat cell with a "three-electrode configuration" with the test sample as the working electrode was used. The Ag/AgCl electrode (with 3 M NaCl electrolyte inside) and platinum grid were the reference and counter electrodes, respectively. The area exposed to the solution was about 1 cm^2. Scanning was performed at a rate of 1 mV/s in the scanning range: −1.50 V vs. E_{OC}; +0.700 V, where E_{OC} is the open circuit potential. The time for determining the open circuit potential was 60 min.

To assess the impact of HPT processing on the biocompatibility of 316LVM SS, the level of hemolysis and cytotoxicity relative to non-transformed immunocompetent blood cells was evaluated. For the evaluation of hemolytic activity, the 316LVM steel samples were incubated in 2 mL of Hank's solution containing 14.8×10^6 human erythrocytes at 37 °C in an atmosphere with 5% carbon dioxide. The incubation was carried out for 2, 4, and 24 h. The level of hemolysis was determined according to a previously described method [34].

To study cytotoxicity, the steel samples were incubated in 2 mL of complete growth medium containing 540,000 mononuclear leukocytes isolated from the venous blood of a healthy donor using a Ficoll gradient. The incubation was performed at 37 °C in an atmosphere containing 5% carbon dioxide for 24 h. In the control group, cells were incubated in complete growth medium under the same conditions. The influence of the steel on tumor cell viability was assessed by calculating the ratio of cell activity on the surface of the samples to the control in the MTT assay, as described previously [35].

For cell colonization experiments, a human MMSC culture from the cell culture collection of the N.N. Blokhin National Medical Research Center of Oncology consisting of 20,000 cells was applied to the surface of the alloy samples after forging and after HPT, with a volume of 15 μL. The samples were incubated for 30 min at 37 °C in an atmosphere with 5% carbon dioxide. After incubation, 1 mL of complete growth medium was added to the samples containing cells, and they were then incubated for 21 days under the same conditions. In the control group, cells were incubated on the bottom of the well plate under the same conditions. The growth medium was changed every 2 days.

Cell colonization was studied using fluorescence microscopy with a LionHeart LX digital microscope (Perkin Elmer, Waltham, MA, USA) after staining the cells with Calcein AM (Sigma, St. Louis, MO, USA) following the manufacturer's instructions. At the end of the experiment, the alloy samples were removed from the well plate in which incubation took place and treated with 1 mL of trypsin. In the control group, the growth medium was removed and the cells were treated with trypsin. Cell viability was assessed using The Muse Count & Viability Kit (Muse, Thermo Scientific, Waltham, MA, USA) and the Muse cell analyzer (Millipore, Darmstadt, Germany). To study the expression of adhesion molecules by the cells, they were stained with antibodies against CD44 (BD Bioscience, San Diego, CA 92121, USA) and CD11b (BD Bioscience, San Diego, CA, USA), and the number of CD44(+), CD11b(+), and CD44(+)CD11b(+) cells was determined using a NovoCyte flow cytometer (ACEA Bioscience Inc., San Diego, CA 92121, USA) based on the analysis of at least 10,000 cells in the sample.

The results of the research were presented as the mean value and standard deviation (mean ± SD). In order to perform comparative analysis with the control groups, a *t*-test was used. Differences were considered statistically significant at $p < 0.05$. The experiments and procedures with cells and animals were assessed and approved by the Local Ethics Committee of "N.N. Blokhin National Medical Research Center of Oncology" of the Health Ministry of Russia (#8-03, 1 March 2023).

3. Results

3.1. Microstructural Characterization of the 316LVM Stainless Steel in the Initial State and after HPT

Figure 2 shows the initial structure of the 316LVM stainless steel in the forged state (Figure 2a) and after solution heat treatment (Figure 2b).

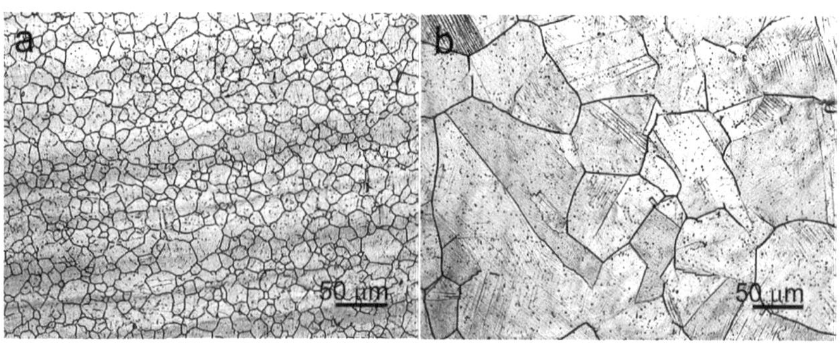

Figure 2. Optical micrographs of the 316LVM stainless steel after forging (**a**) and solution heat treatment (**b**).

After HPT at room temperature and at 400 °C, a nanocrystalline structure with predominantly high-angle grain boundaries is formed in the 316LVM steel (Figure 3). The presence of a grain structure was determined by the circular electron diffraction pattern with point reflections. HPT at room temperature resulted in a microstructure, with an average size of the structural elements of about 46 ± 1.8 nm and deformation twins in the austenitic matrix (Figure 3a,b). With an increase in the HPT temperature up to 400 °C, the character of the forming structure remains the same, and the average size of the structural elements increases slightly to 50 ± 1.6 nm (Figure 3c,d). Moreover, the density of twins after HPT at 400 °C, according to the TEM analysis (Figure 3b,d), is four times higher than at 20 °C. Deformation twinning typically occurs when slip is hindered for various reasons.

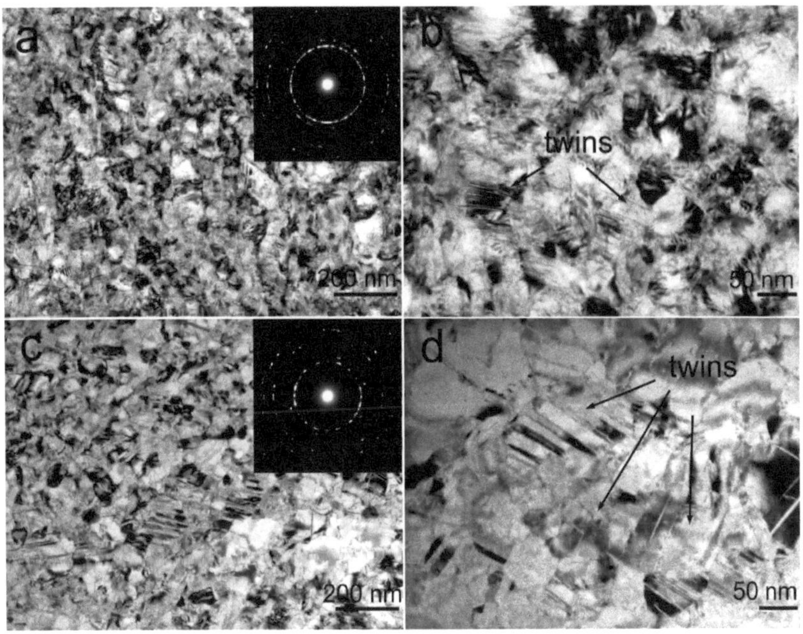

Figure 3. TEM images of the 316LVM stainless steel after HPT at 20 °C (**a**,**b**) and 400 °C, (**c**,**d**) with SAED pattern seen in the inset of (**a**,**c**).

3.2. X-ray Analysis of the 316LVM Stainless Steel after HPT

The X-ray structural quantitative phase analysis revealed the presence of a fully austenitic structure in the steel, deformed under all processing regimes (Figure 4, Table 2). By analyzing the width of the X-ray lines of the austenite (111) and (222) reflections, the parameters of the fine crystalline structure of the 316LVM stainless steel after HPT were determined (Table 3). This included the crystallite size (d) and the root mean square microstrain, which further allowed for the calculation of the dislocation density (Table 3). The line profiles were fitted using Gaussian functions. The X-ray structural analysis did not reveal any significant differences in the sizes of the CSRs (30–40 nm) and showed a twofold decrease in the dislocation density after increasing the deformation temperature from room temperature to 400 °C. The values of microstrains decrease with an increase in the HPT temperature. The highest value of microstrain was obtained after HPT at room temperature. The decrease in the CSR values corresponds to a reduction in the grain size observed through the electron microscopy analysis. However, there is only a slight difference between the average size of the structural elements determined by electron microscopy and the CSR size determined by the X-ray structural analysis. This discrepancy may be attributed to the fact that X-ray structural analysis takes into account subgrain structures that are not always detectable through electron microscopy [36].

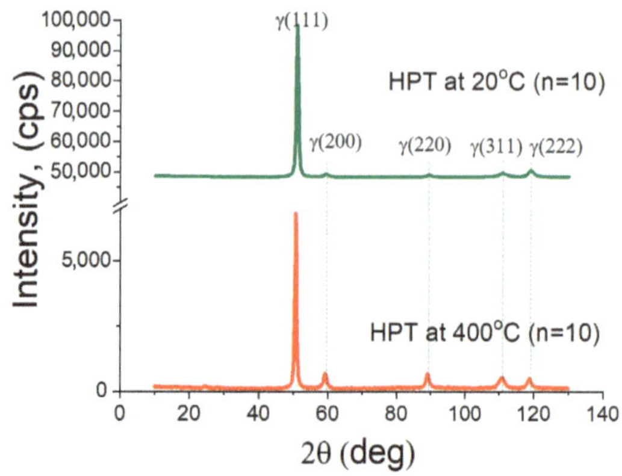

Figure 4. XRD patterns of the 316LVM stainless steel after HPT (n—number of revolutions).

Table 2. Results of the X-ray phase analysis of the 316LVM stainless steel after HPT.

Processing	Space Group	Phase	a, Å	Content, wt.%
HPT, at 20 °C	225: Fm-3m	γ	3.596	100.0 ± 0.0
HPT, at 400 °C	225: Fm-3m	γ	3.599	100.0 ± 0.0

Table 3. Crystallite size, microstrain, and dislocation density of the 316LVM stainless steel after HPT determined by X-ray line profile analysis.

Processing	β_{111}, °	β_{222}, °	d, nm [1]	ε, % [2]	ρ, 10^{15} (m^{-2}) [3]
HPT, at 20 °C	0.82 ± 0.08	2.56 ± 0.25	31.4 ± 1.52	0.33 ± 0.10	143.2
HPT, at 400 °C	0.60 ± 0.06	1.83 ± 0.18	41.5 ± 1.89	0.23 ± 0.07	75.4

[1] d—crystallite size, [2] ε—microstrain, [3] ρ—dislocation density.

3.3. Microhardness of the 316LVM Stainless Steel after HPT

The degree of deformation during HPT increases with the distance from the center along the radius, resulting in smaller sizes of the structural elements towards the edge of the sample and higher microhardness values. All investigated samples after HPT exhibited lower microhardness values in the center of the sample (4.2–5.0 GPa) and an increase in microhardness at the edge of the sample, reaching 5.0–5.5 GPa (Figure 5a). The samples subjected to HPT at room temperature showed more uniform deformation with the smallest difference in microhardness between the center and the edge of the sample.

Figure 5 illustrates the distribution of microhardness over the surface of an HPT sample after HPT from the side of the plunger (Figure 5b,d) and support (Figure 5c,e). The microhardness values from the support side are higher and more uniformly distributed across the surface. The microhardness measurements across the sample surface visually demonstrated not only a significant difference in the microhardness values between the center and edges of the samples after HPT at 400 °C, but also higher microhardness values in these samples at the established stage with the same microhardness (1/2 R—R from the center of the sample).

The investigation of the thermal stability of the steel samples after severe plastic deformation (SPD) revealed that the deformation strengthening for all modes decreases at temperatures above six hundred and fifty degrees due to grain growth (Figure 6). In the temperature range of 400–650 °C, there is a slight increase in microhardness, with the growth in microhardness starting slightly earlier and being more pronounced at a deformation temperature of 400 °C.

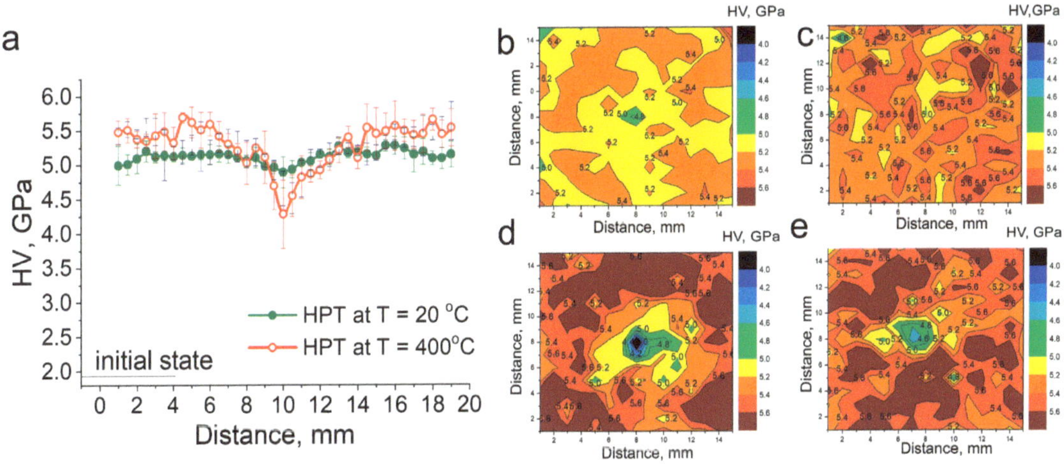

Figure 5. Microhardness of the 316LVM stainless steel produced by HPT: (**a**) along the diameter; (**b–d**) distribution over surface of an HPT sample after HPT at 20 °C (**b**,**c**) and after HPT at 400 °C (**d**,**e**); from the side of the plunger (**b**,**d**) and support (**c**,**e**).

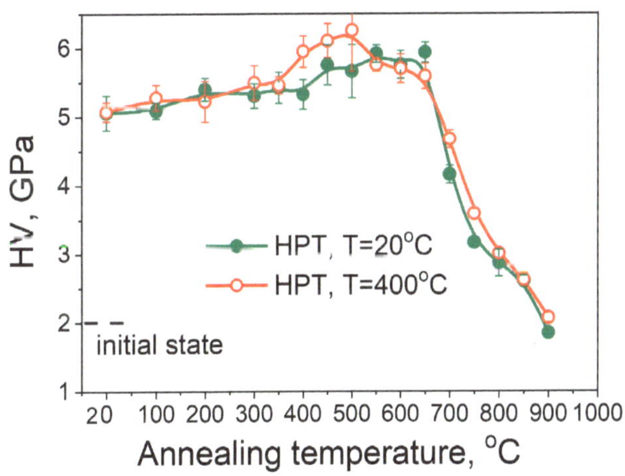

Figure 6. Microhardness of the 316LVM stainless steel produced by HPT as a function of annealing temperature.

3.4. Mechanical Properties under Static and Cyclic Loading

HPT significantly enhances the strength characteristics of the 316LVM steel (Figure 7, Table 4). The yield strength increases by more than nine times, while the ultimate tensile strength increases by four times. However, the ductility decreases after HPT at 20 °C to $\varepsilon = 11\%$, and sharply drops to $\varepsilon = 1\%$ at a deformation temperature of 400 °C. At approximately similar average sizes of the structural elements, an increase in the HPT temperature results in a twofold decrease in the density of free dislocations and an increase in the density of twins, which should not lead to a loss of ductility compared to the deformation level at room temperature.

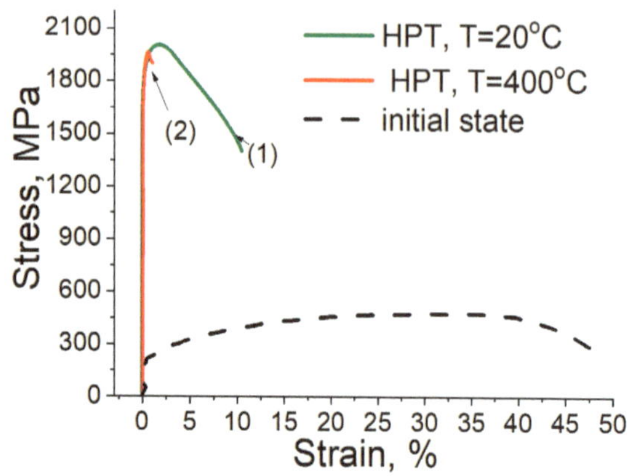

Figure 7. Stress–strain curves of the 316LVM stainless steel after solution heat treatment and HPT.

Table 4. Mechanical properties of the 316LVM stainless steel after solution heat treatment and HPT.

Processing	σ_{UTS} [MPa] [1]	σ_{YS} [MPa] [2]	ε [%] [3]
HPT, at 20 °C	2005 ± 24	1890 ± 6	11 ± 0.45
HPT, at 400 °C	1960± 18	1820 ± 12	1 ± 0.3
Quenching	500 ± 4	200 ± 6.5	47.6 ± 1.5

[1] Ultimate tensile strength; [2] yield strength; [3] total elongation.

The fatigue limit of the samples after HPT at room temperature (550 MPa) exceeds the fatigue limit of the as-quenched sample (250 MPa) by more than twofold (Figure 8). Previously, the authors of this study observed a significant increase in the fatigue strength due to grain refinement and twinning in austenite during severe plastic deformation (SPD) and subsequent cyclic deformation [13].

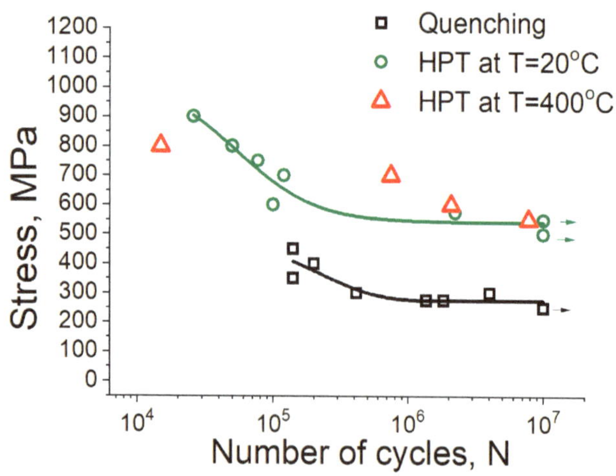

Figure 8. The Woehler (S-N) curves for cyclic deformation of the 316LMV stainless steel after solution heat treatment and HPT.

The results of the study on the samples after HPT at 400 °C did not reveal a decrease in high-cycle fatigue, but showed a slight decrease in stress under low-cycle loading compared to the samples after HPT at 20 °C.

The fractography of the surface of the as-quenched sample after fatigue failure indicates that, in the fatigue crack initiation and propagation zone (Figure 9a), a predominantly ductile nature of the surface relief is observed. In certain areas, the relief appears quasi-ductile with signs of fatigue striations (Figure 9b). In static fractures, typical ductile dimple fractures are observed (Figure 9).

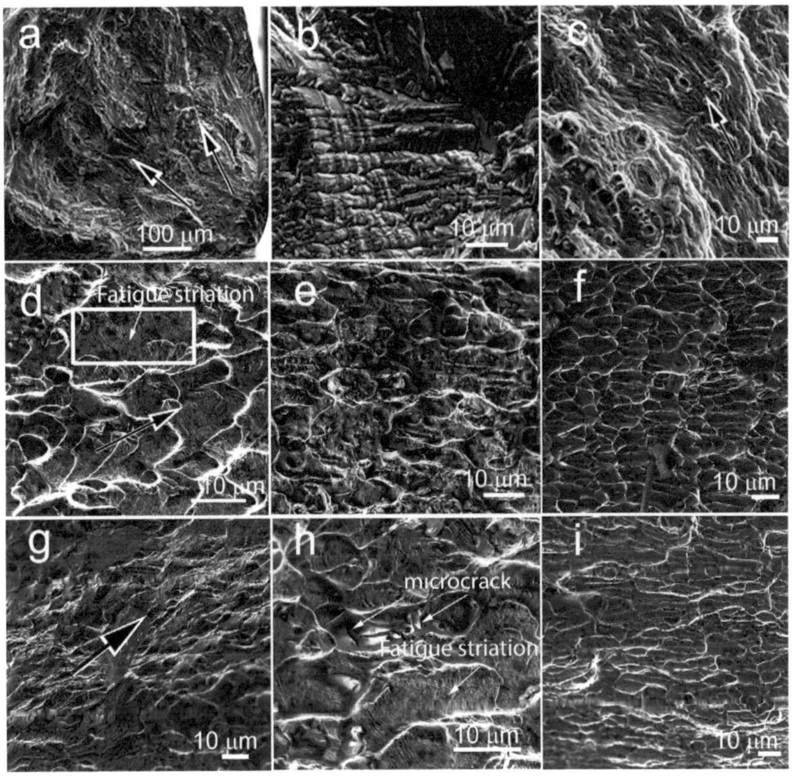

Figure 9. Fracture surfaces of fatigue failure of the 316LVM stainless steel samples after solution heat treatment (**a**–**c**), HPT at RT (**d**–**f**) and HPT at 400 °C (**g**–**i**).

After HPT, the relief in the fatigue crack initiation and propagation zone is less developed compared to the case of samples in the as-quenched state. A distinct zone of accelerated crack growth before the static fracture can be clearly identified (Figure 9d,g). The main crack propagation direction along the striations is shown at the figures. Signs of fatigue striations are observed (Figure 9d,h). The static fracture, as in the as-quenched samples, is associated with a ductile dimple fracture mechanism (Figure 9f,i). Deeper dimples on the fracture surface of the sample after HPT at room temperature indicate higher material ductility compared to the samples after HPT at 400 °C. The presence of microcracks within the dimples (Figure 9h) after HPT at 400 °C suggests the presence of grain boundary precipitates, which could have contributed to their initiation.

3.5. Corrosion Resistance of the 316LVM Stainless Steel in the Initial State and after HPT

The results of the potentiodynamic analysis on the 316LMV stainless steel samples in the as-quenched state and after HPT are presented in Figure 10. HPT at room temperature, based on the polarization curves, shows minimal changes in the corrosion resistance (Figure 10a).

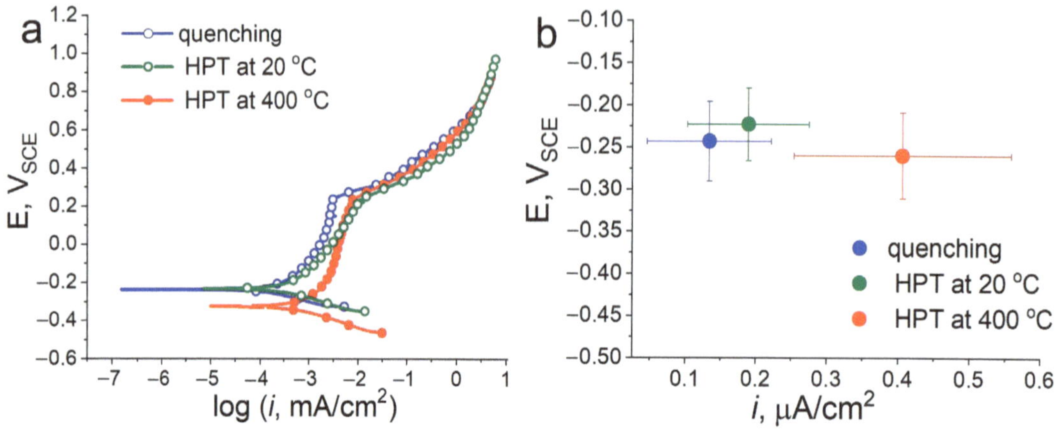

Figure 10. PDP curves in 0.1 M NaCl (pH = 6) electrolytes at scan rate of 1 mV/s (**a**) and E vs. corrosion current density (*i*) (**b**).

The lower corrosion potential (E_{corr} = −260 ± 51 mV) and higher corrosion current density (i_{corr} = 0.406 ± 0.152 μA/cm^2) compared to the as-quenched state (E_{corr} = −243 ± 47 mV, i_{corr} = 0.134 ± 0.088 μA/cm^2) and the state after HPT at 20 °C (E_{corr} = −223 ± 48 mV, i_{corr} = 0.189 ± 0.081 μA/cm^2) indicate a lower corrosion resistance of the samples after HPT at 400 °C (Figure 10b, Table 5).

Table 5. Potentiodynamic polarization parameters of the 316LVM stainless steel.

Processing	E_{corr} [1], mV	i_{corr} [2], μA/cm^2
Quenching, at 1050 °C (1 h)	−243 ± 47	0.134 ± 0.088
HPT at 20 °C	−223 ± 48	0.189 ± 0.081
HPT at 400 °C	−260 ± 51	0.406 ± 0.152

[1] Corrosion potential; [2] corrosion current density.

3.6. Biocompatibility In Vitro of the 316LVM Stainless Steel in the Initial State and after HPT

In order to assess the effect of deformation using the HPT method on the biocompatibility of the 316LVM stainless steel, the hemolysis levels were examined after incubation with the samples, as well as their cytotoxicity relative to non-transformed immune-competent blood cells. It is important to note that, in this study, materials are considered biocompatible with low hemolytic activity if they induce a degree of hemolysis of no more than 5%, in accordance with recommendations [37]. The hemolysis levels were evaluated after 2, 4, and 24 h of incubation relative to the spontaneous hemolysis level in the control with intact cells.

The conducted studies showed that the hemolytic activity of all the examined samples did not exceed 2% throughout the entire observation period (Figure 11a). Furthermore, no significant difference in the impact on the hemolysis level was observed between the alloy in its as-received state and after HPT.

The assessment of the cytotoxicity of the steel (Figure 11b) through measuring the level of cell activity using the MTT (3-[4,5-dimethylthiazol-2-yl]-2,5 diphenyl tetrazolium bromide) assay after 1 day of incubation, relative to the control with intact cells, was performed. The results showed that the presence of the examined samples, both in their as- quenched state and after HPT, did not lead to a significant change in cell activity level ($p > 0.05$). Thus, no signs of the cytotoxic effects of the 316LVM SS were found in this cell model.

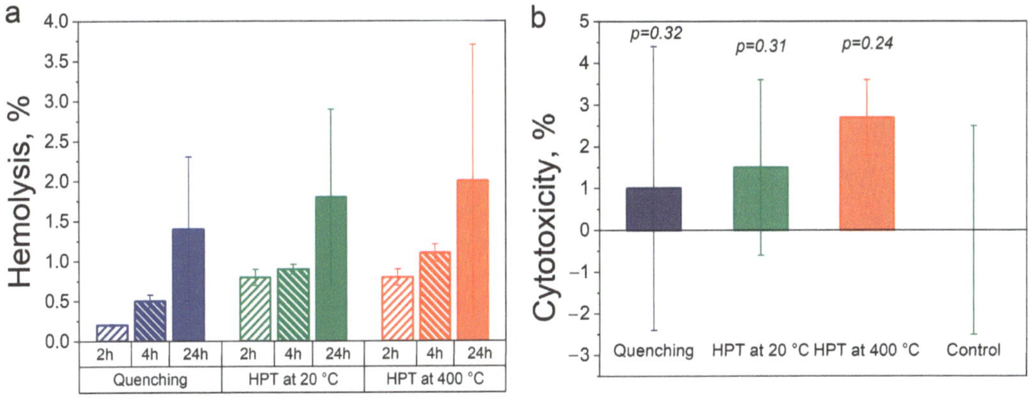

Figure 11. The hemolysis (**a**) and cytotoxicity (**b**) of the 316LVM stainless steel samples after solution heat treatment and HPT calculated as a percentage relative to control.

Since the examined samples of the 316LVM SS in their as-quenched state and after HPT did not induce hemolysis exceeding 2% and did not exhibit cytotoxicity, these samples can be considered biocompatible. HPT did not lead to a significant deterioration in this parameter.

In order to investigate the impact of the processing of the 316LVM SS on the colonization of mesenchymal stromal/stem cells (MSCs), the cells were incubated on the surface of the samples in their as-quenched state, after HPT at room temperature, and after HPT at 400 °C for a period of 21 days. The assessment of their viability on the surface of the samples did not reveal any significant differences in the proportion of viable cells compared to the control ($p < 0.05$) on the surface of the samples in their as-quenched state and after HPT at 20 °C (Figure 12). However, on the surface of the samples after HPT at 400 °C, there was a 1.2-fold decrease in the concentration of adhered viable cells compared to the control ($p = 0.04$).

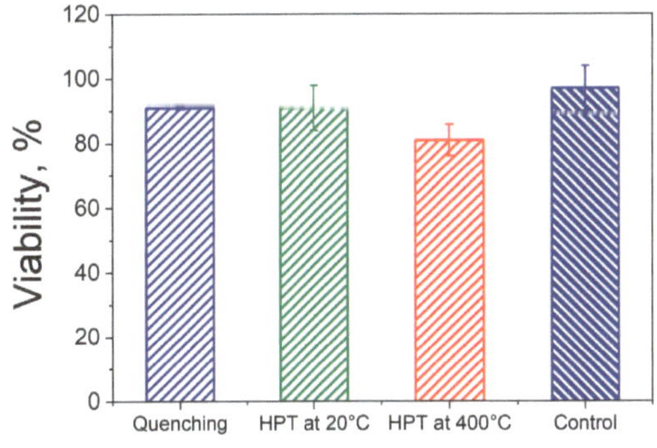

Figure 12. Cell viability on the surface of the 316LVM stainless steel samples after solution heat treatment and HPT relative to control.

The images allowing for the evaluation of the number of live and dead cells colonizing the 316LVM SS samples at the end of the incubation period indicate not only a decrease in the ratio of live/dead cells on the surface of the 316LVM SS sample after HPT at 400 °C but also a suppression of cellular colonization overall (Figure 13). It is important to note that the concentration of live cells on the surface of the 316LVM SS samples after HPT at 20 °C was approximately comparable to or even slightly higher than that of the as-quenched state sample. At the same time, it was observed that the majority of cells were not arranged individually but rather in groups, forming colonies. This can be interpreted as evidence of an actively ongoing cell division process, accompanying the colonization of the surface of the samples.

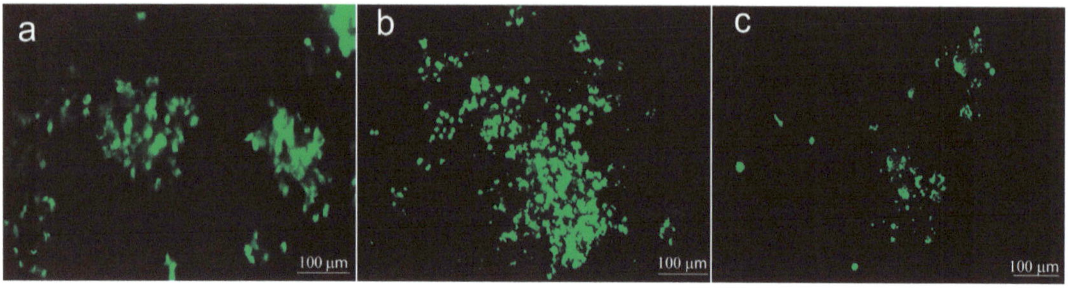

Figure 13. Surface colonization of the 316LVM stainless steel samples after solution heat treatment (**a**), HPT at 20 °C (**b**), and HPT at 400 °C (**c**) by MMSC after 21 days of incubation. Cell staining Calcein AM.

The next step was to further investigate the mechanism underlying the observed effect of 316LVM SS processing on the cell colonization. In order to do this, the change in the concentration of cells expressing the adhesion molecules CD11b and CD44 on the cell membranes, colonizing the surface of the steel, was studied in comparison to the control (Figure 14). The transmembrane molecules of integrin CD11b are involved in various aspects of cellular activity and influence cell attachment to the extracellular matrix. The transmembrane glycoprotein molecules CD44, integrated into the cell membrane, influence both cell adhesion to the extracellular matrix and intercellular interactions, promoting cell migration.

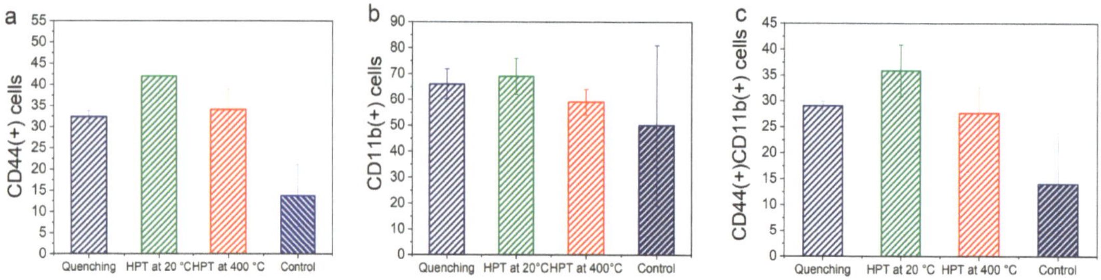

Figure 14. The expression of adhesion molecules CD44 and CD11b on the membranes of MSCs colonizing the surface of the 316LVM SS samples in the initial state and after HPT was compared to the control.

According to the obtained data, contact with the 316LVM SS samples stimulated the expression of the surface adhesion molecules CD44 and CD11b in MSCs compared to the control. Moreover, the impact of 316LVM SS after HPT at 20 °C was more pronounced compared to the as-quenched sample and the sample after HPT at 400 °C. This effect was determined to a lesser extent by an increase in the expression of CD11b integrins on the cell surface and, to a greater extent, by an increase in the expression of CD44. Thus, it can be concluded that the cell colonization of the 316LVM SS samples was caused by the stimulation of synthesis and expression of transmembrane adhesion molecules by the cells. Moreover, the samples after HPT at 20 °C stimulated colonization to the greatest extent through the involvement of this mechanism.

4. Discussion

In this study, surgical grade 316LVM stainless steel was investigated, which undergoes vacuum arc remelting in order to increase its purity (lower phosphorus and sulfur content), making it more corrosion-resistant and biocompatible compared to 316L. The chemical composition of 316LVM stainless steel differs slightly from 316L. It has an increased lower limit for molybdenum, reduced silicon content (maximum allowable value of 0.75% vs. 1.0%), and a higher range for allowable chromium and nickel content according to [38]. In addition, the microstructure should not contain molybdenum-enriched chi and sigma intermetallic compounds, which can reduce corrosion resistance and potential embrittlement.

In this study, two HPT modes, performed at room temperature and at 400 °C, resulted in a nanocrystalline state in the 316LVM steel, with an average size of the structural elements of 46 ± 1.8 nm and 50 ± 1.6 nm, respectively. The structure differed only in the density of the twinning in the austenite (four times higher after HPT at 400 °C, as observed using TEM) and the density of the dislocations (twice higher after HPT at 20 °C, as measured by X-ray). No precipitates were detected in the microstructure of the 316LVM steel using TEM. The above-mentioned distinguishing features of the microstructure do not explain the similar levels of strength properties with a significant difference in ductility between the two states after HPT at room temperature and at 400 °C (ε = 11 ± 0.45% and ε = 1 ± 0.3%, respectively). Only the increased density of twins indicates that something hindered the dislocation glide during the HPT process at 400 °C.

When comparing with the above data from the literature, it should be noted that in this work, after HPT at 20 °C, higher strength characteristics were obtained (σ_{UTS} = 2005 ± 24 MPa) with satisfactory ductility (ε = 11%). At the same time, the best result of the ultimate strength presented in the literature was achieved in [25] and amounted to 1865 MPa. This result was obtained after quasi-constrained high-pressure torsion under pressure for 15 revolutions. After 10 revolutions of HPT in [23], the tensile strength was about 1790 ± 50 MPa. The same amount of revolutions in the case of the paper presented here resulted in a higher level of strength and can be explained by a greater grain refinement, 46 ± 1.8 nm versus 62 nm, in [23]. However, subsequent annealing after HPT led to a further increase in strength to 2390 MPa in [23] and to 2226 MPa in [25]. This increase in strength after annealing was accompanied by a significant increase in fatigue strength up to 1 GPa. In this work, heating during HPT did not increase the strength of the steel compared to HPT at room temperature, but only worsened the ductility.

The mechanism reported in [22–25] of the segregation of Mo–Cr–Si precipitation, allow us to suppose that the increased content of Mo and Cr in the investigated 316LVM steel could have led to more pronounced segregation. This had a negative impact not only on the mechanical properties (Figure 7, Table 4) but also on the operational characteristics. The results of the fatigue tests showed a decrease in fatigue strength under low-cycle loading (Figure 8). Moreover, the fractographic analysis of the fracture surface of the sample after HPT at 400 °C revealed the presence of microcracks (Figure 9h), indicating possible segregation that could have been the cause of their initiation. The corrosion studies using potentiodynamic polarization analysis revealed that HPT at 400 °C reduces the corrosion potential of the surgical steel, and increases the corrosion current density of the samples (Figure 10, Table 5). The in vitro studies (Figures 11–14) of the 316LVM stainless steel revealed lower biocompatibility indicators for samples after HPT at 400 °C. However, while the difference in the hemolysis and cytotoxicity levels of the samples (Figure 11) was not significant (p > 0.05), the colonization of the MMSCs on the surface of the steel after HPT at 400 °C (Figure 12) showed a decrease in the concentration of adhered living cells compared to the control (p = 0.04). Additionally, there was an overall suppression of cell colonization (Figure 13).

Thus, this study has established that the previously observed segregation of Mo–Cr–Si in 316L stainless steel can significantly deteriorate the operational characteristics in surgical grade 316LVM steel with higher Mo and Cr content, negating the advantages of its high-strength state.

Simultaneously, the 316LVM SS samples after HPT at 20 °C demonstrated a high combination of properties. Not only were high strength characteristics observed (σ_{UTS} = 2005 MPa and σ_{YS} = 1890 MPa) with satisfactory ductility for such a high-strength state (ε = 11%), but also high levels of fatigue strength, improved resistance to pitting corrosion, and enhanced in vitro biocompatibility compared to the as-received state. It was determined that the cell colonization on the 316LVM SS samples was induced by the stimulation of cell synthesis and expression of transmembrane adhesion molecules. This mechanism was primarily stimulated by the 316LVM steel samples after HPT at 20 °C. The colonization of 316LVM SS after such processing could be influenced by the surface morphology of the nanocrystalline material and its chemical composition. Both of these factors can promote or inhibit changes in the local membrane potential of the cell in the contact area, which stimulates protein synthesis and the formation of covalent bonds with the cell membrane. As a result, the increase in the concentration of membrane-bound molecules affects the formation of a tighter intercellular contact. This, in turn, enhances the effectiveness of co-stimulation, leading to an increase in the functional potential of the MMSCs. It should also be noted that the inhibition of changes in the local membrane potential of the cell may have been influenced by the segregation of Mo–Cr–Si, which resulted in a deterioration of cell colonization in the steel after HPT at 400 °C.

The obtained data suggest that the 316LVM SS after HPT at 20 °C can be considered as a material for the development of submerged implants and metal structures for osteosynthesis. These implants will effectively stimulate cell colonization on their surface and promote accelerated fixation in the area of bone defect repair, leading to the enhanced cell differentiation of pluripotent stromal cells. Additionally, they will contribute to the increase in periosteal formation and local neo-osteogenesis. The utilization of HPT processing at 20 °C will not only significantly improve the operational characteristics but also reduce the weight of medical devices. This is especially important in the development of implantable prostheses for skeletal defect replacement, as it enables a significant increase in the specific strength of the steels. Furthermore, the miniaturization of medical devices will result in a reduced negative impact on the human body from the presence of foreign implants.

5. Conclusions

1. HPT at room temperature and at 400 °C forms a nanocrystalline austenitic structure in 316LVM steel.
2. HPT at both temperatures increases the microhardness of steel by more than 2.5 times.
3. After HPT at 400 °C, the 316LVM steel is less ductile (ε = 1%) than after HPT at room temperature (ε = 11%) at the same level of strength.
4. HPT increases the endurance limit of the 316LVM steel by more than two times (up to 550 MPa) compared to the initial state. At the same time, HPT at 400 °C reduces low-cycle fatigue.
5. Nanostructuring by HPT at 400 °C worsens the corrosion resistance of steel, in contrast to HPT at 20 °C.
6. HPT at 20 °C of the 316LVM SS stimulates cell synthesis and the expression of transmembrane adhesion molecules, leading to improved cell colonization on the surface of the samples.

These results indicate the potential of nanocrystalline surgical grade 316LVM stainless steel after HPT at 20 °C for medical use. An attempt to additionally strengthen 316LVM steel by the precipitation of segregations during HPT at 400 °C was not successful. It was shown that nanostructured 316LVM steel after HPT at 400 C does not have sufficient ductility, corrosion resistance, and biocompatibility, which does not allow it to be recommended for widespread use.

Author Contributions: Conceptualization, O.R., M.K. and S.D.; methodology, N.A., N.M., G.R. and M.K.; software, G.R., A.T., D.P. and M.G.; validation, O.R., N.A. and N.M.; formal analysis, O.R. and N.M.; investigation, O.R., N.A., N.M., G.R., A.T., E.L., D.P. and M.G.; resources, A.T.; data curation, A.T., E.L. and S.D.; writing—original draft preparation, O.R. and N.A.; writing—review and editing, G.R., S.D., E.L., M.K. and N.M.; visualization, O.R., G.R., M.G., N.A. and E.L.; supervision, S.D. and M.K.; project administration, O.R.; funding acquisition, S.D. All authors have read and agreed to the published version of the manuscript.

Funding: This research was carried out with the support of the Russian Federation state assignment of A.A. Baikov Institute of Metallurgy and Materials Science of the Russian Academy of Science (IMET RAS), Russia (Theme No. 075-01176-23-00).

Data Availability Statement: All the data required to reproduce these experiments are present in the article.

Acknowledgments: The study of fractography was carried out using the research equipment of the Shared Facility Center of the P.N. Lebedev Physical Institute of RAS "Center for the Study of High-Temperature Superconductors and Other Strongly Correlated Electronic Systems".

Conflicts of Interest: The authors declare no conflict of interest.

References

1. Marshall, P. *Austenitic Stainless Steel: Microstructure and Mechanical Properties*; Elsevier Applied Science Publishers: London, UK; New York, NY, USA, 1984.
2. Davis, J.R. *Handbook of Materials for Medical Devices*; Metallic Materials; ASM International: Materials Park, OH, USA, 2003; Chapter 3.
3. Pilliar, R.M. Metallic Biomaterials. In *Biomedical Materials*; Narayan, R., Ed.; Springer: Boston, MA, USA, 2009; pp. 41–81. [CrossRef]
4. Teoh, S.H. Fatigue of biomaterials: A review. *Int. J. Fatigue* **2000**, *22*, 825–837. [CrossRef]
5. *ISO 5838-1:2013*; Implants for Surgery Metallic Skeletal Pins and Wires. Part 1: General Requirements. International Organization for Standardization: Geneva, Switzerland, 2022.
6. Kashyap, B.P.; Tangri, K. On the Hall-Petch relationship and substructural evolution in type 316L stainless steel. *Acta Metall. Mater.* **1995**, *43*, 3971–3981. [CrossRef]
7. Valiev, R.Z.; Langdon, T.G. Principles of equal-channel angular pressing as a processing tool for grain refinement. *Prog. Mater. Sci.* **2006**, *51*, 881–981. [CrossRef]
8. Zhilyaev, A.P.; Langdon, T.G. Using high-pressure torsion for metal processing: Fundamentals and applications. *Prog. Mater. Sci.* **2008**, *53*, 893–979. [CrossRef]
9. Dong, F.Y.; Zhang, P.; Pang, J.C.; Chen, D.M.; Yang, K.; Zhang, Z.F. Optimizing strength and ductility of austenitic stainless steels through equal-channel angular pressing and adding nitrogen element. *Mater. Sci. Eng. A* **2013**, *587*, 185–1917. [CrossRef]
10. Ueno, H.; Kakihata, K.; Kaneko, Y.; Hashimoto, S.; Vinogradov, A. Enhanced fatigue properties of nanostructured austenitic SUS 316L stainless steel. *Acta Mater.* **2011**, *59*, 7060–7069. [CrossRef]
11. Huang, C.X.; Yang, G.; Gao, Y.L.; Wu, S.D.; Zhang, Z.F. Influence of processing temperature on the microstructures and tensile properties of 304L stainless steel by ECAP. *Mater. Sci. Eng. A* **2008**, *485*, 643–650. [CrossRef]
12. Dobatkin, S.V.; Rybal'chenko, O.V.; Raab, G.I. Structure Formation, Phase Transformations and Properties in Cr–Ni Austenitic Steel after Equal-Channel Angular Pressing and Heating. *Mater. Sci. Eng. A* **2007**, *463*, 41–45. [CrossRef]
13. Rybalchenko, O.V.; Anisimova, N.Y.; Kiselevskiy, M.V.; Belyakov, A.N.; Tokar, A.A.; Terent'ev, V.F.; Prosvirnin, D.V.; Rybalchenko, G.V.; Raab, G.I.; Dobatkin, S.V. The influence of ultrafine-grained structure on the mechanical properties and biocompatibility of austenitic stainless steels. *J. Biomed. Mater. Res.* **2020**, *108*, 1460–1468. [CrossRef]
14. Koohdar, H.; Nili-Ahmadabadi, M.; Kalahroudi, F.J.; Jafarian, H.R.; Langdon, T.G. Effect of post-deformation annealing on the microstructure and mechanical behavior of an Fe–Ni–Mn steel processed by high-pressure torsion. *J. Mater. Res. Technol.* **2021**, *15*, 1537–1546. [CrossRef]
15. Jang, G.; Kim, J.N.; Lee, H.; Lee, T.; Enikeev, N.; Abramova, M.; Valiev, R.Z.; Kim, H.S.; Lee, C.S. Microstructural evolution and mechanical properties of nanocrystalline Fe–Mn–Al–C steel processed by high-pressure torsion. *Mater. Sci. Eng. A* **2021**, *827*, 142073. [CrossRef]
16. Biserova-Tahchieva, A.; Chatterjee, D.; van Helvoort, A.T.J.; Llorca-Isern, N.; Cabrera, J.M. Effect of the nanostructuring by high-pressure torsion process on the secondary phase precipitation in UNS S32750 Superduplex stainless steel. *Mater. Charact.* **2021**, *183*, 111639. [CrossRef]
17. Jacob, K.; Yadav, D.; Dixit, S.; Hohenwarter, A.; Jaya, B.N. High pressure torsion processing of maraging steel 250: Microstructure and mechanical behaviour evolution. *Mater. Sci. Eng. A* **2021**, *802*, 140665. [CrossRef]

18. Anijdan, S.H.M.; Koohdar, H.; Nili-Ahmadabadi, M.; Jafarian, H.R.; Langdon, T.G. Evaluation of texture weakening and microstructural evolution in an Fe–10Ni–7Mn martensitic steel severely deformed by six turns of high-pressure torsion. *Mater. Sci. Eng. A* **2022**, *851*, 143660. [CrossRef]
19. Pakieła, Z.; Garbacz, H.; Lewandowska, M.; Drużycka-Wiencek, A.; Suś-Ryszkowska, M.; Zieliński, W.; Kurzydłowski, K. Structure and properties of nanomaterials produced by severe plastic deformation. *Nukleonika* **2006**, *51*, 19–25.
20. Mine, Y.; Horita, Z.; Murakami, Y. Effect of hydrogen on martensite formation in austenitic stainless steels in high-pressure torsion. *Acta Mater.* **2009**, *57*, 2993–3002. [CrossRef]
21. Gubicza, J.; El-Tahawy, M.; Huang, Y.; Choi, H.; Choe, H.; Lábár, J.L.; Langdon, T.G. Microstructure, phase composition and hardness evolution in 316L stainless steel processed by high-pressure torsion. *Mater. Sci. Eng. A* **2016**, *657*, 215–223. [CrossRef]
22. Scheriau, S.; Zhang, Z.; Kleber, S.; Pippan, R. Deformation mechanisms of a modified 316L austenitic steel subjected to high pressure torsion. *Mater. Sci. Eng. A* **2011**, *528*, 2776–2786. [CrossRef]
23. Wang, H.; Shuro, I.; Umemoto, M.; Kuo, H.H.; Todaka, Y. Annealing behavior of nano-crystalline austenitic SUS316L produced by HPT. *Mater. Sci. Eng. A* **2012**, *556*, 906–910. [CrossRef]
24. Renk, O.; Hohenwarter, A.; Eder, K.; Kormout, K.S.; Cairney, J.M.; Pippan, R. Increasing the strength of nanocrystalline steels by annealing: Is segregation necessary? *Scr. Mater.* **2015**, *95*, 27–30. [CrossRef]
25. Renk, O.; Hohenwarter, A.; Gammer, C.; Eckert, J.; Pippan, R. Achieving 1 GPa fatigue strength in nanocrystalline 316L steel through recovery annealing. *Scr. Mater.* **2022**, *217*, 114773. [CrossRef]
26. Abramova, M.M.; Enikeev, N.A.; Valiev, R.Z.; Etienne, A.; Radiguet, B.; Ivanisenko, Y.; Sauvage, X. Grain boundary segregation induced strengthening of an ultrafine-grained austenitic stainless steel. *Mater. Lett.* **2014**, *136*, 349–352. [CrossRef]
27. Etienne, A.; Radiguet, B.; Cunningham, N.; Odette, G.; Valiev, R.; Pareige, P. Comparison of radiation-induced segregation in ultrafine-grained and conventional 316 austenitic stainless steels. *Ultramicroscopy* **2011**, *111*, 659–663. [CrossRef]
28. Dobatkin, S.V.; Rybalchenko, O.V.; Enikeev, N.A.; Tokar, A.A.; Abramova, M.M. Formation of fully austenitic ultrafine-grained high strength state in metastable Cr–Ni–Ti stainless steel by severe plastic deformation. *Mater. Lett.* **2016**, *166*, 276–279. [CrossRef]
29. Han, J.-K.; Liu, X.; Lee, I.; Kuzminova, Y.O.; Evlashin, S.A.; Liss, K.-D.; Kawasaki, M. Structural evolution during nanostructuring of additive manufactured 316L stainless steel by high-pressure torsion. *Mater. Lett.* **2021**, *302*, 130364. [CrossRef]
30. Lee, D.-H.; Gao, Z.; Park, J.-M.; Zhao, Y.; Suh, J.-Y.; Jägle, E.A.; Tsuchiya, K.; Ramamurty, U.; Jang, J. Nanomechanical and microstructural characterization on the synergetic strengthening in selectively laser melted austenitic stainless steel. *Scr. Mater.* **2022**, *209*, 114359. [CrossRef]
31. Liu, M.; Gong, W.; Zheng, R.; Li, J.; Zhang, Z.; Gao, S.; Ma, C.; Tsuji, N. Achieving excellent mechanical properties in type 316 stainless steel by tailoring grain size in homogeneously recovered or recrystallized nanostructures. *Acta Mater.* **2022**, *226*, 117629. [CrossRef]
32. *ISO 6892-1:2019*; Metallic Materials—Tensile Testing—Part 1: Method of Test at Room Temperature. International Organization for Standardization: Geneva, Switzerland, 2019.
33. *ASTM G59-23*; Standard Test Method for Conducting Potentiodynamic Polarization Resistance Measurements. ASTM International: West Conshohocken, PA, USA, 2023.
34. Choudhary, R.; Venkatraman, S.K.; Bulygina, I.; Senatov, F.; Kaloshkin, S.; Anisimova, N.; Kiselevskiy, M.; Knyazeva, M.; Kukui, D.; Walther, F.; et al. Biomineralization, dissolution and cellular studies of silicate bioceramics prepared from eggshell and rice husk. *Mater. Sci. Eng. C* **2021**, *118*, 111456. [CrossRef]
35. Anisimova, N.; Kiselevskiy, M.; Martynenko, N.; Straumal, B.; Willumeit-Römer, R.; Dobatkin, S.; Estrin, Y. Cytotoxicity of biodegradable magnesium alloy WE43 to tumor cells in vitro: Bioresorbable implants with antitumor activity? *J. Biomed. Mater. Res. B Appl. Biomater.* **2020**, *108*, 167–173. [CrossRef]
36. Yagodkin, Y.D.; Dobatkin, S.V. Application of electron microscopy and X-ray structural analysis for the determination of sizes of structural elements in nanocrystalline materials (review). *Inorg. Mater.* **2008**, *44*, 1520–1530. [CrossRef]
37. *ASTM F756-17*; Standard Practice for Assessment of Hemolytic Properties of Materials. ASTM International: West Conshohocken, PA, USA, 2017.
38. *ISO/DIS 5832-1*; Implants for Surgery Metallic Materials. Part 1: Wrought Stainless Steel. International Organization for Standardization: Geneva, Switzerland, 2016.

metals

Article

Fatigue Reliability Assessment for Orthotropic Steel Decks: Considering Multicrack Coupling Effects

Jing Liu [1], Yang Liu [1], Guodong Wang [1,2,*], Naiwei Lu [1], Jian Cui [1] and Honghao Wang [1]

1 Key Laboratory of Bridge Engineering Safety Control by Department of Education, School of Civil Engineering, Changsha University of Science & Technology, Changsha 410114, China; ljtbdr@stu.csust.edu.cn (J.L.); liuyangbridge@163.com (Y.L.); lunaiwei@csust.edu.cn (N.L.); cuijian@stu.csust.edu.cn (J.C.); wanghonghao@stu.csust.edu.cn (H.W.)

2 National-Local Joint Laboratory of Engineering Technology for Long-Term Performance Enhancement of Bridges in Southern District, Changsha University of Science & Technology, Changsha 410114, China

* Correspondence: guodongw@csust.edu.cn; Tel.: +86-18774891894

Abstract: Multiple fatigue cracks are generally present in practical engineering due to the existence of welding; the size and number of cracks of orthotropic steel bridge decks are greatly uncertain. The component failure conditions caused by these cracks may have correlations. Currently, it is still a challenging issue to develop a physical model of multiple fatigue crack propagation in bridge decks and perform a fatigue reliability assessment, which is also the motivation that drives the innovation of this study. A fatigue reliability evaluation method is presented for orthotropic steel bridge decks, considering the coupling effect of multiple cracks and the randomness of vehicle loading. A numerical simulation method for multifatigue crack growth is developed by combining the ABAQUS and FRANC3D programs. The equivalent crack depth under different spacing and depths of collinear cracks is calculated by using numerical simulation and the multicrack equivalent characterization method. The critical damage accumulation function of multiple fatigue cracks is established using linear elastic fracture mechanics. Subsequently, the critical damage accumulation function of multiple fatigue cracks is established based on linear elastic fracture mechanics. In order to solve the time-consuming problem of traditional Monte Carlo method, the iHL-RF method and AK-MCS method are developed for fatigue reliability analysis. The results show that compared with the single-crack model, the fatigue reliability of orthotropic steel deck will be crucially reduced considering the coupling effect of double cracks. The MCS, iHL-RF and AK-MCS methods can effectively solve the fatigue reliability analysis problem. Compared with the MCS method, the reliability calculation time based on AK-MCS method is significantly reduced. The AK-MCS method-based method reduces the time for calculating the reliability of orthotropic steel decks by 50% compared with the iHL-RF method. The reliability analysis of orthotropic steel deck bridge based on AK-MCS method is proved to be efficient and accurate.

Keywords: orthotropic steel deck; multiple cracks; AK-MCS method; fatigue reliability

Citation: Liu, J.; Liu, Y.; Wang, G.; Lu, N.; Cui, J.; Wang, H. Fatigue Reliability Assessment for Orthotropic Steel Decks: Considering Multicrack Coupling Effects. *Metals* **2024**, *14*, 272. https://doi.org/10.3390/met14030272

Academic Editor: Alireza Akhavan-Safar

Received: 24 January 2024
Revised: 23 February 2024
Accepted: 23 February 2024
Published: 25 February 2024

1. Introduction

An orthotropic steel deck bridge has outstanding advantages, such as light weight, high strength, factory production, and convenient construction and assembly [1–3], which is a landmark achievement in modern steel structure bridges. There are many welds on the steel bridge plate, and the welding defects are difficult to detect and maintain effective control of, which leads to crack initiation [4]. Under the long-term action of heavy vehicles, the welding structure details of the top plate and longitudinal ribs are prone to fatigue cracks and rapid growth. Many steel deck bridges, both domestically and internationally, experienced fatigue cracks shortly after their opening to traffic, such as the Severn Bridge in the UK, the Akashi Strait Bridge in Japan, and the Junshan Yangtze River Bridge in China [5,6]. The fatigue properties of single cracks in precracked structures have been extensively studied [7,8]. However, with the increase in bridge service time, the number and size of fatigue cracks in the steel deck bridge crucially increase, and the interaction between dense cracks will accelerate the initiation and propagation of cracks. After multiple crack tips come into contact, fusion occurs, resulting in larger cracks that significantly affect the crack propagation speed, leading to a sharp decrease in the fatigue life of steel bridge decks [9]. Therefore, it is necessary and urgent to conduct a reliability analysis on steel bridge decks under the influence of multiple fatigue cracks in welds.

An orthotropic steel deck bridge is a structure composed of longitudinal and transverse stiffeners (longitudinal and transverse ribs) that are perpendicular to each other, along with the bridge deck cover plate, to jointly bear wheel loads. The stiffness of this structure is different in the two perpendicular directions, resulting in structural anisotropy. The reliability analysis of the orthotropic bridge panel should be carried out to ensure its sufficient strength. In the traditional reliability calculation methods, most studies are based on the known function [10]. However, when calculating the reliability of actual engineering, the engineering structure is very complex, and the calculation is cumbersome. The performance function is a highly nonlinear implicit function, and traditional methods such as first-order and second-order moments can easily lead to low computational efficiency and poor accuracy in the results [11]. Therefore, the Monte Carlo method based on finite element software is usually used to calculate the structure reliability [12], except for the problem of small failure probability due to the limitation of the Monte Carlo simulation method itself. The accuracy of the calculation results depends on the number of calculation times of finite element software. The lower the failure probability of actual engineering structures, the higher the reliability. The number of calculations will increase by tens or hundreds of times, leading the calculation cost to be unacceptable.

To solve the above problems, it is an effective method to construct a proxy model to replace the implicit function of the actual complex structure. The proxy model method is essentially a fitting technique that can discover implicit relationships between input and output variables and make predictions [13]. At present, a variety of proxy model methods have been developed, including polynomial response surface [14], radial basis function [15], neural network [16] and Kriging [17]. The Kriging proxy model only considers the relationship between variable values, which can be combined with other reliability calculation methods to further improve the calculation efficiency and accuracy. Kriging is widely used in reliability calculation due to its good nonlinear fitting ability and unique error evaluation function [18]. Fan et al. [19] adopted the Kriging model to optimize the reliability design of crane bridges. Du et al. [20] used parallel subset simulation and the Kriging model to analyze the reliability of a cantilever tube. The results showed that the efficiency was improved under the condition of meeting the accuracy. Lv et al. [21] proposed an active-learning reliability analysis method (AK-LS) combining the Kriging model and linear sampling, screened out the best sample points for improving model accuracy through the constructed active-learning function and constructed a high-precision Kriging model with fewer sample points. Echard et al. [22] proposed an active-learning reliability analysis method (AK-MCS) combining the Kriging model and Monte Carlo simulation method, screened out the points near the failure surface and the points with large prediction errors and realized the high fitting of the limit state boundary with fewer sample points.

This paper is organized as follows: First, the fatigue crack growth model of the orthotropic steel deck is established. Next, a reliability calculation framework for multiple fatigue cracks in orthotropic bridge decks based on finite element models is proposed. Then, the equivalent crack depth under different spacing and depth of collinear cracks is calculated by using numerical simulation and multicrack equivalent characterization method. Following this, the iHL-RF method and AK-MCS method for fatigue reliability analysis of multicrack orthotropic decks are developed. Finally, the accuracy and timeliness of multiple methods under different working conditions are compared.

2. Materials and Methods

The method based on the S-N curve AASHTO (American Association of State Highway Transportation Officials) is a typical method for evaluating steel bridge components [23]. However, the AASHTO method requires a large number of fatigue tests to obtain the relevant parameters and cannot consider the crack size information in the fatigue evaluation process. Based on the Paris crack growth model, John et al. [24] proposed a linear elastic fracture mechanics method for fatigue reliability assessment. The Paris equation of crack growth:

$$\frac{d\alpha}{dN} = C(\Delta K)^m \tag{1}$$

where α is the fatigue crack size, specifically, the crack depth, in this paper; N is the number of loading cycles; C and m are the fatigue growth correlation coefficients; and ΔK is the amplitude of the stress intensity factor. According to the linear elastic fracture mechanics (LEFM) theory [25], it can be estimated as follows:

$$\Delta K = K_{\max} - K_{\min} = YS_{eq}\sqrt{\pi\alpha} \tag{2}$$

where S_{eq} is the equivalent stress amplitude under varying amplitude load; Y is a geometric function considering the crack shape of the member.

The fatigue crack size corresponds to the number of load cycles. The number of load cycles is defined as N_1 and N_2, then the integral of Equation (1) can be obtained as follows:

$$\int_{\alpha_1}^{\alpha_2} \frac{1}{\left(Y\sqrt{\pi\alpha}\right)^m} d\alpha = \int_{N_1}^{N_2} C S_{eq}{}^m dN \tag{3}$$

A damage accumulation function is proposed to reflect the fatigue crack size from α_1 to α_2 [26], which is defined as follows:

$$\psi(\alpha_2, \alpha_1) = \int_{\alpha_1}^{\alpha_2} \frac{1}{\left(Y\sqrt{\pi\alpha}\right)^m} d\alpha \tag{4}$$

The relationship between damage accumulation function and load accumulation is as follows:

$$\psi(\alpha_2, \alpha_1) = C S_{eq}{}^m (N_2 - N_1) \tag{5}$$

When the critical crack size is specified, the fatigue failure criterion of the structure subjected to $(N_2 - N_1)$ stress cycle can be defined as:

$$\alpha_C - \alpha_N \leq 0 \tag{6}$$

where α_N is the crack size of the in-service structure after N stress cycles, which can be redefined as the crack evolving from the initial size of α_0 (the N_0 stress cycle) to the α_2 size (the N stress cycle). Once α_N exceeds the critical crack size α_C, a failure problem can be considered to have occurred.

For fatigue reliability analysis of bridge components, Equation (6) can be considered the limit state. Since damage accumulation function $\psi(\alpha_2, \alpha_1)$ is monotonically increasing with crack size, the limit state function of Equation (6) can be redefined as:

$$g(\boldsymbol{x}) = \psi(\alpha_C, \alpha_0) - \psi(\alpha_N, \alpha_0) \leq 0 \tag{7}$$

$$g(\boldsymbol{x}) = \int_{\alpha_0}^{\alpha_c} \frac{1}{\left(Y\sqrt{\pi\alpha}\right)^m} d\alpha - C S_{eq}{}^m (N - N_0) \leq 0 \tag{8}$$

where $\psi(\alpha_c, \alpha_0)$ is the fatigue damage accumulation function from the initial crack size to the critical crack size, namely, the critical threshold of the limit state equation. $\Psi(\alpha_N, \alpha_0)$ is the damage accumulation function from the initial crack size α_0 through N stress cycles to α_N, namely, the load-effect part of the limit state function.

It is considered that the initial depth of the double crack α_0 is equivalent to a single-crack depth α_e after the extended coupling effect, and the coupling equivalent to a single crack continues to expand to the critical depth α_c. According to the recommendation of the IIW (International Institute of Welding) [27], when the crack propagation depth reaches half of the thickness of the roof plate, the component is considered to have failed. During this process, Y changes with the crack size; the expression on the right of Equation (8) is processed with piecewise integral.

$$g(\boldsymbol{x}) = \int_{\alpha_0}^{\alpha_e} \frac{1}{(Y\sqrt{\pi\alpha})^m} \mathrm{d}\alpha + \int_{\alpha_e}^{\alpha_c} \frac{1}{(Y\sqrt{\pi\alpha})^m} \mathrm{d}\alpha - C{S_{eq}}^m (N - N_0) \quad (9)$$

The stress cycle number is defined as follows: $N = 365 \times n \times N_d$. Equation (9) can be converted to

$$g(\boldsymbol{x}) = \int_{\alpha_0}^{\alpha_e} \frac{1}{(Y_0\sqrt{\pi\alpha})^m} \mathrm{d}\alpha + \int_{\alpha_e}^{\alpha_c} \frac{1}{(Y_e\sqrt{\pi\alpha})^m} \mathrm{d}\alpha - 365nN_dC{S_{eq}}^m \quad (10)$$

where N_d is the number of daily cycles of stress; n is the service life of the bridge. Y_0 and Y_e are the boundary correction factors for the reference stress intensity factors, respectively [28].

$$Y_0 = 0.4804 - 0.5116(\alpha_0/T) + 0.2072(\alpha_0/T)^2 - 1.0116(\alpha_0/T)^3 + 0.9562(\alpha_0/T)^4 \quad (11)$$

$$Y_e = 0.5459 + 0.2219(\alpha_e/T) + 0.7656(\alpha_e/T)^2 - 1.5501(\alpha_e/T)^3 + 0.9562(\alpha_e/T)^4 \quad (12)$$

where α_e is the crack depth of the collinear double crack considering the coupling effect equivalent to a single crack, which is calculated using the ABAQUS-FRANC3D interactive technique; T is the plate thickness; and R_s is the collinear double-crack spacing ratio.

Considering the actual random variation of load, the lateral distribution coefficient of the wheel track at the bridge panel e and the annual traffic volume growth coefficient α_y are further added to the limit state function.

$$g(\boldsymbol{x}) = \int_{\alpha_0}^{\alpha_e} \frac{1}{(Y_0\sqrt{\pi\alpha})^m} \mathrm{d}\alpha + \int_{\alpha_e}^{\alpha_c} \frac{1}{(Y_e\sqrt{\pi\alpha})^m} \mathrm{d}\alpha - e365nN_dC{S_{eq}}^m (1 + \frac{n+1}{2}\alpha_y) \quad (13)$$

where the double crack spacing ratio, initial crack depth, crack propagation coefficient, wheel track transverse distribution coefficient, equivalent stress amplitude and daily stress cycles were random distribution variables. Table 1 shows the random variable distribution.

Table 1. Random variable distribution.

Variable	Distribution	Mean	Variation Coefficient
R_s	Uniform [29]	0.1	3
α_0	lognormal [30]	0.5	0.2
C	lognormal [31]	5.21	0.6
e	lognormal [32]	0.78	0.1
S_{eq}	normal (driving lane) [33] normal (passing lane) [33]	17.67 16.87	0.44 1.53
N_d	normal (driving lane) [34] normal (passing lane) [34]	5685 1140	480 52

3. Structural Reliability Analysis Based on iHL-RF Method

For the fatigue reliability problem described in performance function Equation (13), the failure probability is defined as follows:

$$P_f = P[g(\boldsymbol{x}) \leq 0] \approx \Phi(-\beta_u) \tag{14}$$

where β_u is the reliability index corresponding to the failure probability; $\Phi(\cdot)$ is the standard normal cumulative distribution function. In the standard normal space, the reliability index β_u can be calculated as:

$$\beta_u = ||\boldsymbol{u}^*||_2 \tag{15}$$

where $\boldsymbol{u}^*$ is a design point in standard normal space. In a geometric sense, β_u is the point corresponding to the minimum Euclidean distance from the limit state plane to the origin in standard space. The solution of design point u^* involves a constrained optimization solution problem, which is defined as

$$\beta_u = \begin{cases} \min\sqrt{u_1^2 + u_2^2 + \ldots + u_n^2} = ||\boldsymbol{u}||_2 \\ \text{s.t.} g(\boldsymbol{u}) = 0 \end{cases} \tag{16}$$

where $\boldsymbol{u} = [u_1, u_2, \ldots, u_n]$ is a random variable in a standard normal space, which corresponds to random variable $\boldsymbol{x}$ one by one and can be calculated from variable $\boldsymbol{x}$ by equal probability transformation method to obtain $\boldsymbol{u}$.

The iHL-RF method is a gradient-based line search method proposed by Hohenbichler [35], which can effectively solve the above-constrained optimization problems. iHL-RF uses the following iterative equation to search design points:

$$\boldsymbol{u}^{(k+1)} = \boldsymbol{u}^{(k)} + \lambda \boldsymbol{d} \tag{17}$$

where k is the number of iterations; d is the search direction; λ is the search step, which defined as:

$$\boldsymbol{d} = [\frac{g(\boldsymbol{u})}{\nabla g(\boldsymbol{u})} + \boldsymbol{\alpha}_u^T \boldsymbol{u}]\boldsymbol{\alpha}_u - \boldsymbol{u} \tag{18}$$

where $\nabla g(\boldsymbol{u})$ is the gradient vector of the function. $\nabla g(\boldsymbol{u})$ and $\boldsymbol{\alpha}_{\text{u}}$ are calculated as:

$$\nabla g(\boldsymbol{u}) = [\frac{\partial g}{\partial u_1}, \frac{\partial g}{\partial u_2}, \ldots, \frac{\partial g}{\partial u_n}]^T \tag{19}$$

$$\boldsymbol{\alpha}_u = \frac{-\nabla g(\boldsymbol{u})}{||\nabla g(\boldsymbol{u})||} \tag{20}$$

According to the Armijo criterion, the search step λ must meet the following conditions [36]:

$$m(\boldsymbol{u}^{(k+1)}) - m(\boldsymbol{u}^{(k)}) \leq \frac{1}{2}\lambda(\nabla g(\boldsymbol{u})^T \boldsymbol{d}) \tag{21}$$

where $m(\cdot)$ is the value function. When Equation (16) reaches the optimal value, the value function will also reach its minimum value.

$$m(\boldsymbol{u}) = \frac{1}{2}\boldsymbol{u}^2 + c|g(\boldsymbol{u})| \tag{22}$$

where c is the penalty parameter and must meet the following conditions:

$$c > \frac{||\boldsymbol{u}||}{||\nabla g(\boldsymbol{u})||} \tag{23}$$

Take a two-dimensional example as an example; its performance function is expressed as follows [37]:

$$g(\boldsymbol{x}) = x_1 - 1.7x_2 + 1.5(x_1 + 1.7x_2)^2 + 5 \tag{24}$$

where x_1 and x_2 are standard normal random variables. The iHL-RF method is used to solve the example, and the reliability of the example is 2.873. The iterative route in the process of solving the iHL-RF method is shown in Figure 1.

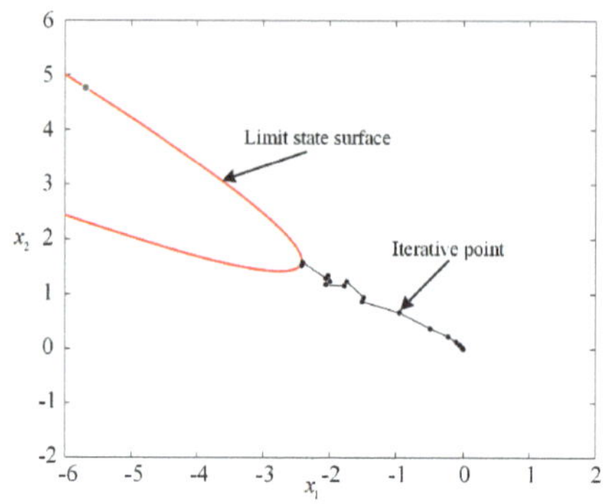

Figure 1. iHL-RF method iteration process diagram.

4. Structural Reliability Analysis Method Based on AK-MCS Method

The AK-MCS method combining active-learning Kriging (AK) and MCS has been a research hotspot in recent years, which can effectively balance the computational efficiency and precision of structural reliability analysis [22].

4.1. Kriging Modeling

The Kriging model regards the performance function as a random field model, which is mainly composed of polynomials and random processes. For a given initial sample set $\boldsymbol{X} = [\boldsymbol{x}_1, \boldsymbol{x}_2, \ldots, \boldsymbol{x}_m]^T$, $\boldsymbol{X} \in R^n$ and corresponding response set $\boldsymbol{G} = [g(\boldsymbol{x}_1), g(\boldsymbol{x}_2), \ldots g(\boldsymbol{x}_m)]^T$, the Kriging model can be expressed as follows:

$$g(\boldsymbol{x}) = \boldsymbol{f}(\boldsymbol{x})^T \boldsymbol{\beta} + \xi(\boldsymbol{x}) \tag{25}$$

where $f(x)$ is the polynomial function variable; β is the regression coefficient vector. $\xi(x)$ is a random process with mean 0 and variance σ^2, whose covariance function is defined as follows:

$$\mathrm{cov}[\xi(\boldsymbol{x}_i), \xi(\boldsymbol{x}_j)] = \sigma^2 R(\boldsymbol{\theta}, \boldsymbol{x}_i, \boldsymbol{x}_j) \tag{26}$$

$$R(\boldsymbol{\theta}, \boldsymbol{x}_i, \boldsymbol{x}_j) = \exp \sum_{k=1}^{n} \left[-\boldsymbol{\theta}^k (\boldsymbol{x}_i^k - \boldsymbol{x}_j^k) \right], \boldsymbol{\theta}^k > 0 \tag{27}$$

where $R(\boldsymbol{\theta}, \boldsymbol{x}_i, \boldsymbol{x}_j)$ is the correlation between samples $\boldsymbol{x}_i$ and $\boldsymbol{x}_j$; this paper adopts the form of the Gaussian function. θ is a parameter variable of $1 \times n$. The superscript k represents the k-th component of sample $\boldsymbol{x}_i$.

Based on sample set $\boldsymbol{X}$ and response set $\boldsymbol{G}$, the regression coefficient $\boldsymbol{\beta}$ and variance σ^2 are estimated as follows:

$$\hat{\beta} = (\boldsymbol{F}^T \boldsymbol{R}^{-1} \boldsymbol{F})^{-1} \boldsymbol{F}^T \boldsymbol{R}^{-1} \boldsymbol{G} \tag{28}$$

$$\hat{\sigma}^2 = \frac{1}{m} (\boldsymbol{G} - \boldsymbol{F}\hat{\beta})^T \boldsymbol{R}^{-1} (\boldsymbol{G} - \boldsymbol{F}\hat{\beta}) \tag{29}$$

where F is the regression coefficient matrix of the training sample; R is the regression coefficient matrix of the training sample: $R = [R_{ij}]_{m\times m}$, $R_{ij} = R(\theta, x_i, x_j)$.

The estimated value and prediction variance at unknown point x is calculated as follows:

$$\mu_{\hat{g}} = f(x)^T\beta + r(x)R^{-1}(G - F\hat{\beta}) \tag{30}$$

$$\sigma_{\hat{g}} = \hat{\sigma}^2\left[1 + u(x)^T(F^TR^{-1}F)^{-1}u(x) - r(x)^TR^{-1}r(x)\right] \tag{31}$$

$$u(x) = F^TR^{-1}r(x) - f(x) \tag{32}$$

$$r(x) = [R(\theta, x_1, x), R(\theta, x_2, x), \ldots R(\theta, x_m, x)]^T \tag{33}$$

Compared with proxy models such as NN (Neural Network) and SVM (Support Vector Machine), the prediction variance Equation (31) provided by the Kriging model can be used to measure the prediction uncertainty of the model, which is one of the basic prerequisites for carrying out the active-learning Kriging technique.

4.2. AK-MCS Method

There is a fatigue reliability analysis problem for performance function $g(x)$ and the input random variable x. The direct MCS method requires a large sample sampling of input random variables and a calculation of output response values of each group of samples in turn. The whole calculation process of the MCS method is lengthy in duration, leading the proxy model to not be adopted into the field of structural reliability analysis.

In essence, the AK-MCS method uses the Kriging proxy model to approximate the actual function. In addition, it uses the MCS method to estimate the failure probability, which can effectively avoid a lot of calls to the actual function. This is different from the traditional one-time sampling directly counting sample sets. AK-MCS will first sample a small initial sample set and then gradually sample the key areas through the active-learning function and gradually improve the global and local approximate accuracy of the Kriging model.

According to the principle of MCS calculation of failure probability, as long as the positive and negative coincidence of each sample can be correctly identified, the failure probability can be accurately calculated. For those samples with a high risk of crossing the limit state surface $\hat{g}(x)$, it must be added to the training sample set to improve the approximate accuracy of the model near the limit state surface. Samples selected for the Kriging model construction generally have three characteristics: (1) they are close to the limit state surface; (2) they have large model prediction variance, which means that the symbol of the sample is prone to be misestimated; (3) and simultaneously possessing the above two points. Therefore, the learning function is constructed as follows:

$$|\mu_{\hat{g}}(x)| - U(x)\sigma_{\hat{g}}(x) = 0 \tag{34}$$

The risk that indicates whether the sign of the Kriging model's predicted value $U(x)$ is consistent with the sign of the actual function $\hat{g}(x)$ is related to the low confidence bounding function (LCB), and Equation (34) can be redescribed as:

$$U(x) = \frac{|\mu_{\hat{g}}(x)|}{\sigma_{\hat{g}}(x)} \tag{35}$$

For U function, when $U = 2$, the probability that its symbol is incorrectly estimated at an unknown point is $\Phi(-2) = 0.023$, so $\min(U(x)) \geq 2$ can be used as the convergence criterion for the active-learning process.

Taking a two-dimensional four-series system as an example, its performance functions are defined as follows [38]:

$$g(\boldsymbol{x}) = \begin{cases} 3 + \dfrac{(x_1 - x_2)^2}{10} - \dfrac{(x_1 + x_2)}{\sqrt{2}} \\ 3 + \dfrac{(x_1 - x_2)^2}{10} + \dfrac{(x_1 + x_2)}{\sqrt{2}} \\ (x_1 - x_2) + \dfrac{7}{\sqrt{2}} \\ (x_2 - x_1) + \dfrac{7}{\sqrt{2}} \end{cases} \tag{36}$$

where x_1 and x_2 are random variables subject to a standard normal distribution, respectively. AK-MCS is used to solve the reliability problem of the above series system, and the reliability index is 2.845. The point selection process of the $\boldsymbol{U}$ function is shown in Figure 2. The selected sample points are uniformly distributed near the limit state surface. Due to the $\boldsymbol{U}$ function, we can balance the searchability of the region near the limit state surface and the global region and effectively prevent the agglomeration phenomenon of the selected sample points while ensuring the selection of points near the limit state surface.

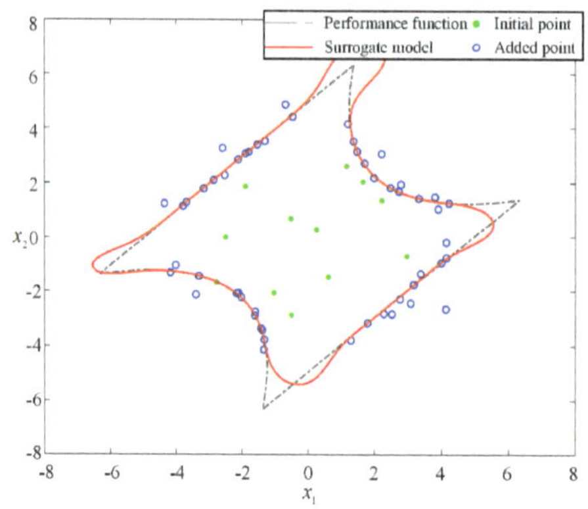

Figure 2. AK-MCS solution process.

4.3. AK-MCS Calculation Process

According to the calculation principle of AK-MCS, its specific calculation process is shown in Figure 3.

(1) A candidate sample set $\boldsymbol{S}_c$ is generated. According to the distribution of input variables, a candidate sample set $\boldsymbol{S}_c$ with size $\boldsymbol{N}_c$ is obtained with sampling.

(2) The initial training sample set $\boldsymbol{X}$ and the corresponding response value $\boldsymbol{G}$ are generated. To construct an initial Kriging model, it is necessary to obtain N initial samples $\boldsymbol{X} = [\boldsymbol{x}_1, \boldsymbol{x}_2, \ldots, \boldsymbol{x}_N]^T$ first and then calculate its true performance function response value $\boldsymbol{G} = [g(\boldsymbol{x}_1), g(\boldsymbol{x}_2), \ldots g(\boldsymbol{x}_N)]^T$. The size of the initial sample set $\boldsymbol{X}$ defines $N = \max(n+2, 12)$.

(3) Construct the Kriging model. According to the sample set $\boldsymbol{X}$ and its response value $\boldsymbol{G}$, the Kriging model is constructed based on the DACE toolbox.

(4) The predicted response value and failure probability of Kriging were calculated. Based on the Kriging model, the predicted response value $\hat{g}(\boldsymbol{x}_i)$, $i = 1, 2, \ldots N_c$ of all samples in candidate sample set $\boldsymbol{S}_c$ is calculated, and its failure probability is calculated as follows:

$$\hat{P}_f = \frac{N_{\hat{g}(\mathrm{x}_i)<0}}{N_c} \tag{37}$$

(5) Determine the next best update point x_b. Based on the Kriging method, the prediction response values and prediction variance of all samples in candidate sample set S_c are estimated, respectively. Then, function U values in all candidate samples are calculated according to Equation (35), and the sample corresponding to the minimum value is selected as the best update point x_b.

(6) Determine whether it is convergent. When $U(x_b) \geq 2$, the convergence condition is met, go to step (8); otherwise, go to step (7).

(7) Update the Kriging agent model. When the convergence criterion is not satisfied, calculate the true response value $g(x_b)$ of the best update point x_b and add it to the sample set $\boldsymbol{X}$ and its response value $\boldsymbol{G}$ to reconstruct the Kriging model and return to step (4).

(8) Calculate the coefficient of variation of failure probability. If the convergence criterion in step (6) is satisfied, the coefficient of variation for calculating the failure probability is as follows:

$$Cov(\hat{P}_f) = \sqrt{\frac{1-\hat{P}_f}{\hat{P}_f(N_c-1)}} \tag{38}$$

If $Cov(\hat{P}_f)$ is less than 0.05, go to step (10); otherwise, go back to step (9).

(9) Update the candidate sample set. If the coefficient of variation of the failure probability is oversized, the candidate sample set Sc is increased, and the update process is restarted by going back to step (4).

(10) End AK-MCS, output $\hat{P}_f$.

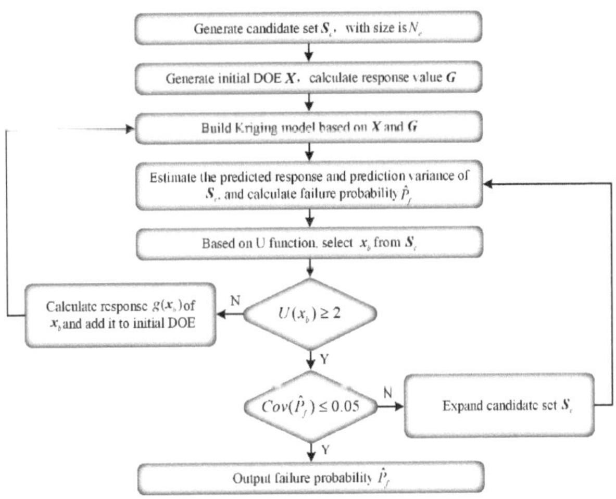

Figure 3. AK-MCS calculation flowchart.

5. Simulation Method of Crack Propagation Based on ABAQUS-FRANC3D Interactive Technique

The ABAQUS-FRANC3D interactive technique was used to analyze the fatigue crack growth. Liu et al. [39] used ABAQUS-FRANC3D interaction technology to calculate the fatigue crack stress intensity factor at the weld toe between the longitudinal rib and the roof of the steel bridge panel and verified its accuracy through tests. Two semi-elliptical initial cracks were inserted at the toe of the joint between the top plate and the U-rib weld base on FRANC3D (version 7.5.5) software. The dimensions were as follows: short half-axis a (depth direction), long half-axis c, spacing s. Two cracks were vertically symmetrically arranged along both sides of the central axis of the welding toe. A crack grid was adopted to repartition the crack, and a solid submodel with a mesh size of 0.2 mm was established. The crack propagation depth was calculated using the finite element method until the crack depth reached 50% of the roof thickness. Figure 4 shows the ABAQUS-FRANC3D interactive workflow. The model is cut along the axis to indicates collinear cracks, and the grid of collinear crack fronts is shown in Figure 4.

A long-span steel box girder suspension bridge was used to establish the finite element model, as shown in Figure 5. The model material was Q345D steel, the elastic modulus was 2.1×10^5 MPa, and the Poisson ratio was 0.3. There were two transverse partitions and two U-ribs. The model length and width were 3200 mm and 1400 mm, respectively. The thicknesses of the top plate and U-rib were 16 mm and 8 mm, respectively. The top plate U-rib upper mouth width was 300 mm, the lower mouth width 170 mm, the height 280 mm, and the transverse partition thickness was 10 mm. The welding seam with 80% was used to connect the U-rib and the top plate of the steel bridge panel. The assembly clearance parameter g between the top plate and the U-rib was 0.5 mm. The finite element model adopted two-point loading; the loading area was 200 mm × 200 mm. The three-way displacement of all nodes at the bottom of the transverse partition of the steel bridge panel was constrained.

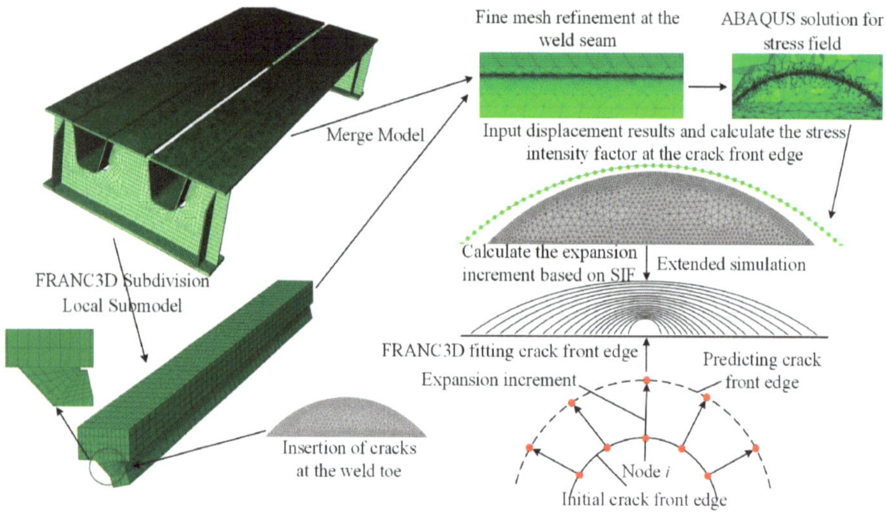

Figure 4. ABAQUS-FRANC3D interactive workflow.

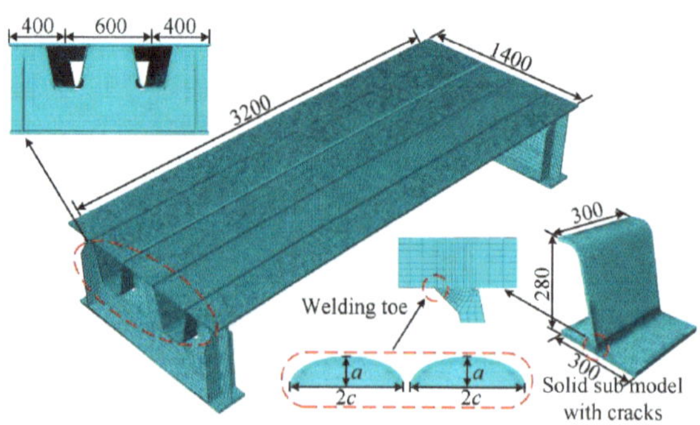

Figure 5. Parameters of U-rib and top plate of steel bridge deck. (unit: mm).

In actual engineering, the coupling effect between multiple cracks in engineering structures results in the structural fracture analysis becoming complicated. At present, the equivalent elliptic criterion is commonly used in multicrack analysis if the two cracks located on the same side are coplanar cracks. When the crack spacing is less than the long axis length of the smaller crack, the semi-elliptical crack that can envelop the two cracks can be equivalent to one crack. Figure 6 shows the multiple crack equivalent characterization method [40].

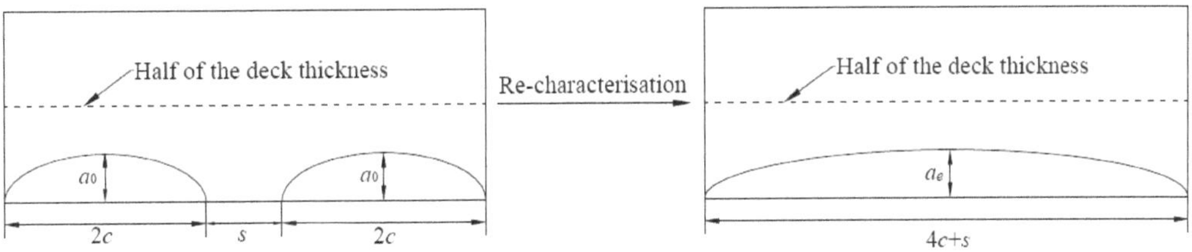

Figure 6. The multiple crack equivalent characterization method.

The stress intensity factor and crack depth at the crack front of a collinear double crack can be calculated with the interaction technique of ABAQUS-FRANC3D. Table 2 shows the equivalent crack depth under a different initial crack depth and tip spacing. The above response surface and the corresponding test data points are represented as shown in Figure 7. Figure 7 shows that there is a certain correlation between α_e, α_0 and R_s. In this study, a polynomial response surface is used to construct the relationship between the three based on the test data. The overall trend of the test data points and the response surface is consistent.

Table 2. Relevant test data between α_e, α_0 and R_s.

α_0/mm	R_s	α_e/mm	α_0/mm	R_s	α_e/mm
1	0.5	1.2948	3	0.33	3.348
1	1	1.6284	3	0.5	3.5868
1	1.5	1.9169	3	0.66	3.8187
1	2	2.2819	3	0.83	4.174
1	2.5	2.6597	3	1	4.41
1	3	3.06	4	0.25	4.271
2	0.5	2.437	4	0.375	4.4913
2	1.5	2.682	4	0.5	4.681
2	1	3	4	0.625	4.8997
2	1.25	3.364	4	0.75	4.8
2	1.5	3.79			

Third order polynomial with cross terms is used to fit the above data

$$\begin{aligned}\alpha_e = 1.814 - 1.519\alpha_0 - 2.184R_s + 0.8039{\alpha_0}^2 + 3.195\alpha_0 R_s + 0.8079{R_s}^2 - 0.0804{\alpha_0}^3 \\ -0.4312{\alpha_0}^2 R_s - 0.8022\alpha_0 {R_s}^2 + 0.00938{R_s}^3\end{aligned} \quad (39)$$

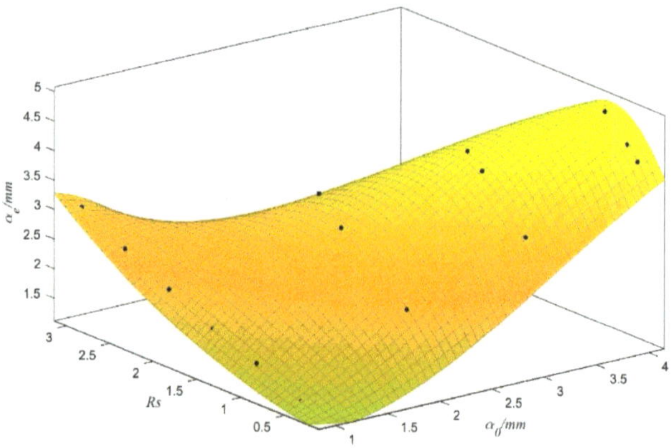

Figure 7. Test data and response surface.

6. Reliability Calculation

To study the influence of annual traffic growth on the fatigue reliability of the steel bridge deck, this section analyzes the cases where annual traffic growth is 0%, 1%, 2% and 3%, respectively, while considering the difference between the driving lane and passing lane on the steel bridge deck. The equivalent stress amplitude S_{eq} and the number of daily stress cycles N_{d} of different lanes are corrected accordingly; namely, different variable distribution parameters are adopted.

Figure 8 shows the fatigue reliability analysis results of the steel deck based on the MCS method (for driving lane state) for both the traditional single-crack model and the multicrack coupling effect. Figure 8 shows that when the single-crack model is adopted, the fatigue reliability indexes of 0%, 1%, 2% and 3% corresponding to the annual traffic increase in the bridge design base period (100 years) are 3.88, 3.19, 2.71 and 2.32, respectively. Considering the coupling effect of double cracks, the fatigue reliability indexes of 0%, 1%, 2% and 3% corresponding to the annual traffic increase are 3.30, 2.63, 2.15 and 1.77, respectively. Compared with the single-crack model, considering the coupling effect of double cracks, the fatigue reliability indexes of steel deck all decrease.

In order to verify the computational efficiency of the iHL-RF method and AK-MCS method in fatigue reliability analysis of steel bridge deck, this study adopts the above methods to solve the fatigue reliability indexes of steel bridge deck driving lanes when the annual traffic growth volume α_{y} is 0%, 1%, 2% and 3%, respectively. Figures 9–12 shows the analysis result. Figures 9–12 shows that both the iHL-RF method and AK-MCS method can efficiently solve the fatigue reliability problem of steel bridge decks, and the errors are within the acceptable range of engineering. To further compare the computational efficiency of the iHL-RF method and the AK-MCS method, the reliability results of the single crack and multicrack under different working conditions are shown in Tables 3 and 4. The AK-MCS method has a great reduction in the number of function calls compared with the iHL-RF method in both single-crack and double-crack cases, and the estimated relative error is less than 2% in all cases.

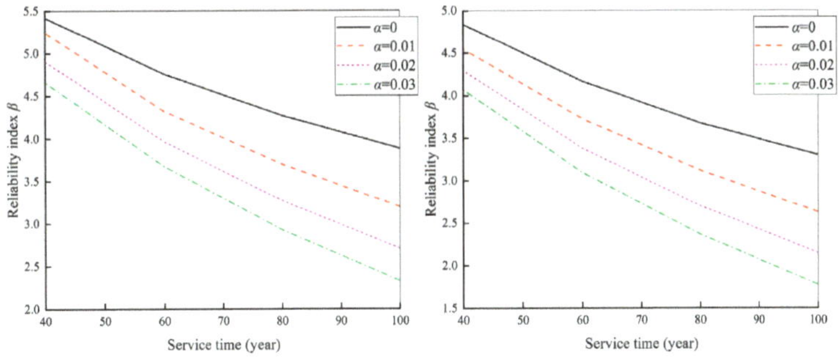

(**a**) (**b**)

Figure 8. Fatigue reliability analysis based on MCS (driving lane). (**a**) Single crack. (**b**) Double crack.

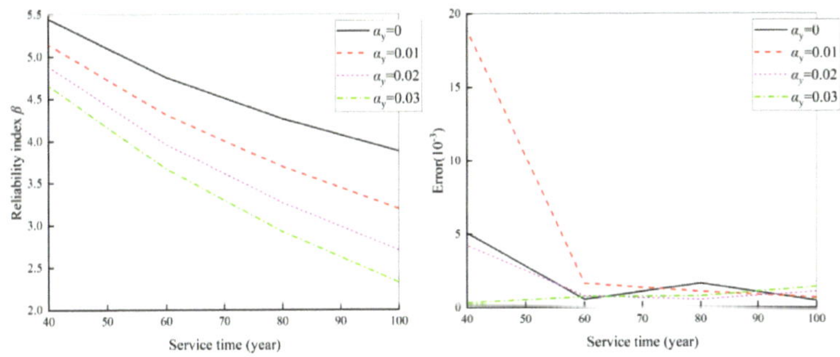

(**a**) (**b**)

Figure 9. Fatigue reliability analysis of single crack based on iHL-RF (driving lane). (**a**) Reliability calculation results. (**b**) Estimation error.

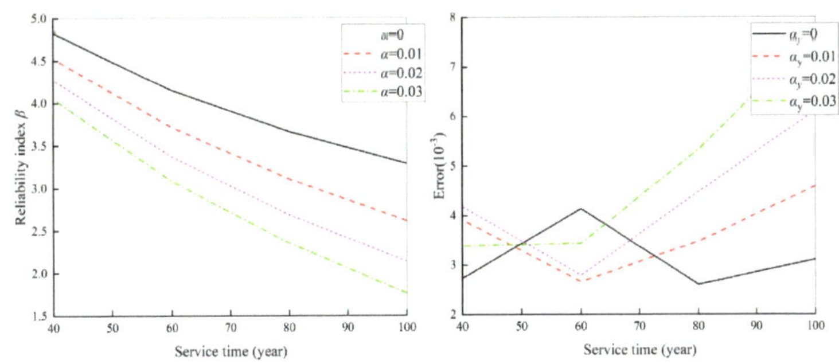

(**a**) (**b**)

Figure 10. Fatigue reliability analysis of double cracks based on iHL-RF (driving lane). (**a**) Reliability calculation results. (**b**) Estimation error.

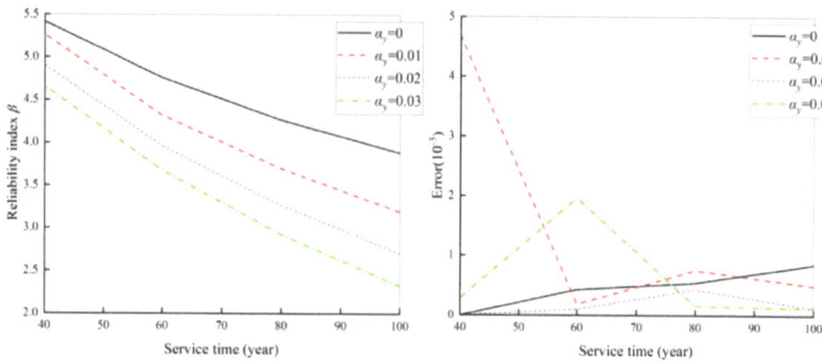

(**a**) (**b**)

Figure 11. Fatigue reliability analysis of single crack based on AK-MCS (driving lane). (**a**) Reliability calculation results. (**b**) Estimation error.

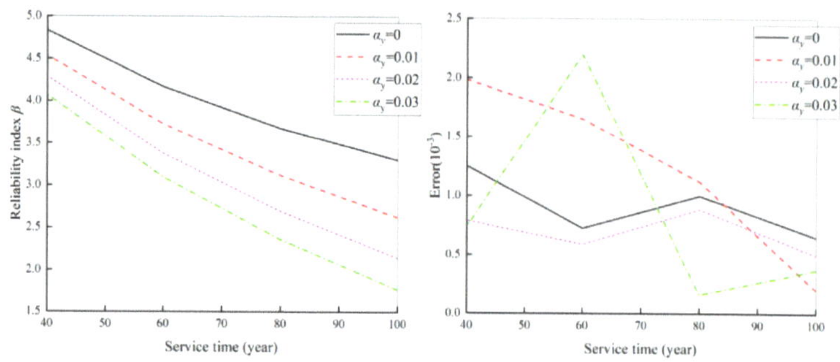

(**a**) (**b**)

Figure 12. Fatigue reliability analysis of double cracks based on AK-MCS (driving lane). (**a**) Reliability calculation results. (**b**) Estimation error.

Figure 13 shows the fatigue reliability analysis results of the steel bridge deck based on the MCS method (for passing lane state). Figure 13 shows the reliability of the single-crack model and the multicrack coupling effect model when the annual flow growth rate is 0%, 1%, 2% and 3%, respectively. The fatigue reliability index when the bridge reaches the design base period (100 years) is superior to the target fatigue reliability index of 1.5. Compared with the single-crack model, considering the coupling effect of double cracks, the fatigue reliability indexes of steel decks all decrease.

In order to verify the computational efficiency of the iHL-RF method and AK-MCS method for steel bridge deck fatigue reliability analysis (passing lane state), this paper adopts the above methods to solve the fatigue reliability indexes of the steel bridge deck passing lane with annual traffic growth of 0%, 1%, 2% and 3%, respectively. Since the equivalent stress amplitude and the number of daily stress cycles of the passing lane are smaller than that of the driving lane, the failure probability of the passing lane within about 100 years is relatively small. Therefore, longer service lives are used to analyze the reliability of passing lanes, which are 180, 200, 220, and 240 years, respectively. Figures 14–17 show the analysis results. It can be seen that the accuracy and effectiveness of the iHL-RF method and the AK-MCS method can be effectively verified under the passing lane conditions. To further compare the computing efficiency of the iHL-RF method and the AK-MCS method, Tables 5 and 6 show the calculation results. The results of the two methods are similar to those under driving lane conditions. Compared with the iHL-RF method, the number of calls to performance functions with the AK-MCS method is crucially reduced, and the estimated relative errors in all cases are less than 2%.

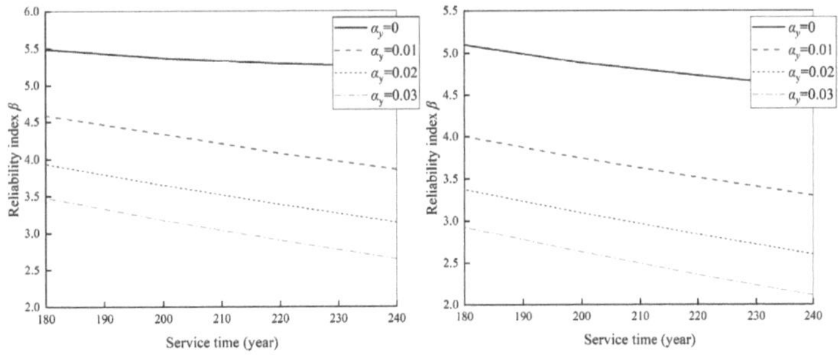

(a) (b)

Figure 13. Fatigue reliability analysis based on MCS (passing lane). (**a**) Single crack. (**b**) Double cracks.

Table 3. Fatigue reliability analysis results of single crack (driving lane).

Conditions	MCS	iHL-RF		AK-MCS	
	β	β	N_{call}	β	N_{call}
$n = 40, \alpha_y = 0\%$	5.42	5.45	328	5.42	121
$n = 40, \alpha_y = 1\%$	5.24	5.14	306	5.27	156
$n = 40, \alpha_y = 2\%$	4.90	4.88	302	4.90	188
$n = 40, \alpha_y = 3\%$	4.66	4.66	286	4.66	156
$n = 60, \alpha_y = 0\%$	4.76	4.75	300	4.76	212
$n = 60, \alpha_y = 1\%$	4.32	4.31	278	4.32	247
$n = 60, \alpha_y = 2\%$	3.96	3.96	274	3.96	259
$n = 60, \alpha_y = 3\%$	3.67	3.67	272	3.68	129
$n = 80, \alpha_y = 0\%$	4.27	4.26	278	4.27	206
$n = 80, \alpha_y = 1\%$	3.70	3.69	272	3.70	186
$n = 80, \alpha_y = 2\%$	3.27	3.27	262	3.27	143
$n = 80, \alpha_y = 3\%$	2.93	2.92	252	2.93	157
$n = 100, \alpha_y = 0\%$	3.88	3.88	274	3.88	125
$n = 100, \alpha_y = 1\%$	3.19	3.19	252	3.19	143
$n = 100, \alpha_y = 2\%$	2.71	2.70	252	2.71	162
$n = 100, \alpha_y = 3\%$	2.33	2.32	254	2.33	179

Table 4. Fatigue reliability analysis results of double cracks (driving lane).

Conditions	MCS	iHL-RF		AK-MCS	
	β	β	N_{call}	β	N_{call}
$n = 40, \alpha_y = 0\%$	4.84	4.83	346	4.83	104
$n = 40, \alpha_y = 1\%$	4.55	4.52	324	4.54	122
$n = 40, \alpha_y = 2\%$	4.29	4.27	324	4.29	127
$n = 40, \alpha_y = 3\%$	4.07	4.05	310	4.07	135
$n = 60, \alpha_y = 0\%$	4.16	4.15	302	4.17	153
$n = 60, \alpha_y = 1\%$	3.72	3.71	294	3.73	125
$n = 60, \alpha_y = 2\%$	3.38	3.36	286	3.38	139
$n = 60, \alpha_y = 3\%$	3.09	3.08	276	3.10	154
$n = 80, \alpha_y = 0\%$	3.67	3.66	282	3.68	129
$n = 80, \alpha_y = 1\%$	3.12	3.10	276	3.12	153
$n = 80, \alpha_y = 2\%$	2.70	2.69	276	2.70	171
$n = 80, \alpha_y = 3\%$	2.36	2.35	270	2.36	185
$n = 100, \alpha_y = 0\%$	3.30	3.29	286	3.30	146
$n = 100, \alpha_y = 1\%$	2.63	2.61	274	2.62	169
$n = 100, \alpha_y = 2\%$	2.15	2.13	266	2.14	193
$n = 100, \alpha_y = 3\%$	1.77	1.76	260	1.77	112

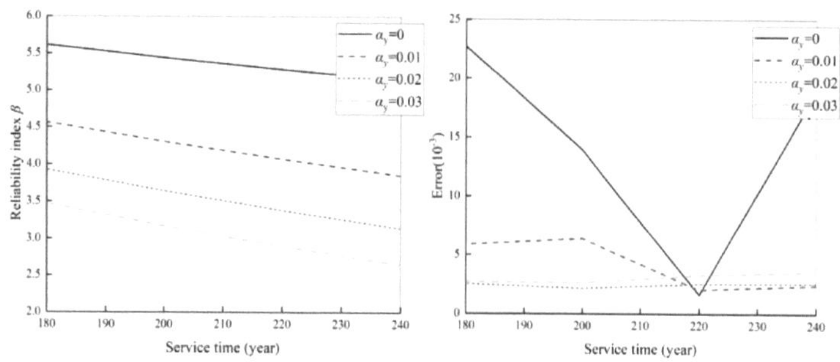

(a) (b)

Figure 14. Single-crack fatigue reliability analysis based on iHL-RF (passing lane). (**a**) Single crack. (**b**) Double cracks.

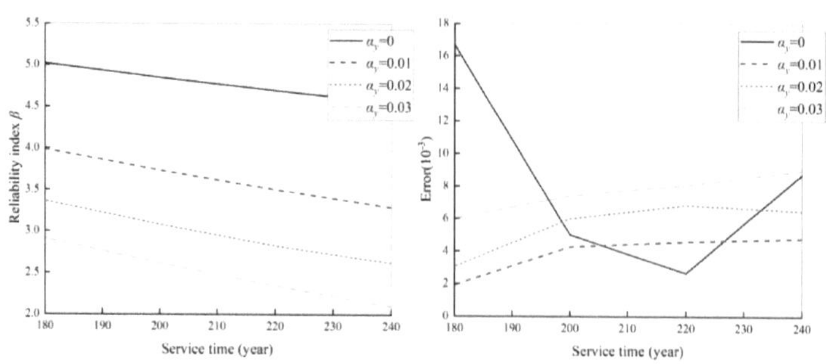

(a) (b)

Figure 15. Double-crack fatigue reliability analysis based on iHL-RF (passing lane). (**a**) Reliability calculation results. (**b**) Estimation error.

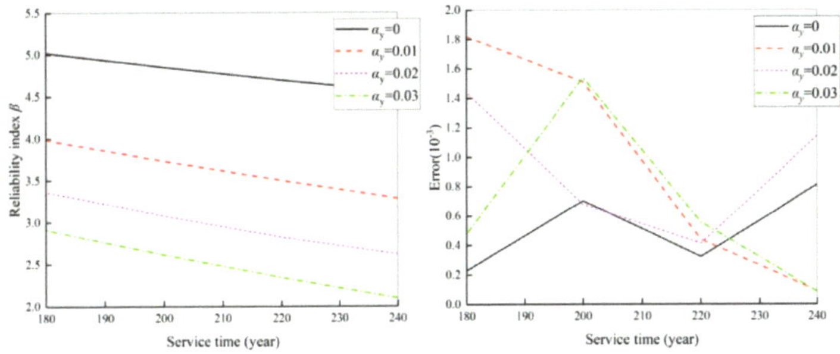

(**a**) (**b**)

Figure 16. Single-crack fatigue reliability analysis based on AK-MCS (passing lane). (**a**) Reliability calculation results, (**b**) Estimation error.

Table 5. Fatigue reliability analysis results of single crack (passing lane).

Conditions	MCS	iHL-RF		AK-MCS	
	β	β	N_{call}	β	N_{call}
$n = 180, \alpha_y = 0\%$	5.49	5.62	346	5.49	156
$n = 180, \alpha_y = 1\%$	4.58	4.56	286	4.58	162
$n = 180, \alpha_y = 2\%$	3.94	3.93	274	3.93	161
$n = 180, \alpha_y = 3\%$	3.48	3.47	270	3.48	138
$n = 200, \alpha_y = 0\%$	5.37	5.44	342	5.37	164
$n = 200, \alpha_y = 1\%$	4.33	4.30	278	4.33	197
$n = 200, \alpha_y = 2\%$	3.65	3.64	272	3.65	136
$n = 200, \alpha_y = 3\%$	3.18	3.17	252	3.17	147
$n = 220, \alpha_y = 0\%$	5.29	5.29	326	5.30	172
$n = 220, \alpha_y = 1\%$	4.08	4.07	274	4.07	189
$n = 220, \alpha_y = 2\%$	3.39	3.38	252	3.39	142
$n = 220, \alpha_y = 3\%$	2.90	2.89	252	2.90	172
$n = 240, \alpha_y = 0\%$	5.24	5.14	322	5.24	139
$n = 240, \alpha_y = 1\%$	3.86	3.85	274	3.86	160
$n = 240, \alpha_y = 2\%$	3.15	3.13	252	3.14	148
$n = 240, \alpha_y = 3\%$	2.65	2.64	252	2.65	176

Table 6. Fatigue reliability analysis results of double cracks (passing lane).

Conditions	MCS	iHL-RF		AK-MCS	
	β	β	N_{call}	β	N_{call}
$n = 180, \alpha_y = 0\%$	5.10	5.02	350	5.10	176
$n = 180, \alpha_y = 1\%$	4.00	3.99	310	3.99	169
$n = 180, \alpha_y = 2\%$	3.38	3.36	286	3.37	152
$n = 180, \alpha_y = 3\%$	2.93	2.91	278	2.93	164
$n = 200, \alpha_y = 0\%$	4.88	4.85	348	4.87	171
$n = 200, \alpha_y = 1\%$	3.75	3.73	304	3.75	184
$n = 200, \alpha_y = 2\%$	3.10	3.08	276	3.10	163
$n = 200, \alpha_y = 3\%$	2.63	2.62	274	2.64	180
$n = 220, \alpha_y = 0\%$	4.73	4.69	346	4.71	178
$n = 220, \alpha_y = 1\%$	3.52	3.50	284	3.52	141
$n = 220, \alpha_y = 2\%$	2.84	2.82	278	2.84	172
$n = 220, \alpha_y = 3\%$	2.36	2.35	272	2.36	193
$n = 240, \alpha_y = 0\%$	4.59	4.55	328	4.59	145
$n = 240, \alpha_y = 1\%$	3.30	3.28	286	3.30	147
$n = 240, \alpha_y = 2\%$	2.60	2.62	258	2.60	185
$n = 240, \alpha_y = 3\%$	2.12	2.10	266	2.12	117

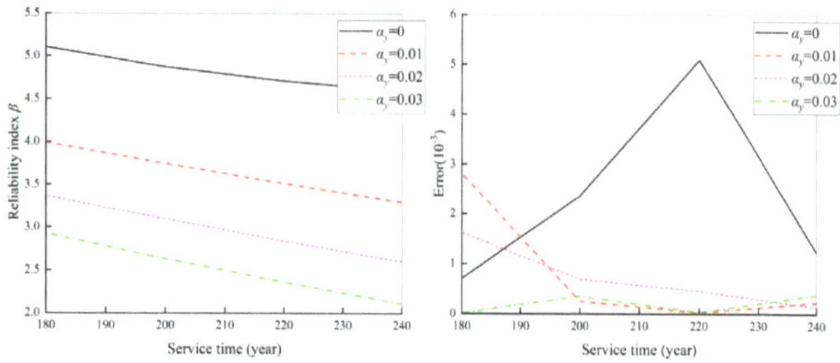

(a) (b)

Figure 17. Double-crack fatigue reliability analysis based on AK-MCS (passing lane). (**a**) Reliability calculation results, (**b**) Estimation error.

Figures 13–17 show the fatigue reliability of steel bridge deck decreases with the increase in annual traffic volume, whether it is the driving lane or passing lane. Compared with the single-crack model, considering the coupling effect of double cracks, the fatigue reliability of the steel deck will be reduced to a certain extent. Both the iHL-RF method and the AK-MCS method can effectively solve the above fatigue reliability analysis problems. The performance function for calculating the reliability of bridge deck panels is quite complex. Using the AK-MCS algorithm to calculate the fatigue reliability of steel bridge decks can effectively reduce the number of calls to the performance function.

7. Conclusions

This study focuses on the fatigue reliability analysis of orthotropic steel decks. A fatigue reliability analysis model considering the coupling effect of multiple cracks and the randomness of vehicle load is established. The iHL-RF method and AK-MCS method for fatigue reliability analysis are developed to solve the problem that the finite element model analysis of random traffic flow takes time consumption. Considering the load difference of steel bridge decks under two working conditions, the fatigue reliability of steel bridge decks is analyzed, respectively. The main conclusions are as follows:

(1) Compared with the traffic lane, the equivalent stress amplitude S_{eq} and the number of daily stress cycles N_d of the passing lane are smaller. The fatigue reliability of the passing lane on the steel bridge deck is higher under the same working conditions (annual traffic increase and service time).

(2) Compared with the single-crack model, the fatigue reliability of steel decks will be significantly reduced, considering the coupling effect of double cracks. The main reason is that when the coupling effect of double cracks is considered, the threshold item $\psi(\alpha_c, \alpha_0)$ in the fatigue reliability function is relatively small, which leads to an increase in the probability of fatigue failure of steel bridge decks. When the design life reaches 100 years and the annual traffic growth amount α_y = 3%, the multicrack fatigue reliability of the steel bridge deck driving lane is lower than 2.

(3) The annual traffic increase α_y has a crucial influence on the fatigue reliability of steel bridge decks. For the driving lanes, the reliability difference between annual traffic growth α_y = 1% and α_y = 3% after 100 years of service is 1.55 (single crack) and 1.53 (double crack). For the passing lane, the difference in reliability between annual traffic growth α_y = 1% and α_y = 3% after 240 years of service is 2.59 (single crack) and 2.46 (double crack).

(4) The MCS, iHL-RF and AK-MCS methods can effectively solve the above fatigue reliability analysis problems. However, considering the calculation efficiency, iHL-RF and AK-MCS can effectively reduce the number of function calls and ensure calculation accuracy. When the performance function involves a complex and time-consuming finite element

analysis process, AK-MCS can well balance the computational efficiency and calculation accuracy of fatigue reliability analysis.

Author Contributions: J.L.—Conceptualization; methodology; validation; formal analysis; resources; writing—original draft preparation; writing—review and editing; supervision; funding acquisition; Y.L.—methodology; data curation; writing—original draft preparation; supervision; project administration; G.W.—Conceptualization; validation; formal analysis; resources; writing—original draft preparation; writing—review and editing; supervision; funding acquisition; N.L.—Conceptualization; methodology; validation; investigation; writing—original draft preparation; writing—review and editing; supervision; J.C.—software; validation; investigation; writing—original draft preparation; H.W.—software; validation; investigation; writing—original draft preparation. All authors have read and agreed to the published version of the manuscript.

Funding: This study was financially supported by the National Science Foundation of China (grant number 52308138), the innovative projects of Key Disciplines of Civil Engineering of Changsha University and Science and Technology (23ZDXK05), and the Hunan Graduate Innovation Project (QL20210186).

Data Availability Statement: The raw data supporting the conclusions of this article will be made available by the authors on request.

Conflicts of Interest: The authors declare that they have no known competing financial interests or personal relationships that could have appeared to influence the work reported in this paper.

Nomenclature

α	fatigue crack size	β_u	reliability index corresponding to the failure probability
N	number loading cycles	$\Phi(\cdot)$	standard normal cumulative distribution function
C	fatigue growth correlation coefficient	u^*	design point in standard normal space
m	fatigue growth correlation coefficient	$\boldsymbol{u}$	random variable in a standard normal space
ΔK	amplitude of stress intensity factor	k	number of iterations
Seq	equivalent stress amplitude under varying amplitude load	d	search direction
Y	geometric function considering the crack shape of the member	λ	search step
α_N	crack size of the in-service structure after N stress cycles	$\nabla g(u)$	gradient vector of the function
$\psi(\alpha_c, \alpha_0)$	fatigue damage accumulation function from the initial crack size to the critical crack size	$m(\cdot)$	value function
$\psi(\alpha_N, \alpha_0)$	damage accumulation function from the initial crack size α0 through N stress cycles to α_N	c	penalty parameter
N_d	number of daily cycles of stress	$f(x)$	polynomial function variable
n	service life of the bridge	β	regression coefficient vector
Y_0	boundary correction factors for the reference stress intensity factor	$\xi(x)$	random process
Y_e	boundary correction factors for the reference stress intensity factor	$R(\theta, x_i, x_j)$	correlation
α_e	crack depth of collinear double crack considering the coupling effect equivalent to a single crack	θ	parameter variable
T	plate thickness	$\boldsymbol{F}$	regression coefficient matrix of the training sample
R_s	collinear double-crack spacing ratio	$\boldsymbol{R}$	regression coefficient matrix of the training sample
R_s	double-crack spacing ratio	$U(x)$	consistent with the sign of the actual function
α_0	initial crack depth	$\hat{g}(x)$	related to the low confidence bounding function
e	wheel track transverse distribution coefficient		

References

1. Ke, L.; Wang, Y.; Li, C.; Chen, Z.; Feng, Z.; Li, Y. Fatigue evaluation and structural optimization of rib-to-diaphragm connection in orthotropic steel decks. *J. Constr. Steel Res.* **2023**, *208*, 107998. [CrossRef]
2. Luo, Y.; Liu, X.; Chen, F.; Zhang, H.; Xiao, X. Numerical simulation on Crack–inclusion interaction for rib-to-deck welded joints in orthotropic steel deck. *Metals* **2023**, *13*, 1402. [CrossRef]
3. Yuan, D.; Cui, C.; Zhang, Q.; Zhang, X.; Li, Z. Influence of resin asphalt pavement on stress behaviors of double-side welded rib-to-deck joints in orthotropic steel decks. *J. Constr. Steel Res.* **2022**, *197*, 107491. [CrossRef]
4. Dang, H.; Liang, A.; Feng, R.; Zhang, J.; Yu, X.; Shao, Y. Experimental study on mechanical properties and low-cycle fatigue behaviour of stainless steels subjected to salt spray and dry/wet cycle. *Int. J. Fatigue* **2023**, *165*, 107187. [CrossRef]
5. Tang, X.; Roberts, G.W.; Li, X.; Hancock, C.M. Real-time kinematic PPP GPS for structure monitoring applied on the Severn Suspension Bridge, UK. *Adv. Space Res.* **2017**, *60*, 925–937. [CrossRef]
6. Megawati, K.; Higashihara, H.; Koketsu, K. Derivation of near-source ground motions of the 1995 Kobe (Hyogo-ken Nanbu) earthquake from vibration records of the Akashi Kaikyo Bridge and its implications. *Eng. Struct.* **2001**, *23*, 1256–1268. [CrossRef]
7. Liao, D.; Zhu, S.-P.; Keshtegar, B.; Qian, G.; Wang, Q. Probabilistic framework for fatigue life assessment of notched components under size effects. *Int. J. Mech. Sci.* **2020**, *181*, 105685. [CrossRef]
8. Niu, X.; Zhu, S.-P.; He, J.-C.; Liao, D.; Correia, J.A.; Berto, F.; Wang, Q. Defect tolerant fatigue assessment of AM materials: Size effect and probabilistic prospects. *Int. J. Fatigue* **2022**, *160*, 106884. [CrossRef]
9. Kharin, V. Crack tip hydrogen diffusion with multiple type traps and implications for hydrogen assisted cracking. *Eng. Fract. Mech.* **2023**, *291*, 109473. [CrossRef]
10. Bai, Z.; Song, S. Structural reliability analysis based on neural networks with physics-informed training samples. *Eng. Appl. Artif. Intell.* **2023**, *126*, 107157. [CrossRef]
11. Conceição António, C.; Hoffbauer, L.N. Reliability-based design optimization and uncertainty quantification for optimal conditions of composite structures with non-linear behavior. *Eng. Struct.* **2017**, *153*, 479–490. [CrossRef]
12. Wang, L.; Ni, B.; Wang, X.; Li, Z. Reliability-based topology optimization for heterogeneous composite structures under interval and convex mixed uncertainties. *Appl. Math. Model.* **2021**, *99*, 628–652. [CrossRef]
13. Peng, X.; Qiu, C.; Li, J.; Wu, H.; Liu, Z.; Jiang, S. Multiple-scale uncertainty optimization design of hybrid composite structures based on neural network and genetic algorithm. *Compos. Struct.* **2021**, *262*, 113371. [CrossRef]
14. Subramanian, R.; Anantharaman, V. Reliability analysis of a complex standby redundant systems. *Reliab. Eng. Syst. Saf.* **1995**, *48*, 57–70. [CrossRef]
15. Souza, O.L.D.C.; Sánchez Filho, E.D.S.; Vaz, L.E.; Silva Filho, J.J.H. Reliability analysis of RC beams strengthened for torsion with carbon fibre composites. *Struct. Concr.* **2014**, *15*, 38–44. [CrossRef]
16. Siddiqui, N.A.; Khan, F.H.; Umar, A. Reliability of underground concrete barriers against normal missile impact. *Comput. Concr.* **2009**, *6*, 79–93. [CrossRef]
17. Jia, D.; Wu, Z. An improved adaptive Kriging model for importance sampling reliability and reliability global sensitivity analysis. *Struct. Saf.* **2024**, *107*, 102427. [CrossRef]
18. Nan, H.; Liang, H.; Di, H.; Li, H. A gradient-assisted learning function of Kriging model for robust design optimization. *Reliab. Eng. Syst. Saf.* **2024**, *244*, 109944. [CrossRef]
19. Fan, X.; Wang, P.; Hao, F. Reliability-based design optimization of crane bridges using Kriging-based surrogate models. *Struct. Multidiscip. Optim.* **2019**, *59*, 993–1005. [CrossRef]
20. Du, W.; Luo, Y.; Wang, Y. Time-variant reliability analysis using the parallel subset simulation. *Reliab. Eng. Syst. Saf.* **2019**, *182*, 250–257. [CrossRef]
21. Lv, Z.; Lu, Z.; Wang, P. A new learning function for Kriging and its applications to solve reliability problems in engineering. *Comput. Math. Appl.* **2015**, *70*, 1182–1197. [CrossRef]
22. Echard, B.; Gayton, N.; Lemaire, M. AK-MCS: An active learning reliability method combining Kriging and Monte Carlo Simulation. *Struct. Saf.* **2011**, *33*, 145–154. [CrossRef]
23. Zhao, Z.; Haldar, A.; Breen, F.L. Fatigue-reliability evaluation of steel bridges. *J. Struct. Eng.* **1994**, *120*, 1608–1623. [CrossRef]
24. Leander, J. Reliability evaluation of the Eurocode model for fatigue assessment of steel bridges. *J. Constr. Steel Res.* **2018**, *141*, 1–8. [CrossRef]
25. Chen, J.; Diao, B.; He, J.; Pang, S.; Guan, X. Equivalent surface defect model for fatigue life prediction of steel reinforcing bars with pitting corrosion. *Int. J. Fatigue* **2018**, *110*, 153–161. [CrossRef]
26. Toyoda-Makino, M. Cost-based optimal history-dependent inspection strategy for random fatigue crack growth. *Probabilistic Eng. Mech.* **1999**, *14*, 339–347. [CrossRef]
27. Internation Institute of Welding. *Recommendations for Fatigue Design of Welded Joints and Components: IIW-1823-07 [R]*; IIW: Paris, France, 2008.
28. Qiang, B.; Qiu, H.; Li, Y.; Wang, X.; Kang, G. Stress intensity factors and weight functions for semi-elliptical cracks at weld toes in U-rib-to-deck joints. *Theor. Appl. Fract. Mech.* **2023**, *123*, 103697. [CrossRef]
29. Cheng, X.; Shi, D.; Liu, C.; Xia, R.; Zhang, Y.; Zhou, J. Double cracks with single-inclusion fatigue propagation of surface-quenched large modulus rack. *Adv. Mech. Eng.* **2019**, *11*, 168781401989574. [CrossRef]
30. Zhang, R.; Mahadevan, S. Reliability-based reassessment of corrosion fatigue life. *Struct. Saf.* **2001**, *23*, 77–91. [CrossRef]

31. Zhang, R.; Mahadevan, S. Fatigue reliability analysis using nondestructive inspection. *J. Struct. Eng.* **2001**, *7*, 957–965. [CrossRef]
32. Liu, Y.; Lu, N.; Deng, Y. Fatigue reliability assessment of steel bridge decks under measured traffic flow. *China J. Highw. Transp.* **2016**, *29*, 58–66.
33. Deng, Y.; Li, A.; Liu, Y. Probabilistic modeling of fatigue loading effects and fatigue reliability evaluation for steel bridges based on monitored data. *China Civ. Eng. J.* **2014**, *47*, 79–87.
34. Deng, Y.; Li, A. Fatigue reliability analysis for welds of U ribs in steel box girders based on fracture mechanics and long-term monitoring data. *J. Southeast Univ. (Nat. Sci. Ed.)* **2019**, *49*, 68–75.
35. Hohenbichler, M.; Rackwitz, R. Improvement of second-order reliability estimates by importance sampling. *J. Eng. Mech.* **1988**, *114*, 2195–2199. [CrossRef]
36. Xia, Y.; Xie, B.; Tang, F.; Yu, Y.Y.; Li, J. An improved approach of Armijo-based adaptive stability transformation method. *Structures* **2023**, *50*, 1827–1837. [CrossRef]
37. Jiang, C.; Han, S.; Ji, M. A new method to solve thestructural reliability index based on homotopy analysis. *Acta Mech.* **2015**, *226*, 1067–1083. [CrossRef]
38. Shi, Z.; Lu, Z.; Zhang, X.; Li, L. A novel adaptive support vector machine method for reliability analysis. *J. Risk Reliab.* **2021**, *235*, 896–908. [CrossRef]
39. Liu, Y.; Chen, F.; Lu, N.; Wang, L.; Wang, B. Fatigue performance of rib-to-deck double-side welded joints in orthotropic steel decks. *Eng. Fail. Anal.* **2019**, *105*, 127–142. [CrossRef]
40. *BS7910-Amendment 1*; Guide to Methods for Assessing the Acceptability of Flaws in Metallic Structures. UK British Standards Institution: London, UK, 2019.

metals

MDPI

Article

Development and Performance Evaluation of a Mechanical Connection for Steel and Shape Memory Alloy Bars

Min-Kyu Song [1,2], Eunsoo Choi [3] and Jong-Han Lee [2,*]

1 Smart Engineering Team, Civil Business Division, DL E&C, Seoul 03181, Republic of Korea; minkyusong@dlenc.co.kr
2 Department of Civil Engineering, Inha University, Incheon 22212, Republic of Korea
3 Department of Civil Engineering, Hongik University, Seoul 04066, Republic of Korea; eunsoochoi@hongik.ac.kr
* Correspondence: jh.lee@inha.ac.kr

Abstract: Shape memory alloys (SMAs) demonstrate a shape memory effect and superelasticity that can provide recovery performance to structural members. In this study, a round SMA bar was designed to replace the conventional deformed steel bar, particularly within the plastic hinge section of structural members. To integrate the SMA bar and the existing steel bar, a mechanical coupler was proposed by utilizing the advantages of both one-touch and threaded couplers. Uniaxial tensile tests were conducted to analyze the performance of the proposed coupler and the mechanical properties of the SMA–steel connected bar. Stress and strain relationships were examined for steel bars mechanically connected with the SMA bar and for SMA bars before and after exhibiting the shape memory effect. To induce the shape memory effect, SMA should be heated above the finished austenite temperature. Due to the difficulty of accurately measuring strain on the heated bar using traditional contact methods, we employed digital image correlation technology for precise strain measurement of the heated SMA bar. The experimental results indicate the effective application of SMA bars within the plastic hinge region of structural members using the proposed mechanical coupler.

Keywords: shape memory alloy (SMA); mechanical coupler; SMA bar; deformed steel bar; uniaxial tensile test; plastic hinge section; digital image correlation; experimental analysis

Citation: Song, M.-K.; Choi, E.; Lee, J.-H. Development and Performance Evaluation of a Mechanical Connection for Steel and Shape Memory Alloy Bars. *Metals* **2024**, *14*, 300. https://doi.org/10.3390/met14030300

Academic Editors: Francesca Borgioli, Denis Benasciutti, Umberto Prisco and Tomasz Tański

Received: 22 January 2024
Revised: 25 February 2024
Accepted: 25 February 2024
Published: 2 March 2024

1. Introduction

Concrete structures are susceptible to deformation and cracking under external loads. In particular, seismic loads cause residual deformation in structural members. This residual deformation triggers a P-Δ second-order effect that has a significant impact on the safety of structures. Moreover, residual deformation poses challenges when repairing and reinforcing structures for reuse. The integration of materials with recovery properties, such as shape memory alloys (SMAs), becomes essential in eliminating residual deformation from structural members.

The discovery of the shape memory effect in SMAs traces back to the 1950s in certain materials [1]. Active investigation into SMAs began with the discovery of Ni–Ti alloys in 1963 [2]. Subsequently, a variety of SMAs emerged, including Fe-based alloys, Cu-based alloys, Ni–Ti alloys, and HTSMAs [3]. Research into displacement and strain recovery performance according to the thermodynamic behavior of diverse SMAs continued until the 2010s [4–8]. The exploration of introducing SMAs into civil engineering structures commenced in the 2000s. Janke et al. [9] examined the potential introduction of SMAs into civil engineering structures, addressing the nonlinear aspects of the shape memory effect and superelasticity related to fatigue. As the feasibility of introducing SMAs to structural members gained attention, research emphasized processing SMAs into fiber forms to achieve self-centering properties by incorporating SMA fibers into cement mortar [10,11]. In addition, research has also pursued stress recovery and crack closure capabilities of cold-drawn and rolled wires [12,13]. Subsequent research has been conducted using SMA wires to prestress structural elements [14,15], and research continues to apply SMAs to structural members using various methods [16–20].

This study utilized an SMA bar fabricated from a nickel–titanium (Ni-Ti) alloy known for its safety, practicality, and excellent thermodynamic performance. The Ni–Ti SMA enables displacement recovery through the shape memory effect, which returns to its original shape above a specific temperature threshold, and the superelastic ability, which returns to its initial state without residual deformation even after surpassing the yield point [21–23]. A recent study was conducted to enhance the seismic performance of concrete walls by considering the introduction of SMA bars to replace conventional steel bars in structural members [24–27]. However, the high production costs associated with SMA bars pose an economic challenge compared to conventional steel bars. Therefore, this study developed a technique for localized SMA bar application in the plastic hinge area—the most vulnerable part of a structural member where significant displacements occur.

The localized application of SMA bars necessitates a connection method with conventional steel bars in structural members. Common joining methods used in the field include overlapping, welding, and mechanical connections. In particular, mechanical connections using couplers can resist tensile forces regardless of attachment strength and are easily implemented in the field due to their excellent constructability. Recent studies have concentrated on assessing the ductility and strength of mechanical couplers, as well as the nonlinear cyclic performance of reinforced concrete members utilizing couplers [28–30]. Some studies have introduced SMA bars into the plastic hinge area of concrete structures using couplers [31–33]. These previous approaches used threaded couplers to connect SMA bars to steel bars, which requires additional machining and labor at the end of the existing steel bar connected to the coupler. We propose a mechanical coupler that combines a one-touch coupler and a threaded coupler to enhance usability while ensuring the connection behavior between steel bars and SMA bars.

A uniaxial tensile test was conducted to evaluate the mechanical properties and structural performance of the round SMA and deformed steel bars connected by the proposed coupler. Additional tests were carried out to evaluate the mechanical performance of martensite SMA bar when heated above the transition temperature in the uniaxial tensile test. Accurate measurement of the strain values is crucial in determining the elasticity and yielding of SMA and steel bars. Therefore, we utilized digital image correlation (DIC), enabling precise measurements for both the SMA and steel bars.

2. Experiment Design

2.1. SMA Bar

Ni–Ti SMA demonstrates two distinct states, austenite and martensite, each are defined by its crystal structure. Typically, Ni–Ti alloys exist in the austenite state at high temperatures. Austenitic properties in Ni–Ti alloys are achieved through a special treatment process involving gradual cooling from a specific high temperature above room temperature. The austenitic state possesses a body-centered crystal structure that allows the material to return to its original shape when no external stress is applied. When the SMA in the high-temperature austenite state is cooled to room temperature without any special treatment, it undergoes a transformation into the martensite state, characterized by a crystal structure belonging to the monoclinic system [34,35]. Consequently, martensite SMA loses superelasticity and exhibits low ductility compared to austenite SMA.

Figure 1a illustrates the stress–strain curve of austenite SMA, which demonstrates its superelastic behavior characterized by the absence of residual deformation and the return to the original state of the material, even after passing through the stress plateau region [36]. On the other hand, the stress–strain curve of martensite SMA, shown in Figure 1b, displays the detwinning process where displacement is partially recovered upon the removal of the applied load. However, complete displacement recovery does not occur, leaving residual deformation. This residual deformation can be recovered using the shape memory effect, which restores the initial shape of the material and transforms the martensite state back into the austenite states when the temperature of martensite increases.

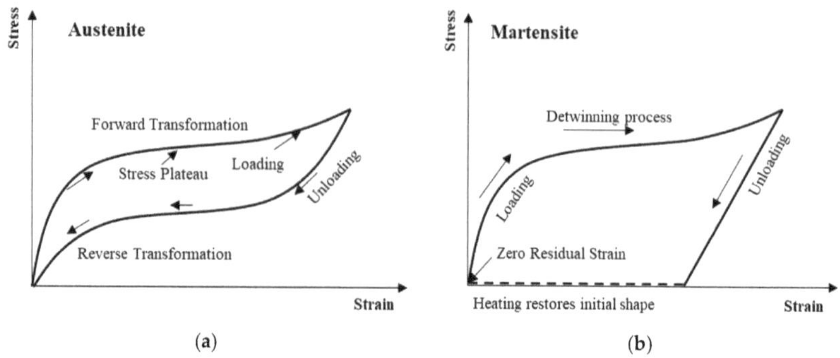

Figure 1. Stress–strain relationship: (**a**) austenite and (**b**) martensite SMAs.

This study encompassed both austenite and martensitic SMA bars, each exhibiting superelastic and shape memory effects, respectively. The two types of SMA bars primarily consist of nickel (Ni) and titanium (Ti) components with trace elements, such as carbon (C), oxygen (O), and iron (Fe). The austenite SMA comprised 55.03% Ni and 44.84% Ti by weight ratio, gradually cooled at a temperature of 90 ± 5 °C. The martensite SMA consisted of 55.74% Ni and 44.12% Ti, cooled at 5 ± 5 °C. For connection to deformed steel bars, the SMA bars were fabricated as depicted in Figure 2. The overall length of the SMA bar was 197 mm, with a central section 110 mm in length and 20 mm in diameter. The two ends connected to the coupler were shaped to 33.6 mm in length and 25 mm in diameter. Threads were implemented at both ends of the SMA bar to minimize slip during the coupler connection and when subjected to potential stress. The spacing and depth of the thread were standardized at 2 mm. To prevent yielding or failure caused by additional stresses, such as external or frictional forces, at the coupling points where the SMA bar interfaces with the coupler, the connection points were designed with a large diameter of D25 compared to the central section diameter of D20. The section transitioned from the coupler attachment to the SMA bar to minimize the concentration of stress spanned 9.9 mm in length with a curvature radius of R20.8 mm.

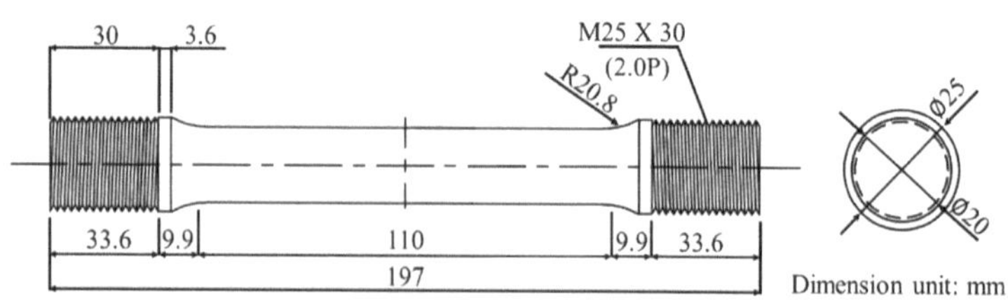

Figure 2. Drawing of SMA bar.

2.2. Mechanical Coupler

For the economic feasibility and practical application of SMA bars in real structures, the connection method with conventional steel bars is crucial. Among the various joining methods, mechanical connection with a coupler offers the advantage of avoiding the need for welding application and highly skilled labor. Therefore, we designed a mechanical coupler to connect SMA bars with conventional deformed steel bars.

Mechanical coupling typically involves a threaded coupler that incorporates threads on both the coupler and the steel bar for connection. However, threaded couplers have the drawback of the need for additional threading of the steel bar, resulting in unnecessary time and resource consumption. To overcome this challenge, we utilized a one-touch coupler that enables a straightforward connection of deformed steel bars without the need for additional machining of the steel bar. Furthermore, the one-touch coupler was combined with a threaded coupler. Specifically, the threaded connection at one end of the coupler is pre-attached to the SMA bar in a factory or another location before being transported to the construction site. Meanwhile, the one-touch connection is designed for use in the field to join deformed steel bars without requiring further machining of the steel bar.

Figure 3 shows the specially designed and manufactured mechanical coupler used in this study. The upper section of the coupler in the figure represents a one-touch coupler, while the lower section depicts a threaded coupler. The one-touch coupler incorporates a spring and a wedge to connect the steel bar and prevent potential separation due to tensile forces. When a steel bar is attached to the one-touch coupler, the embedded spring forces the wedge upwards. The wedge plays a crucial role in securely fastening the coupler and the steel bar when subjected to tensile force. As the tensile force increases, the wedge coupling structure becomes more firmly fixed within the space between the coupler and the steel bar. The lower segment of the coupler is designed as a threaded coupler, ensuring a robust attachment to the SMA bar. The connection between these two types of couplers was purposefully made to be easily interchangeable using threads. This design facilitates the straightforward replacement of the one-touch coupler section, which enables easy adaptation to different sizes of steel bars. Therefore, in this study, the coupling between two types of steel bars, D13 and D25, was executed by simply replacing the one-touch coupler section without the need for additional procedures. The proposed coupler can improve the construction process of the section linked to the steel bar and ensure sufficient attachment strength of the coupler to withstand the tensile behavior of the SMA bar.

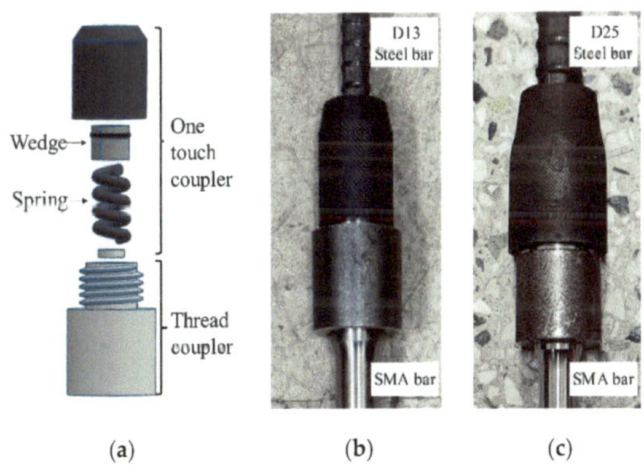

Figure 3. Manufactured coupler details and appearance: (**a**) coupler detail drawing, (**b**) D13 coupler, and (**c**) D25 coupler.

3. Testing and Measurement

3.1. Test Procedure

The uniaxial tensile tests of the specimens followed the guidelines outlined in KS B 0802 [37] and ASTM A370 [38]. The connected bars are required to possess a tensile strength of at least 125% of the specified yield strength of the base bars, as stipulated in KS D 0249 [39]. The yielding point was determined using the offset method. The applied load was controlled by displacement in accordance with ASTM A370 guidelines using a universal testing machine. The maximum stress rate was maintained within the range of 1.2 to 11.5 $N/mm^2 \cdot s$, as per ASTM A370.

The specimens were constructed for two scenarios: when the steel bar yields and when the SMA bar yields. D13 and D25 steel bars represented the yielding of steel and SMA bars, respectively. In addition, this study encompassed both austenite and martensite

SMA bars to examine the stress and deformation states in the SMA connected to a steel bar. Figure 4 illustrates the uniaxial tensile test conducted on the base steel bar and the steel bar connected to the SMA bar using the mechanical coupler. The SMA bar was attached to a 35 cm long steel bar at both ends. The steel bar portion of the specimen was clamped in a hydraulic universal testing machine for the tensile test.

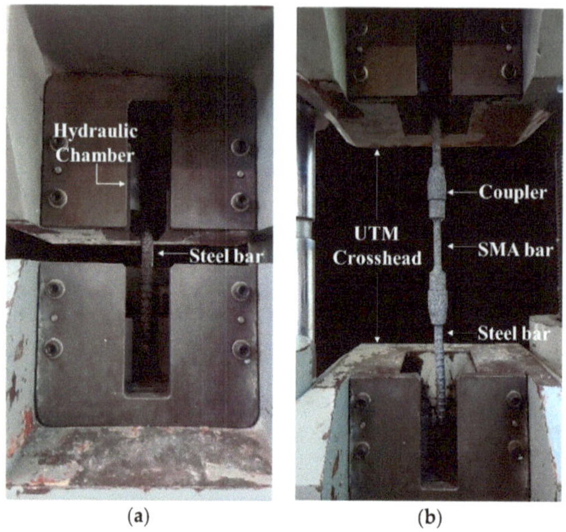

Figure 4. Uniaxial tensile test: (**a**) steel bar and (**b**) SMA bar.

For the D13 steel bar connected to the SMA bar using the proposed mechanical coupler, SD400 grade steel, which represents a yield strength over 400 Mpa, as specified in KS D3504 [40], was utilized to induce the yielding and fracture of the steel bar during the uniaxial tensile test. Thus, the stress–strain relationships in the steel and SMA bars were analyzed to evaluate the stress states at the yielding and fracture points. In addition, the amount of slip at the mechanical coupler was assessed to evaluate the performance of the complete mechanical connection. For the base steel bars, uniaxial tensile tests were also conducted to ensure the reliability of the yielding and tensile strength. In cases where the SMA bar yielded before the steel bar, an SD600 grade D25 steel bar, which represents a yield strength over 660 Mpa, was employed to connect to the SMA bar using the proposed mechanical coupler. Similar to the uniaxial test for the yielding of the steel bar, we examined the stress and strain states at the yielding and fracture of the SMA bar and assessed the amount of slip at the mechanical coupler. In particular, an additional uniaxial test was conducted to evaluate the difference in the yielding of the martensite SMA bar before and after the shape memory effect, which involves heating the SMA bar above the transformation temperature.

3.2. *Design of Heating Equipment*

Martensitic SMA bars exhibit shape recovery when heated above the phase-transformation temperature. The exertion of recovery stress due to heating can change the mechanical properties of martensitic SMA. Therefore, this study conducted a tensile test on martensitic SMA bars subjected to heating above the transformation temperature.

Heating methods commonly include conduction and convection. Electrical resistance for conduction can expedite heating times and enhance temperature reliability, but it can impact surrounding and attached conductive objects and equipment. Although this study considered induction heating, this approach necessitated substantial energy for heating SMA bars and additional cooling equipment. Moreover, experiments involving connectivity with conventional deformed steel bars indicate significantly higher and faster heating in the steel bars than the SMA bar, potentially altering the material properties of the steel bar. Therefore, in this study, a heating apparatus based on a radiative heating method using near-infrared rays was designed and fabricated to minimize the heating time, ensure heating temperature reliability, and exclusively target only the heating of the SMA bar.

Figure 5 illustrates the fabricated heating device designed for heating the martensite SMA bar beyond its transformation temperature above the austenite finish temperature of 101 °C [41]. Operating on a power capacity of 1800 W with an AC 220 V power supply, the device integrates a temperature controller to regulate the power levels and achieve the desired heating temperatures. Positioned at a distance of 50 mm from the SMA specimen, the device heated the specimen up to 350 °C. In actual tests, the SMA bar, positioned 100 mm to 200 mm away from the heating device, attained temperatures ranging from 190 °C to 300 °C. For this study, two heating devices designed for the SMA bar specimens were positioned approximately 100 mm away from the specimens at a 45-degree angle from the rear.

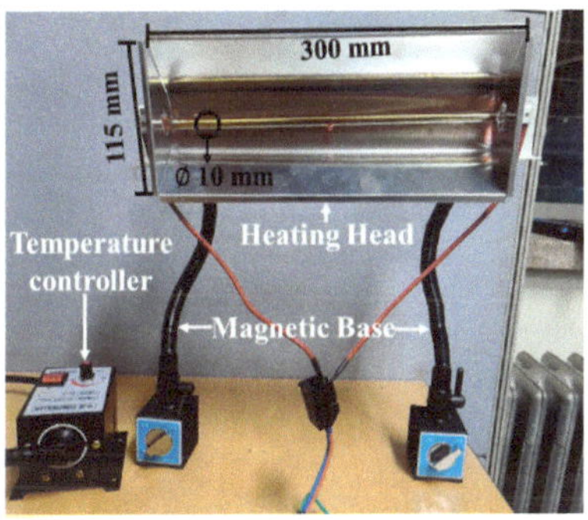

(a)

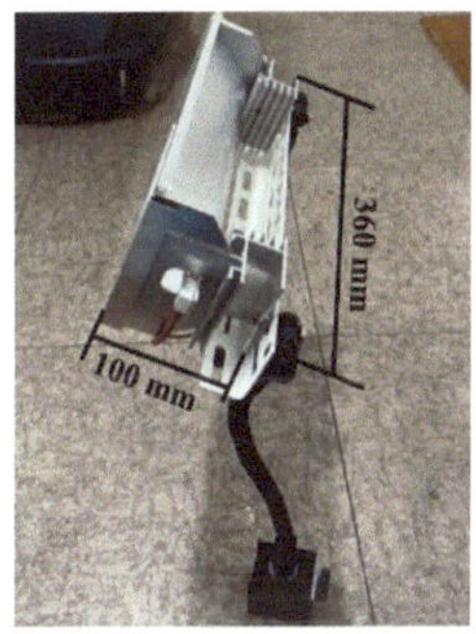

(b)

Figure 5. SMA bar heating equipment: (**a**) front view and (**b**) side view.

Figure 6 depicts the tensile test conducted on the martensite SMA bar using the heating equipment. The temperature was measured on the specimen using Fluke's Tis10 thermal imaging camera with a sensitivity of 0.15 °C (150 mK). The SMA bar was heated to 190 °C for approximately 5 min. The temperature was maintained for an additional 5 min to ensure that the entire cross-section reached the transformation temperature. The temperature of the SMA bar remained above the transformation temperature until the end of the tensile test.

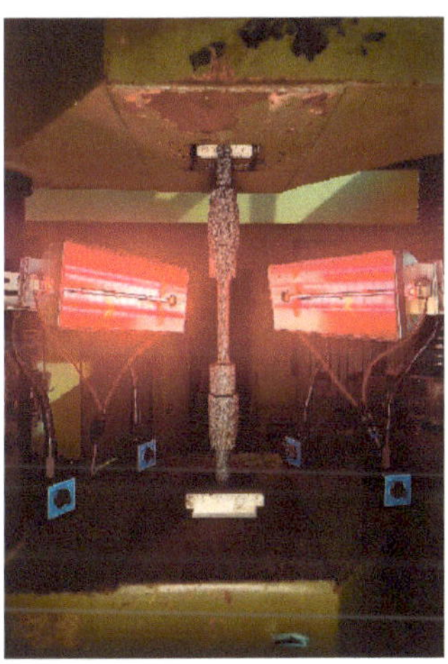

Figure 6. SMA bar heating tensile test.

3.3. *DIC-Based Displacement Measurement*

Conventional measurement techniques using extensometers and strain gauges necessitate direct physical contact. On the other hand, digital image correlation (DIC) offers a non-contact approach for strain measurement [42–44], which is particularly advantageous when multi-point data are required and specimen numbers and test repetitions are limited. Moreover, DIC can analyze strain and displacement at desired locations from immediate recalibration using stored images. In experiments where specimens are subjected to high temperatures, contact sensors can undergo deformation or damage due to temperatures, while DIC can mitigate measurement errors caused by such deformations and temperatures.

DIC consists of four main stages: image acquisition, image processing, image analysis, and post-processing, as illustrated in Figure 7. The image acquisition phase involves capturing the entire test procedure using video capturing equipment and storing these images on a connected digital storage device. In the image-processing phase, the targeted analysis area is identified within the images, and extraneous elements are eliminated. Recognition utilizes a predefined irregular dot pattern applied to the target and analyzes the surrounding color information of the pixels [45]. During the image analysis phase, displacement is calculated using the pixels within the recognized area. Figure 8 shows the process for computing displacement in the images, which involves calculating the pixel distance between sets of pixels from the reference and deformed images. Finally, post-processing is performed to derive stress–strain curves from the time data acquired through DIC and the load data obtained from the universal testing machine.

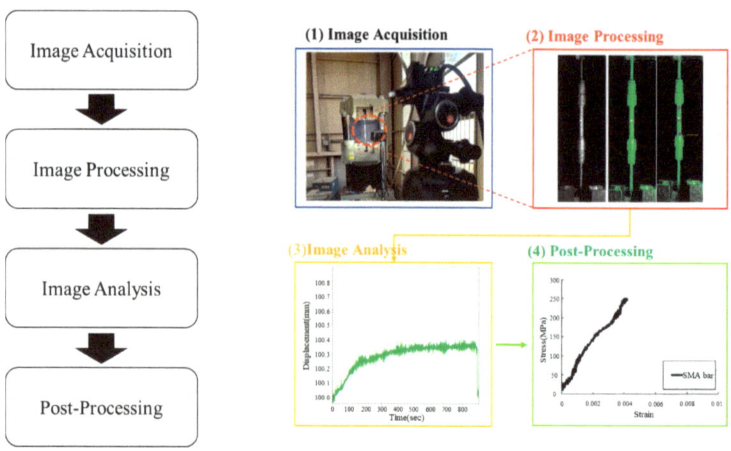

Figure 7. Flow chart of stages, from image acquisition to post-processing.

Figure 9 shows the measurement positions for the steel and SMA bars using DIC during the uniaxial tensile test. Strain measurements were taken over a target length five times the diameter of each specimen. Therefore, for the D20 SMA bar, D13 steel bar, and D25 steel bar, the designated strain measurement distances were set at 10 cm, 6.5 cm, and 12.5 cm, respectively. The gauge length for slip measurements were determined as 2 cm above and below the connection of the coupler and steel bar, respectively. The measurement data were captured at intervals of 0.1 s. All specimens were positioned at a consistent distance from the camera to maintain a scale factor of approximately 0.45 to 0.42 mm per pixel, representing the real-world size per pixel in the image.

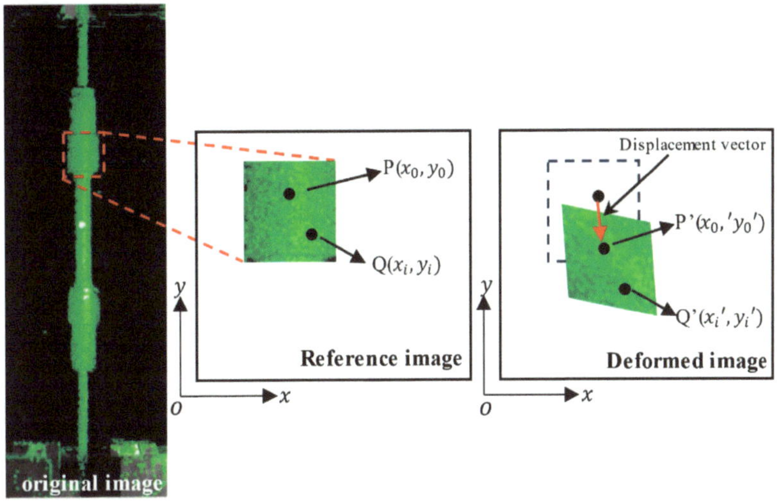

Figure 8. Relationship between the reference and deformed sets.

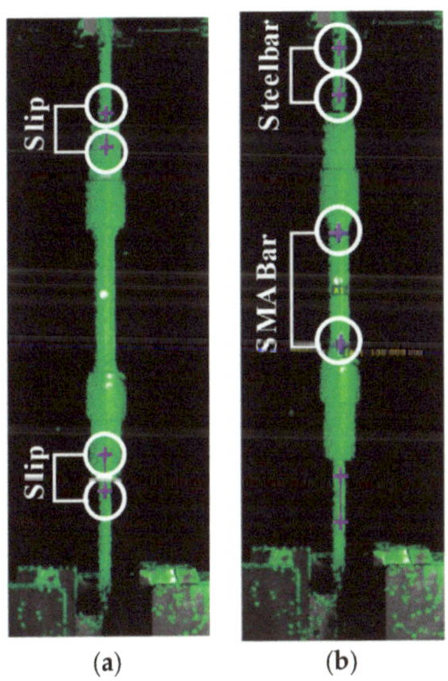

Figure 9. Locations for the stress–strain relationship and slip measurement: (**a**) slip and (**b**) strain.

4. Experimental Results and Discussion

4.1. Tensile Performance of Steel Bars

Tensile tests were conducted on two specimens each for the D13 and D25 base steel bars to examine the mechanical properties of the steel bars. Figure 10 shows the stress–strain curve obtained from one of the D13 and D25 base steel bar specimens. The curve for the D13 base steel bar presents extended perfectly plastic and strain-hardening regions compared to that of the D25 base steel bar. This difference in ductility can be attributed to the difference in strength, as strength and ductility typically have an inverse relationship.

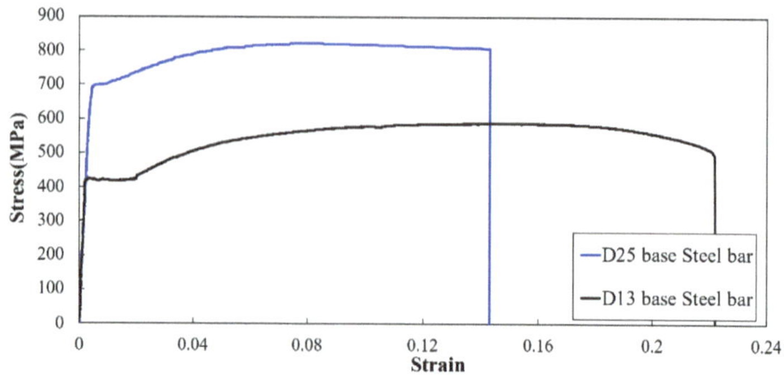

Figure 10. Stress–strain curves of the base steel bars.

The D13 and D25 base steel bars correspond to the SD400 and SD 600 grades, which represent yield strengths of more than 400 MPa and 600 MPa, respectively. The average yield strength of the D25 base steel bar was 691 MPa, which was 1.67 times that of the D13 base steel bar. The D13 base steel bar, which possesses a lower yield strength than the D25 bar, exhibits higher ductility. Table 1 summarizes the mechanical properties of the tested base steel bars, including the yield strength (σ_y), elastic modulus (E), and yield strain (ε_y). The elastic modulus for both types of steel bars obtained from the test ranged from 190 to 203 GPa, with a difference of less than 5%. The average tensile strength (σ_T) of the D25 base steel bar measured 822 MPa, which is 1.40 times higher than the measurement of 587 MPa for the D13 base steel bar. In contrast, the ultimate strain (ε_T) of the D25 base steel bar averaged 0.0812, which was 0.52 times that of the D13 base steel bar.

Table 1. Mechanical properties of the base steel bars.

Specimens		σ_y (MPa)	σ_T (MPa)	E (GPa)	ε_y	ε_T
D25	Specimen 1	696	821	190	0.0037	0.0808
	Specimen 2	686	823	196	0.0035	0.0815
D13	Specimen 1	422	587	201	0.0021	0.1576
	Specimen 2	403	586	203	0.0020	0.1548

4.2. Connected Steel–SMA Bars: Steel Bar Yielding

To assess the tensile capacity of the steel bar when mechanically joined with an SMA bar, a D13 steel bar was connected to the austenite or martensite SMA bars. Figures 11 and 12 show the stress–strain curves obtained from the upper and lower steel bars connected to the martensite and austenite SMA bars, respectively. The strain values, as mentioned previously, were measured over a target length five times the diameter of each rebar. In the graph, the H-steel and L-steel bars represent the steel bars located above and below the connected SMA bar, respectively. All connected steel bars exhibited perfectly plastic and strain-hardening regions similar to the D13 base steel bar.

Table 2 presents the yield strength (σ_y), tensile strength (σ_T), yield strain (ε_y), ultimate strain (ε_T), and elastic modulus (E) of the steel bars coupled with the SMA bars for two specimens. The average yield strength of the D13 steel bar connected to the austenite SMA bar was 429 MPa for the H-steel and 425 MPa for the L-steel. Similarly, the H-steel and L-steel bars connected to the martensite SMA bar also exhibited average yield strengths of 415 MPa and 416 MPa, respectively. The yield strains for the H-steel and L-steel bars connected to both types of SMA bars ranged from 0.0020 to 0.0023. The yield strength and strain of the D13 steel bars connected to the SMA bars using the proposed coupler closely matched those of the D13 base steel bar. The elastic modulus derived from the yield strength and strain of the connected D13 bars was very similar to that of the base steel bar with a difference of less than 5%. Therefore, the D13 steel bars connected to the SMA bars using the proposed mechanical coupler exhibited mechanical behavior similar to the base steel bar in terms of strength and stiffness. Moreover, the fracture strength of these connected D13 bars, ranging from 601 MPa to 658 MPa, satisfied the requirement of more than 125% of the yield strength of the base bar for a fully mechanical connection, as specified in KS D 0249 [39].

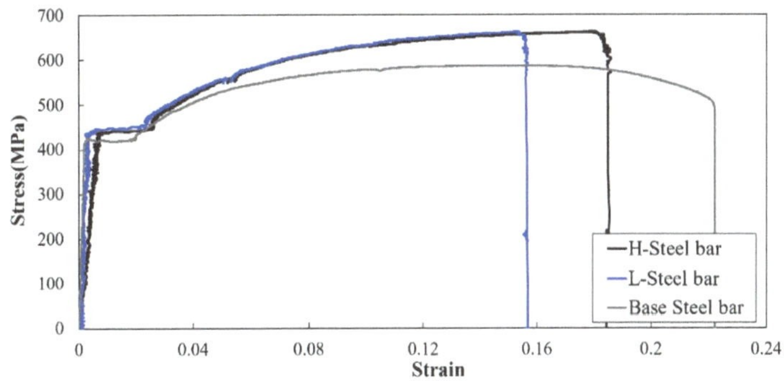

Figure 11. Stress–strain curves of the D13 steel bar coupled with austenite SMA bar.

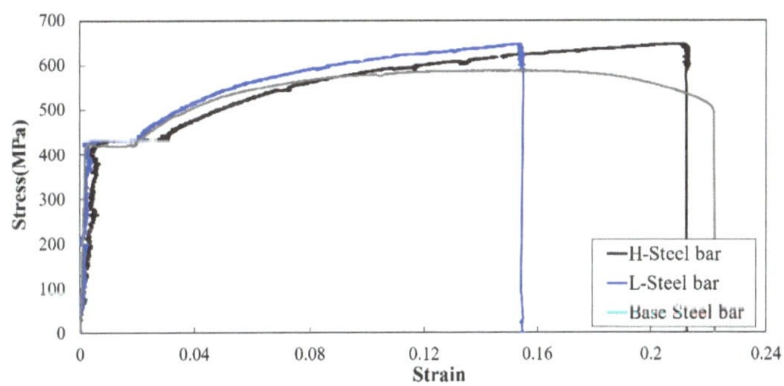

Figure 12. Stress–strain curves of the D13 steel bar coupled with martensite SMA bar.

Table 2. Mechanical properties of the steel bars coupled with SMA bar.

Specimens		σ_y (MPa)	σ_T (MPa)	E (GPa)	ε_y	ε_T
Austenite	H-Steel bars	433	658	190	0.0023	0.1818
		423	610	193	0.0022	0.1230
	L-Steel bars	430	651	196	0.0022	0.1560
		420	614	200	0.0021	0.1220
Martensite	H-Steel bars	421	638	200	0.0021	0.1545
		410	601	205	0.0020	0.1340
	L-Steel bars	420	641	191	0.0022	0.2099
		412	605	187	0.0022	0.1450

Figure 13 shows the fractured specimens. The D13 base steel bar underwent yielding and necking, followed by fracture during the uniaxial tensile test. Similarly, the steel bar connected to the SMA using the coupler exhibited a failure model from yielding to fracture within the steel bar section. The strength and stiffness of the steel bars showed minimal changes due to the proposed coupler connection up to the tensile fracture of the steel bar.

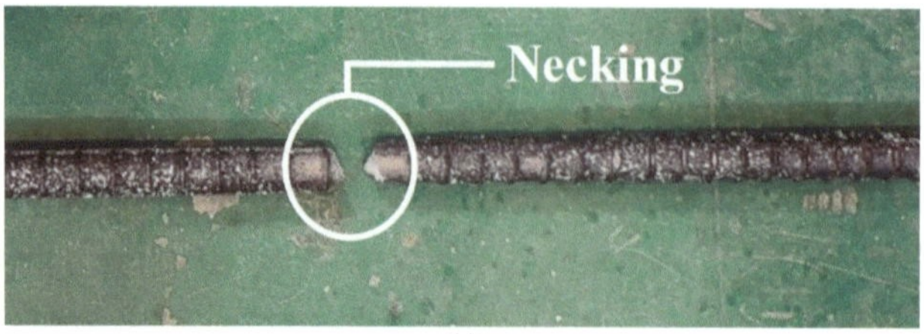

(**a**)

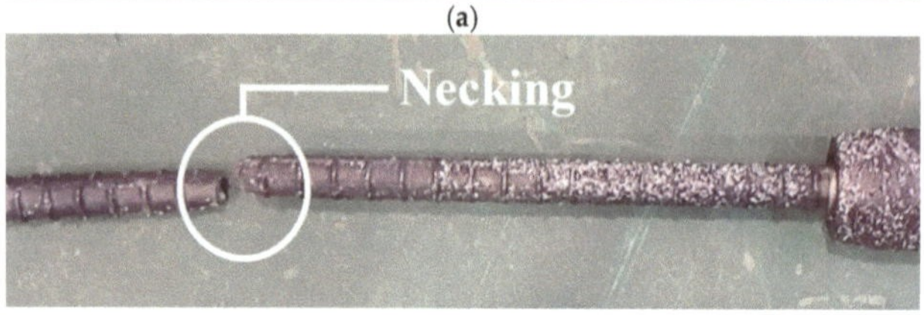

(**b**)

Figure 13. Failure shape: (**a**) D13 base bar and (**b**) D13 bar coupled with SMA bar.

4.3. Connected Steel–SMA Bars: SMA Bar Yielding

Tensile tests were conducted to analyze the mechanical properties of the SMA bar connected to the D25 steel bar. The SMA bar, mechanically joined with the D25 steel bars, exhibited yielding, necking, and eventually fracture in the section of the SMA bar, as shown in Figure 14. On the other hand, the D25 steel bar connected to the SMA bar maintained elastic behavior. Figure 15 shows the stress–strain curves obtained from one specimen of the austenite and martensite SMA bars. Both SMA bars exhibited a linear increase in stress within the elastic region, followed by distinct yielding points and eventual tensile fracture. Upon surpassing the yielding point, the austenite SMA bar demonstrated a linearly increasing strain-hardening region until fracture, while the martensite SMA bar exhibited a perfectly plastic region and transitioned into a strain-hardening phase, leading to fracture.

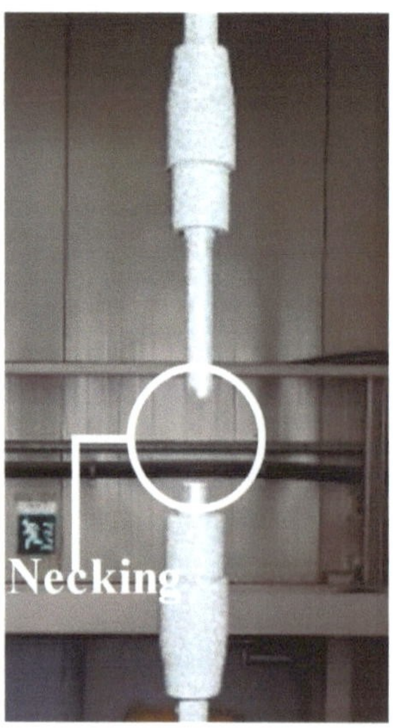

Figure 14. Failure shape of the SMA bar coupled with the D25 steel bar.

Table 3 provides a summary of the yield strength (σ_y), tensile strength (σ_T), yield strain (ε_y), ultimate strain (ε_T), and elastic modulus (E) of the SMA bars. The stress–strain curve reveals that both the austenite and martensite SMA bars exhibited similar yield strain values. However, the austenite SMA bar demonstrated an average yield strength and elastic modulus approximately 2.66 times and 2.54 times those of the martensite SMA bar, respectively. In addition, the austenite SMA bar showed an average tensile strength of 844 MPa and an average ultimate strain of 0.1321, which were 1.34 times and 1.12 times those of the martensite SMA bar.

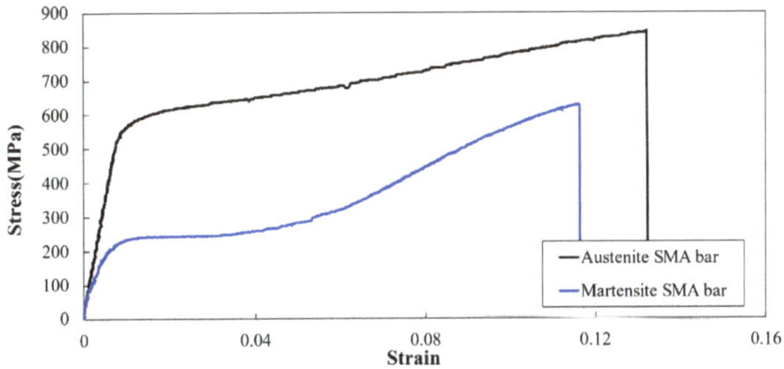

Figure 15. Stress–strain curves of the SMA bars.

Table 3. Mechanical properties of the SMA bar coupled with the D 25 steel bar.

Specimens		σ_y (MPa)	σ_T (MPa)	E (GPa)	ε_y	ε_T
Austenite	Specimen 1	571	844	72	0.0079	0.1321
	Specimen 2	590	845	78	0.0076	0.1286
Martensite	Specimen 1	223	626	30	0.0074	0.1162
	Specimen 2	214	635	29	0.0073	0.1165

4.4. Connected Steel–SMA Bars: Heated SMA Bar Yielding

Heating the SMA bar is essential for the activation of its shape memory ability, which can influence the mechanical properties of the SMA material. Therefore, the fabricated equipment was used to heat the martensite SMA bar to approximately 190 °C above the austenite finish temperature of 101 °C. This temperature was maintained for approximately 5 min to reach the transformation temperature across the entire cross-section. Subsequently, a tensile test was conducted to assess the changes in the mechanical properties of the SMA bar after the introduction of the shape memory effect due to heating. The results of the heating tensile tests were compared with those of the non-heated martensite SMA bar for two specimens.

Figure 16 shows the stress–strain curves for the martensite SMA bar before and after heating above the finish austenite temperature. As previously described, the non-heated martensite SMA bar exhibited a distinct yield region, followed by a perfectly plastic region and subsequent strain hardening until failure. For the heated martensite SMA bar, the perfectly plastic region disappeared, and immediately after the yield region, nonlinear strain hardening appeared and extended up to failure. In particular, a clear yielding point was absent in the heated martensite, necessitating the use of the 0.2% offset method to determine the yield strength.

Table 4 summarizes the yield strength (σ_y), tensile strength (σ_T), yield strain (ε_y), ultimate strain (ε_T), and elastic modulus (E) for the heated and non-heated martensite SMA bars. The heated martensite SMA bars exhibited an average strength of 220 MPa with a yield strain of 0.0074, which are very similar to those of the non-heated SMA bar. The elastic modulus for the two heated two martensite SMA bars ranged from 27 GPa to 33 GPa, which showed a maximum difference of approximately 10% from the non-heated SMA bar. The mechanical properties of the martensite SMA bars up to the yielding point exhibited minimal differences before and after the activation of the shape memory effect. However, after yielding, the ultimate tensile strain for the heated martensite SMA bar showed an increase of approximately 38% to 79% compared to the non-heated martensite SMA bar. The tensile strength of the heated martensite SMA bar was approximately 5% lower than that of the non-heated bar; the difference was deemed negligible. Figure 17 shows the tensile fracture of the heated specimens. Both specimens fractured near the center of the SMA bar. In particular, the heated martensite SMA bar showed a high elongation and significant deformation.

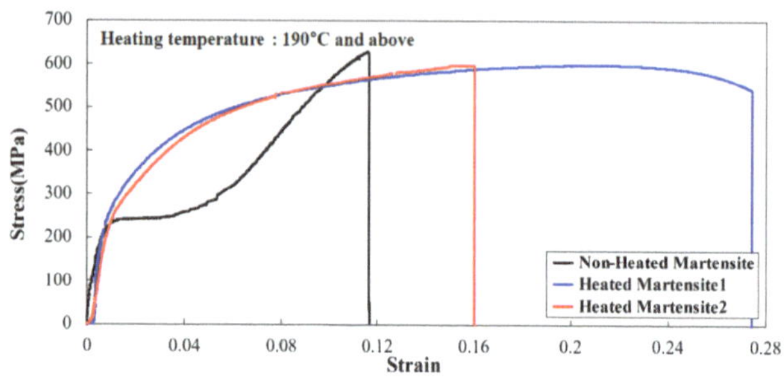

Figure 16. Stress–strain curves of the heated martensite SMA bars.

Table 4. Mechanical properties of the heated martensite SMA bars.

Specimen		σ_y (MPa)	σ_T (MPa)	E (GPa)	ε_y	ε_T
Non-Heated Martensite		223	626	30	0.0074	0.1162
Heated Martensite	Specimen 1	230	598	33	0.0069	0.2082
	Specimen 2	210	596	27	0.0078	0.1600
	Average	220	597	30	0.0074	0.1841

(a) (b)

Figure 17. Failure shape of the heated martensite SMA bar coupled with the D25 steel bar: (**a**) heated martensite 1 and (**b**) heated martensite 2.

4.5. Slip Evaluation of Connected Steel–SMA Bars

Slip in a mechanical joint can affect the usability of the mechanical coupler in concrete structures. Therefore, we evaluated the slip that may occur at mechanical connections during tensile testing. Figure 18 shows the load–displacement relationship at the connection between the D13 steel and SMA bars. Slip measurements were obtained from a 4 cm segment of the connection between the one-touch coupler and the D13 steel bar, divided into 2 cm intervals extending toward both the steel bar and the coupler from the connection point. The A-H-coupled and A-L-coupled parts in the graph represent the regions between the steel bar and one-touch coupler at the top and bottom of the austenite SMA bar, respectively. Similarly, the M-H-coupled and M-L-coupled parts denote variations in displacement at the top and bottom of the martensite SMA bar during the tensile test. The absence of an abrupt increase or change in displacement at the outset of the tensile load indicates no initial slip in the proposed coupler used for the D13 steel bar.

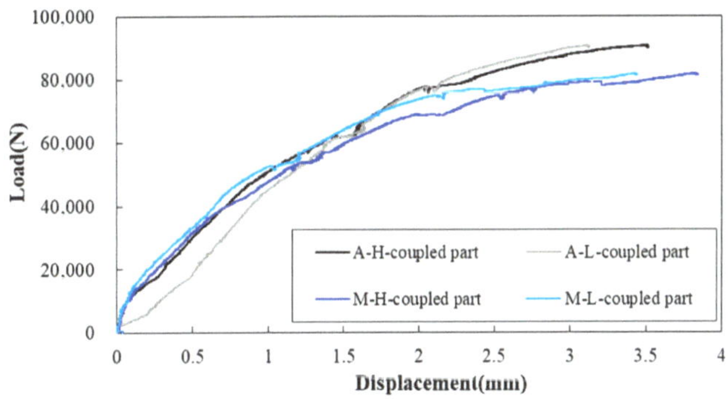

Figure 18. Load–displacement curves between the D13 and SMA bars.

Figures 19 and 20 show the load–displacement relationship at the connection between the D25 steel and SMA bars. In the D25 steel and austenite SMA bar specimen, initial slips of 0.9 mm and 1.2 mm occurred at the top and bottom of the SMA bar, respectively. The martensite SMA bar also showed initial slips of 0.6 mm and 1.2 mm at the top and bottom when joined with the D25 steel bar using the one-touch coupler, respectively. When considering the use of the one-touch coupler with large steel bars, caution should be exercised during field connections, or the SMA bar should be extended adequately beyond the plastic region of the structure to accommodate the coupler.

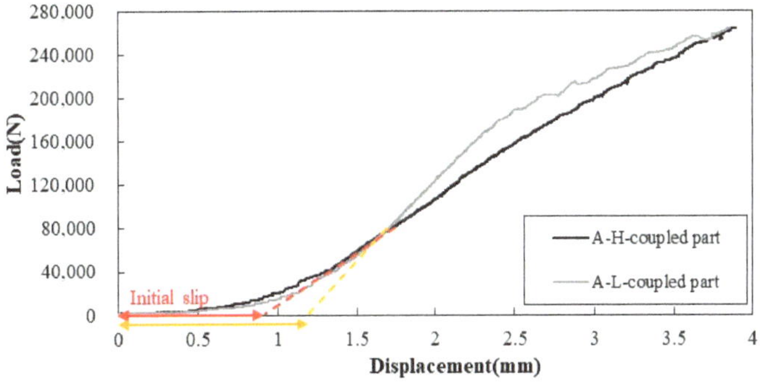

Figure 19. Load–displacement curves at the connection of the D25 and austenite SMA bars.

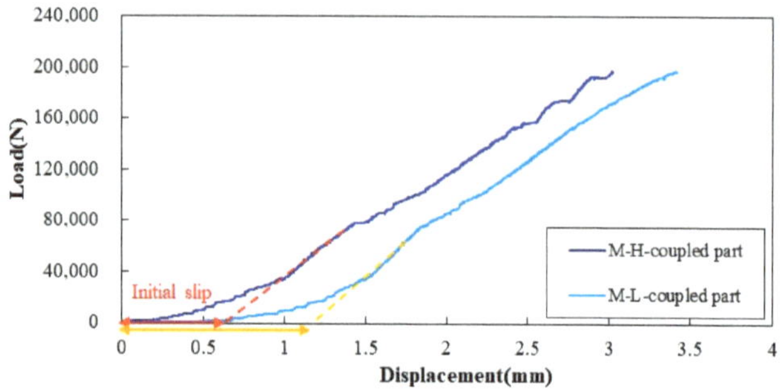

Figure 20. Load–displacement curves at the connection of the D25 and martensite SMA bars.

5. Conclusions

In this study, a mechanical connection method was proposed to locally reinforce round SMA bars in the plastic hinge zone. Uniaxial tests were then conducted to evaluate the performance of the SMA and steel bars connected using the proposed coupler. For precise multi-point measurements, in particular, for the heated section of the SMA bar, the DIC technique was utilized to obtain variations in the strains and displacements of the connected SMA and steel bars during tensile testing. The main outcomes of the study are as follows:

The connected steel bar, when yielding during tensile testing, exhibited mechanical behavior similar to the base steel bar concerning strength and stiffness. The yield strength, elastic modulus, and yield strain of the connected steel bar closely resembled those of the base steel bars. Moreover, the fracture strength of the connected steel bar satisfied the requirement of exceeding 125% of the yield strength of the base bar for a fully mechanical connection.

The austenite SMA bars demonstrated a significantly high yield strength and elastic modulus, approximately 2.66 times and 2.54 times those of the martensite SMA bars on average, respectively. The average tensile strength and ultimate strain were also 1.34 times and 1.12 times higher than the martensite SMA bars, respectively. However, two types of SMA bars exhibited similar yield strain values.

The martensite SMA bars displayed no distinct yielding or perfectly plastic regions after the activation of the shape memory effect due to heating. The mechanical properties of the heated martensite SMA bar, including the yield strength and strain, tensile strength, and elastic modulus, showed minimal differences. However, the ultimate tensile strain increased significantly by approximately 38% to 79%, resulting in considerable elongation and significant deformation.

The one-touch coupler section of the proposed coupler showed no abrupt increase or change in the initial slip for the D13 steel bar, whereas the D25 steel bar exhibited a maximum initial slip of 1.2 mm. When utilizing the one-touch coupler with large steel bars, cautious field connection and adequate extension of the SMA bar beyond the plastic region would be necessary.

Author Contributions: Conceptualization, J.-H.L. and E.C.; methodology, J.-H.L. and E.C.; validation, J.-H.L., M.-K.S. and E.C.; formal analysis, J.-H.L., M.-K.S. and E.C.; writing—original draft preparation, M.-K.S.; writing—review and editing, J.-H.L.; funding acquisition, J.-H.L. and E.C. All authors have read and agreed to the published version of the manuscript.

Funding: This research was funded by Inha University.

Data Availability Statement: The raw data supporting the conclusions of this article will be made available by the authors on request.

Acknowledgments: The authors acknowledge the National Research Foundation of Korea (NRF) (Project No. NRF 2020R1A4A-1018826) for providing the samples and SMA bars.

Conflicts of Interest: The authors declare no conflicts of interest.

References

1. Chang, L.C.; Read, T.A. Plastic deformation and diffusionless phase changes in metals—The gold-cadmium beta phase. *J. Miner. Met. Mater. Soc.* **1951**, *3*, 47–52. [CrossRef]
2. Buehler, W.J.; Gilfrich, J.V.; Wiley, R.C. Effect of low-temperature phase changes on the mechanical properties of alloys near composition TiNi. *J. Appl. Phys.* **1963**, *34*, 1475–1477. [CrossRef]
3. Van Humbeeck, J.; Stalmans, R. Shape memory alloys, types and functionalities. In *Encyclopedia of Smart Materials*, 2nd ed.; Schwartz, M., Ed.; John Wiley and Sons: New York, NY, USA, 2002.
4. Dong, Z.; Klotz, U.E.; Leinenbach, C.; Bergamini, A.; Czaderski, C.; Motavalli, M. A novel Fe-Mn-Si shape memory alloy with improved shape recovery properties by VC precipitation. *Adv. Eng. Mater.* **2009**, *11*, 40–44. [CrossRef]
5. Sadiq, H.; Wong, M.B.; Al-Mahaidi, R.; Zhao, X.L. The effects of heat treatment on the recovery stresses of shape memory alloys. *Smart Mater. Struct.* **2010**, *19*, 035021. [CrossRef]
6. Liu, Y.; Li, Y.; Ramesh, K.T.; Van Humbeeck, J. High strain rate deformation of martensitic NiTi shape memory alloy. *Scr. Mater.* **1999**, *41*, 89–95. [CrossRef]
7. Huang, X.; Liu, Y. Effect of annealing on the transformation behavior and superelasticity of NiTi shape memory alloy. *Scr. Mater.* **2001**, *45*, 153–160. [CrossRef]
8. Mahmud, A.S.; Yang, H.; Tee, S.; Rio, G.; Liu, Y. Effect of annealing on deformation-induced martensite stabilisation of NiTi. *Intermetallics* **2008**, *16*, 209–214. [CrossRef]
9. Janke, L.; Czaderski, C.; Motavalli, M.; Ruth, J. Applications of shape memory alloys in civil engineering structures—Overview, limits and new ideas. *Mater. Struct.* **2005**, *38*, 578–592.
10. Shajil, N.; Srinivasan, S.M.; Santhanam, M. Self-centering of shape memory alloy fiber reinforced cement mortar members subjected to strong cyclic loading. *Mater. Struct.* **2013**, *46*, 651–661. [CrossRef]
11. Jung, C.Y.; Lee, J.H. Crack closure and flexural tensile capacity with SMA fibers randomly embedded on tensile side of mortar beams. *Nanotechnol. Rev.* **2020**, *9*, 369–381. [CrossRef]
12. Choi, E.; Ostadrahimi, A.; Lee, J.H. Pullout resistance of crimped reinforcing fibers using cold-drawn NiTi SMA wires. *Constr. Build. Mater.* **2020**, *265*, 120858. [CrossRef]
13. Choi, E.; Kim, H.S.; Nam, T.H. Effect of crimped SMA fiber geometry on recovery stress and pullout resistance. *Compos. Struct.* **2020**, *247*, 112466. [CrossRef]
14. Dębska, A.; Gwoździewicz, P.; Seruga, A.; Balandraud, X.; Destrebecq, J.F. The application of Ni–Ti SMA wires in the external prestressing of concrete hollow cylinders. *Materials* **2021**, *14*, 1354. [CrossRef]
15. Schleiting, M.; Wetzel, A.; Bauer, A.; Frenck, J.M.; Niendorf, T.; Middendorf, B. Potential of Fe-Mn-Al-Ni Shape Memory Alloys for Internal Prestressing of Ultra-High Performance Concrete. *Materials* **2023**, *16*, 3816. [CrossRef] [PubMed]
16. Beßling, M.; Czaderski, C.; Orlowsky, J. Prestressing effect of shape memory alloy reinforcements under serviceability tensile loads. *Buildings* **2021**, *11*, 101. [CrossRef]
17. Qian, H.; Zhang, Q.; Zhang, X.; Deng, E., Gao, J. Experimental investigation on bending behavior of existing RC beam retrofitted with SMA-ECC composites materials. *Materials* **2021**, *15*, 12. [CrossRef] [PubMed]
18. Sung, M.; Andrawes, B. Innovative local prestressing system for concrete crossties using shape memory alloys. *Eng. Struct.* **2021**, *247*, 113048. [CrossRef]
19. Raza, S.; Shafei, B.; Saiidi, M.S.; Motavalli, M.; Shahverdi, M. Shape memory alloy reinforcement for strengthening and self-centering of concrete structures—State of the art. *Constr. Build. Mater.* **2022**, *324*, 126628. [CrossRef]
20. Alshannag, M.J.; Alqarni, A.S.; Higazey, M.M. Superelastic Nickel–Titanium (NiTi)-Based Smart Alloys for Enhancing the Performance of Concrete Structures. *Materials* **2023**, *16*, 4333. [CrossRef]
21. Pogrebnjak, A.D.; Bratushka, S.N.; Beresnev, V.M.; Levintant-Zayonts, N. Shape memory effect and superelasticity of titanium nickelide alloys implanted with high ion doses. *Russ. Chem. Rev.* **2013**, *82*, 1135. [CrossRef]
22. Ebrahimi, M.; Attarilar, S.; Gode, C.; Kandavalli, S.R.; Shamsborhan, M.; Wang, Q. Conceptual Analysis on Severe Plastic Deformation Processes of Shape Memory Alloys: Mechanical Properties and Microstructure Characterization. *Metals* **2023**, *13*, 447. [CrossRef]
23. Nespoli, A.; Ninarello, D.; Fanciulli, C. A Review on Shape Memory Alloys with Martensitic Transition at Cryogenic Temperatures. *Metals* **2023**, *13*, 1311. [CrossRef]
24. Abraik, E.; El-Fitiany, S.F.; Youssef, M.A. Seismic performance of concrete core walls reinforced with shape memory alloy bars. *Structures* **2020**, *27*, 1479–1489. [CrossRef]
25. Abraik, E.; Youssef, M.A. Ductility and overstrength of shape-memory-alloy reinforced-concrete shear walls. *Eng. Struct.* **2021**, *239*, 112236. [CrossRef]
26. Siddiquee, K.N.; Billah, A.M.; Issa, A. Seismic collapse safety and response modification factor of concrete frame buildings reinforced with superelastic shape memory alloy (SMA) rebar. *J. Build. Eng.* **2021**, *42*, 102468. [CrossRef]
27. Ferraioli, M.; Concilio, A.; Molitierno, C. Seismic performance of a reinforced concrete building retrofitted with self-centering shape memory alloy braces. *Earthq. Eng. Eng. Vib.* **2022**, *21*, 785–809. [CrossRef]
28. Bompa, D.V.; Elghazouli, A.Y. Ductility considerations for mechanical reinforcement couplers. *Structures* **2017**, *12*, 115–119. [CrossRef]
29. Bompa, D.V.; Elghazouli, A.Y. Inelastic cyclic behaviour of RC members incorporating threaded reinforcement couplers. *Eng. Struct.* **2019**, *180*, 468–483. [CrossRef]

30. Ben-dahou, A.; Ferrier, E.; Gabor, A.; Michel, L.; Gardes, R.; Boisson, R.; Poissonnet, C.; Dolo, J.M. Influence of rebar couplers on the cracking behavior of reinforced concrete beams. *Nucl. Eng. Des.* **2024**, *416*, 112801. [CrossRef]
31. Pareek, S.; Suzuki, Y.; Araki, Y.; Youssef, M.A.; Meshaly, M. Plastic hinge relocation in reinforced concrete beams using Cu-Al-Mn SMA bars. *Eng. Struct.* **2018**, *175*, 765–775. [CrossRef]
32. Billah, A.M.; Alam, M.S. Plastic hinge length of shape memory alloy (SMA) reinforced concrete bridge pier. *Eng. Struct.* **2016**, *117*, 321–331. [CrossRef]
33. Molod, M.A.; Spyridis, P.; Barthold, F.J. Applications of shape memory alloys in structural engineering with a focus on concrete construction—A comprehensive review. *Constr. Build. Mater.* **2022**, *337*, 127565. [CrossRef]
34. Otsuka, K.; Sawamura, T.; Shimizu, K. Crystal structure and internal defects of equiatomic TiNi martensite. *Phys. Status Solidi (A)* **1971**, *5*, 457–470. [CrossRef]
35. Chowdhury, P.; Sehitoglu, H. Deformation physics of shape memory alloys-fundamentals at atomistic frontier. *Prog. Mater. Sci.* **2017**, *88*, 49–88. [CrossRef]
36. Dolce, M.; Cardone, D. Mechanical behaviour of shape memory alloys for seismic applications 2. Austenite NiTi wires subjected to tension. *Int. J. Mech. Sci.* **2001**, *43*, 2657–2677. [CrossRef]
37. *KS B 0802: Korean Standard (KS)*; Method of Tensile Test for Metallic Materials. Korean Standards Association: Seoul, Republic of Korea, 2003.
38. *ASTM A370-23*; Standard Test Methods and Definitions for Mechanical Testing of Steel Products. ASTM International: West Conshohocken, PA, USA, 2020.
39. *KS D 0249: Korean Standard (KS)*; Method of Inspection for Mechanical Splicing Joint of Bars for Concrete Reinforcement. Korean Standards Association: Seoul, Republic of Korea, 2019.
40. *KS D 3504: Korean Standard (KS)*; Steel Bars for Concrete Reinforcement. Korean Standards Association: Seoul, Republic of Korea, 2021.
41. Choi, E.; Jeon, J.S.; Lee, J.H. Self-centering capacity of RC columns with smart plastic hinges of martensitic NiTi SMA bars. *Smart Mater. Struct.* **2023**, *32*, 115015. [CrossRef]
42. Quanjin, M.; Rejab, M.R.M.; Halim, Q.; Merzuki, M.N.M.; Darus, M.A.H. Experimental investigation of the tensile test using digital image correlation (DIC) method. *Mater. Today Proc.* **2020**, *27*, 757–763. [CrossRef]
43. Chu, T.C.; Ranson, W.F.; Sutton, M.A. Applications of digital-image-correlation techniques to experimental mechanics. *Exp. Mech.* **1985**, *25*, 232–244. [CrossRef]
44. Bruck, H.A.; McNeill, S.R.; Sutton, M.A.; Peters, W.H. Digital image correlation using Newton-Raphson method of partial differential correction. *Exp. Mech.* **1989**, *29*, 261–267. [CrossRef]
45. Pan, B. Recent progress in digital image correlation. *Exp. Mech.* **2011**, *51*, 1223–1235. [CrossRef]

metals

MDPI

Review

Very High Cycle Fatigue of Welds: A Review

Andrew England *, **Athanasios Toumpis** and **Yevgen Gorash** *

Department of Mechanical & Aerospace Engineering, University of Strathclyde, 75 Montrose Street, Glasgow G1 1XJ, UK; athanasios.toumpis@strath.ac.uk

* Correspondence: andrew.england@strath.ac.uk (A.E.); yevgen.gorash@strath.ac.uk (Y.G.)

Abstract: The design life of welded structures and components extends into the very high cycle fatigue (VHCF) regime across various applications. However, the availability of data on the fatigue behaviour of welded joints in the VHCF regime is limited, particularly when compared to the low and high cycle fatigue regimes. The development of ultrasonic fatigue testing equipment has accelerated fatigue testing and allowed for the VHCF properties of welds to be investigated in a feasible timeframe. In the present review, the emerging research concerning the VHCF behaviour of welds of various steels and non-ferrous alloys are individually explored. Overall, it is observed that welded joints have significantly lower fatigue strength than the base metal in the VHCF regime and that welding defects have a considerable influence on fatigue strength. Through the discussion of the relevant literature, important findings concerning the effects of specimen geometry and fatigue improvement methods are underlined. Furthermore, the guidance provided within design standards is compared, and some examples of VHCF failures of in-service components are highlighted. Finally, perspectives on future directions of investigation are put forward with the aim of encouraging further research in the field of VHCF of welds.

Keywords: very high cycle fatigue; welded joint; ultrasonic fatigue testing; fatigue; fractography

1. Introduction

Fatigue failure of metals is a challenge faced across almost all industries; it is commonly estimated that fatigue damage is responsible for 90% of all in-service mechanical failures [1]. Repeated loading at stresses less than the yield strength of the material can lead to the initiation and propagation of cracks, resulting in a complete fracture [2]. Welded joints are generally the critical location in the structural integrity assessments of cyclically loaded components [3]. The fatigue strength of fusion-welded joints is almost always lower than the corresponding base metal due to several factors [4]. Unless it is removed by machining, the abrupt change in geometry caused by the weld toe creates a stress concentration region [5]. The thermal contraction during cooling generates tensile residual stresses of a magnitude similar to the yield strength of the base metal, significantly reducing the fatigue strength of the welded joint [6]. Additionally, the inhomogeneous microstructures and defects commonly found in welded joints make them susceptible to fatigue failure during service [7].

Fatigue regimes are generally classified by the number of cycles to failure: low cycle fatigue (LCF) as up to 10^4 cycles; high cycle fatigue (HCF) as 10^4 to 10^7 cycles; and very high cycle fatigue (VHCF) as over 10^7 cycles [8]. These fatigue regimes are displayed schematically in the form of an *S–N* curve in Figure 1. In the LCF regime, applied stresses are greater than the yield strength of the material; therefore, significant plastic deformation occurs in each loading cycle [9]. The HCF regime is generally characterised by stresses less than the yield strength and purely elastic deformation in each loading cycle [8]. It is a classical assumption that below a certain constant amplitude stress, fatigue failure will never occur, a concept known as the fatigue limit [10]. The fatigue limit appears on an *S–N* curve as a horizontal asymptote (Figure 1). The existence of a fatigue limit between 10^6 and

check for updates

Citation: England, A.; Toumpis, A.; Gorash, Y. Very High Cycle Fatigue of Welds: A Review. *Metals* **2023**, *13*, 1860. https://doi.org/10.3390/met13111860

Academic Editor: Antonio Mateo

Received: 10 October 2023
Revised: 30 October 2023
Accepted: 3 November 2023
Published: 7 November 2023

10^7 cycles has been proven invalid [11], and subsequently, the VHCF of metals has become a well-established research field.

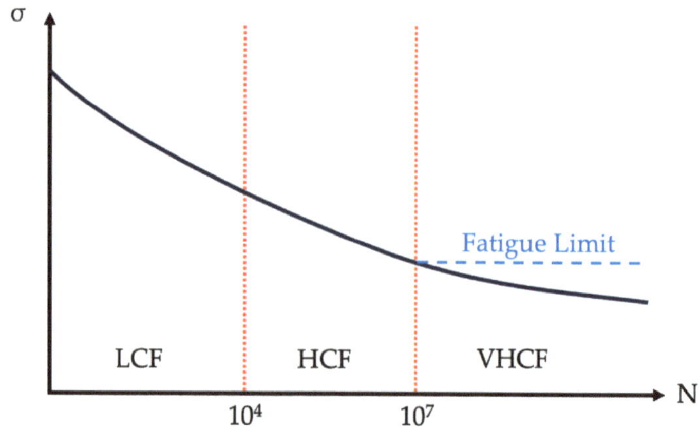

Figure 1. Fatigue regime schematic diagram.

Welded structures have service lives in the VHCF regime across a range of applications, including bridges, automotive components, and heavy industry equipment [12–14]. To ensure the safe operation of engineering components, machines, and structures, the VHCF behaviour of welds has become an emerging field of research following the first published paper on the topic in 2010 [15]. Figure 2 reveals that only a few papers have been published each year, with a relatively steady interest since 2012. For this review, publications were gathered from an extensive range of databases, including Scopus, Web of Science, and Google Scholar.

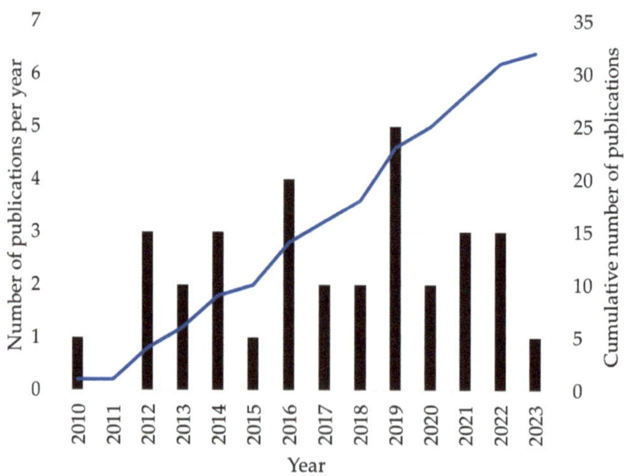

Figure 2. Number of publications related to the VHCF of welds, 2010–2023.

The present paper represents the first review of the topic of VHCF of welds, with the primary aim of acting as a comprehensive introduction to the field. The subjects of ultrasonic fatigue testing (UFT) and the VHCF of metals are also briefly discussed so that the main topic can be better understood. This work also serves to encourage further research in the field of VHCF of welds by identifying crucial knowledge gaps.

2. Ultrasonic Fatigue Testing and Very High Cycle Fatigue of Metals

2.1. The Ultrasonic Fatigue Testing Method

Conducting laboratory-based fatigue testing to the full extent of the VHCF regime in a realistic and economical timeframe is only feasible using UFT, where the loading frequency is typically 20 kHz [16]. Achieving 10^9 loading cycles using a servo-hydraulic fatigue testing machine at 20 Hz requires over 500 days of continuous loading, but this can be reduced to just 14 h using UFT [17]. Undertaking resonance fatigue testing at 100 Hz allows for the early portion of the VHCF regime to be explored; as an illustration, 5×10^7 cycles can be reached in less than 6 days. Fatigue tests conducted at low frequencies (<150 Hz) will hereafter be referred to as conventional fatigue tests. The frequency of 20 kHz is used for UFT as it is outside of the audible range of the human ear, but greater frequencies than this result in increased cycle counting errors and a reduction in the length of the specimen [18]. It is possible to employ UFT for cyclic bending [19], torsional [20], or multiaxial [21] loading, but only uniaxial tension-compression loading will be considered in this paper.

UFT machines (Figure 3) use ultrasonic transducers (usually a piezoelectric actuator) to generate mechanical vibration from an electrical signal [22]. The longitudinal vibration is then amplified as it is transmitted through the booster and horn to the specimen, typically attached by threads [22]. Specimens are excited at their natural frequency; therefore, the maximum stress, but minimum displacement, is experienced at the centre of the specimen. VHCF tests have been demonstrated with both bespoke UFT apparatuses constructed at research institutions and equipment from commercial manufacturers [16].

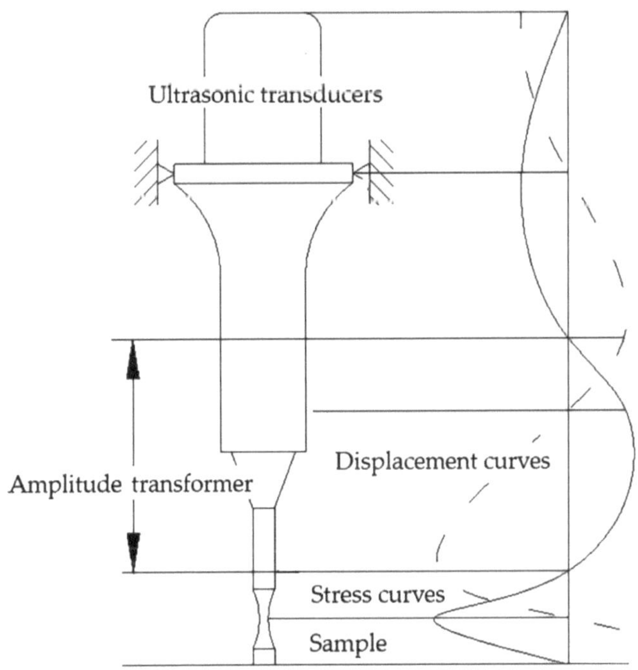

Figure 3. Ultrasonic fatigue testing machine schematic [23] (used under Creative Commons CC-BY license).

The UFT method was standardised by the Japan Welding Engineering Society in WES 1112 [24]. Within this standard, the procedure for determining dimensions for standard-shaped specimens, e.g., cylindrical hourglass (Figure 4), is specified. Additionally, aspects of the test procedure, such as the tolerance of the resonant frequency and the maximum specimen surface temperature, are detailed [25]. Similar to conventional fatigue testing, UFT can be used to construct an *S–N* curve of a particular material by testing multiple identical specimens at a range of constant amplitude stress levels.

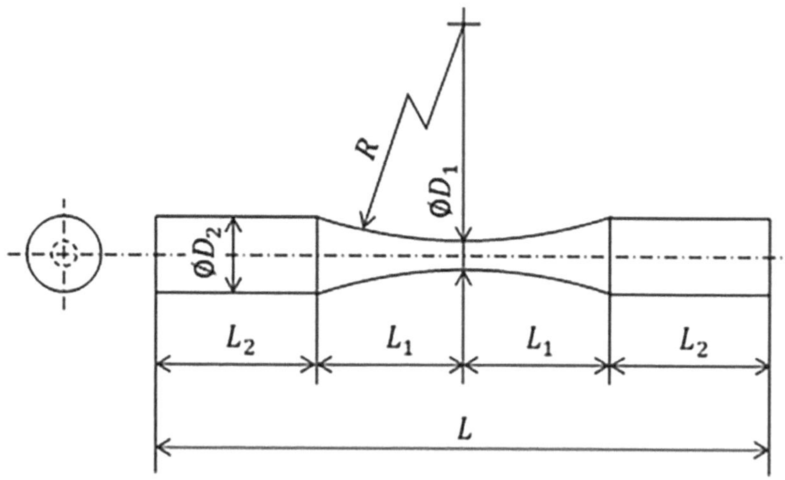

Figure 4. Cylindrical hourglass-shaped ultrasonic fatigue testing specimen showing main dimensions [25] (with permission from John Wiley and Sons, 2023).

It has been reported that the fatigue data obtained using UFT are often not comparable to that obtained at conventional frequencies, a phenomenon known as the "frequency effect" [25]. For instance, Bach et al. [26] found a 50% increase in fatigue strength at 20 kHz compared to 120 Hz for C15E low-carbon steel. A wide-ranging review of the frequency effect is provided by Hong et al. [27]. The primary cause of the frequency effect is the sensitivity of flow stress to strain rate for certain crystal lattice structures, especially the body-centred cubic (BCC) structure [28]. Low-carbon steels, which consist primarily of ferrite with a BCC structure, exhibit a significant frequency effect [29,30]. Recently, Guennec et al. [31] proposed a model based on dislocation motion to predict the fatigue limit of carbon steels, depending on the alloy's wt.% C and the applied loading frequency. The model, validated from results from the literature, showed that as the wt.% C reduces, and consequently, the ferrite grain fraction increases, the severity of the frequency effect increases [31]. A limitation of the model is that it requires the measurement of the fatigue limit, the value of which depends on the selection of the number of cycles defined as a runout, and the number of tests conducted. The frequency effect is not an issue with high-strength steels, since it has been found to diminish as tensile strength increases [32]. Further, the fatigue strength of metallic materials with a face-centred cubic (FCC) lattice structure, e.g., stainless steel or aluminium alloys, demonstrates minimal sensitivity to the loading frequency [27].

2.2. Fatigue Behaviour of Metals at Very High Cycles

The course of the *S–N* curve in the VHCF regime has been shown to vary between different metals. Mughrabi [33] proposed that the VHCF behaviour of metals can be classified into two different types. "Type I" metallic materials show a relatively little decrease in fatigue strength in the VHCF domain and generally only fail from cracks initiated at the specimen surface [33]. Metals that fall under this category are typically ductile and have homogenous microstructures, for instance, copper alloys and low-carbon steels [8]. It is more likely to observe a fatigue limit for Type I metals due to the slow decline in fatigue strength in the VHCF domain [34].

Conversely, the *S–N* curve of "Type II" metallic materials shows a continuous decrease in the VHCF domain [33]. The fatigue behaviour of this class of materials is influenced by the presence of pores, non-metallic inclusions or secondary phases within their microstructure [8]. A duplex-shaped *S–N* curve (Figure 5) represents the VHCF behaviour of Type II materials, examples of which include high-strength steels and titanium alloys [33]. Additionally, aluminium alloys do not typically display a fatigue limit [35].

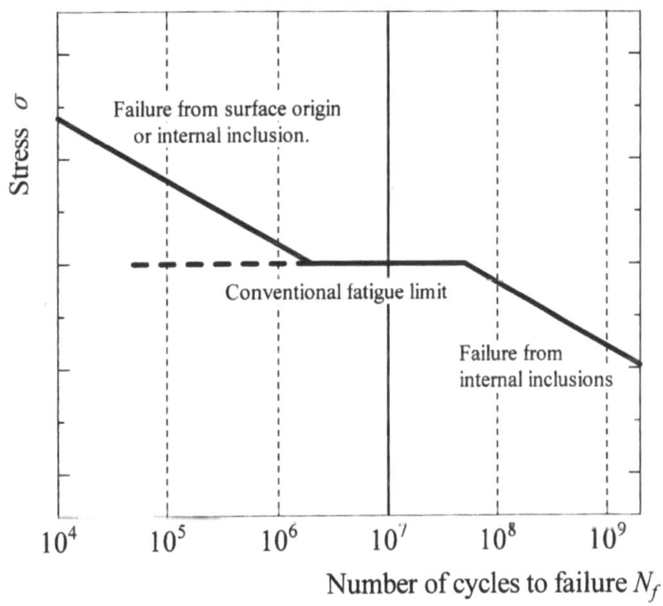

Figure 5. Typical *S–N* curve of Type II materials in VHCF domain [36] (with permission from John Wiley and Sons, 2023).

Fatigue cracks tend to initiate in the VHCF domain from internal inclusions for Type II metals, as indicated in Figure 5. The surface of internal inclusion-initiated fractures of Type II metals (Figure 6a) typically shows a distinctive bright region (Figure 6b), known as the fine granular area (FGA) [37]. The formation mechanisms of FGAs are not fully understood, but a thorough review of the current models has been put forward by Li et al. [38]. However, it is agreed that the FGA forms early in cyclic loading and remains the same approximate size throughout most of the fatigue life [39]. Crack growth occurs outward from the FGA to form the fish-eye propagation zone, which is characteristic of VHCF fracture surfaces [40].

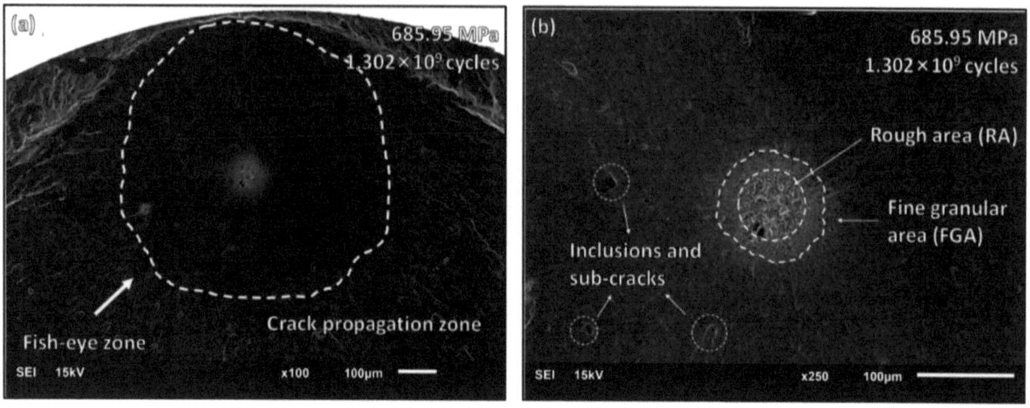

Figure 6. VHCF fracture surface of Inconel 625: (**a**) fish-eye zone; (**b**) inclusions and fine granular area [40] (used under Creative Commons CC-BY license).

3. Very High Cycle Fatigue Performance of Welds

Investigating the VHCF performance of welded joints of metals has become feasible due to the maturity of the UFT method. Compared to the testing of unwelded metals, the study of welded joints is less explored, although there are a limited number of published papers concerning this subject area. This section will summarise the research conducted concerning the VHCF properties of welded joints, categorised by material type.

3.1. Low-Carbon Steels

Low-carbon steels are typically categorised as Type I materials, showing a low decrease in fatigue strength in the VHCF domain due to their homogenous microstructure and ductility [8,33]. However, welded joints of low-carbon steels cannot be assumed to be Type I materials. This is because they frequently contain inclusions and pores that are characteristic of Type II materials, which exhibit a significant reduction in fatigue strength in the VHCF regime [7,8].

Weldments of low-carbon steel are commonly used in applications that experience VHCF, such as bridges [12] and heavy industry equipment [14]. In this context, they have been the subject of some of the first research on the VHCF of welds. Zhao et al. [41] investigated the VHCF performance of EH36 structural steel base metal and butt welds fabricated using flux-core arc welding (FCAW), with the welds ground flush before testing. Figure 7 displays the test results of the base metal and welded specimens under fully reversed axial loading at 20 kHz at room temperature. Although the slope of the *S*–*N* curve reduces in the VHCF regime, no fatigue limit was observed, and multiple fatigue failures occurred above 10^9 cycles for both specimen types. The fatigue strength of the welded joints was less than that of the base metal, especially as the number of cycles to failure increased. Fractography revealed that the fatigue crack initiation sites in welded specimens were inclusions and pores; hence, these were the principal reasons for the reduction in fatigue strength [41]. One limitation of this study is the absence of recorded runout specimens, which would have aided in the statistical evaluation of the results [42].

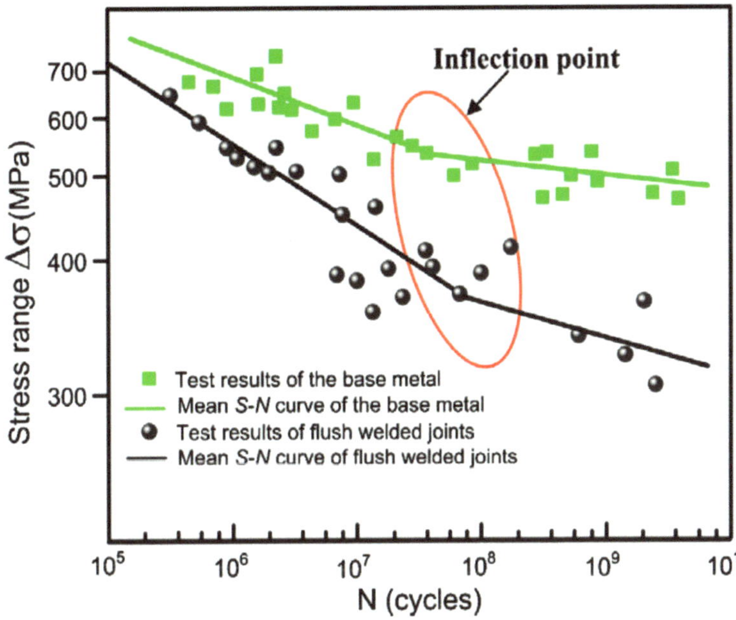

Figure 7. *S–N* curves for EH36 base metal and welded joint [41] (with permission from Elsevier, 2023).

The VHCF properties of cruciform arc-welded joints of Q235 and Q345 structural steels were studied by Yin et al. [15]. The cruciform joints, shown in Figure 8, were comprised of four non-load-carrying fillet welds and assessed using UFT. A continually decreasing *S–N* curve in the VHCF regime was observed for welded specimens of both Q235 and Q345. Failures occurred above 10^9 cycles, and no runout specimens were reported; hence, a fatigue limit was not observed for the cruciform joints and cracks initiated at the weld toes from slag inclusions or machining marks. Additionally, the effect of ultrasonic peening treatment (UPT) was investigated; it was found that the VHCF strength was significantly increased for welded joints of both base metals due to grain refinement and the generation of beneficial compressive residual stresses [15]. Cylindrical, hourglass-shaped specimens of Q345 base metal and arc-welded joints were tested using UFT by He et al. [12], and a fatigue limit was not observed for either specimen type. The fatigue strength of the welded joints was found to be significantly lower in the VHCF domain when compared to the base metal due to cracks initiating owing to lack of penetration, slag inclusion, and gas pore defects [12].

UFT of as-welded butt joints of 16Mn structural steel was conducted by Liu et al. [43]. A similar geometry to the standard hourglass specimen recommended by WES 1112 [24] was used, with a circumferential gas tungsten arc weld (GTAW) in the central gauge section. Fatigue failures were observed in the VHCF regime, with cracks initiating at either the welding stop/start point or the weld toe due to the geometric stress concentration in these regions. Welded joints subjected to UPT were also tested, but showed only a small increase in fatigue strength, in contrast to the results of Yin et al. [15].

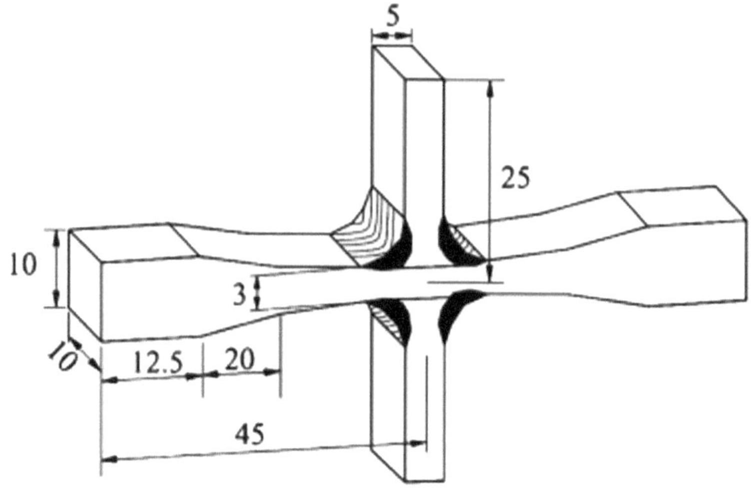

Figure 8. Cruciform welded joints ultrasonic fatigue testing specimens with main dimensions shown (mm) [15] (with permission from Elsevier, 2023).

He et al. [44] investigated the VHCF behaviour of Q345 welds joined by shielded metal arc welding (SMAW). A specially designed semi-circular notched UFT specimen (Figure 9) was used to discretely evaluate the fatigue behaviour of the base metal, heat-affected zone (HAZ), and fusion zone (FZ). The base metal had significantly greater fatigue strength than the HAZ and FZ in the VHCF regime, with a decrease of approx. 55% and 60%, respectively (Figure 9d). The onset of fatigue failure was assessed by monitoring the natural frequency of specimens during testing since crack propagation reduces the stiffness of a specimen, hence the natural frequency [45]. Through this method, it was revealed that the fatigue crack initiation stage was significantly shorter for the FZ when compared to the HAZ, as cracks readily propagated from pores and inclusions early in the fatigue life of specimens [44]. While the study [44] examined the effect of internal defects, the notched specimen design omitted the weld cap and root. Thus, the potential influence of external defects, such as undercuts or inadequate root penetration [46], was overlooked.

The dependence of the microstructure on the VHCF behaviour of low-carbon steel welds was shown by Nishikawa and Furuya [47]. Heat treatments were used to simulate the temperature cycle experienced by different regions of the HAZ of low-carbon steel (0.15% wt.%C). Specimens consisting of either fine- or coarse-grained microstructures were produced by altering the maximum temperature and cooling rate. The fine-grained specimens revealed a distinct fatigue limit, with no failures occurring above 10^6 cycles. In contrast, the coarse-grained specimens showed a continually decreasing *S–N* curve in the VHCF regime due to internally originated fractures [47].

Schaumann and Steppeler [48] used a resonance-based fatigue testing machine to assess the VHCF behaviour of gas metal arc welding (GMAW) butt joints of S355J2+N structural steel. Fatigue testing was conducted at 390 Hz, and the machine configuration (Figure 10) allowed for the testing of welded joints with a 25 mm $\times$ 4 mm cross-section in the gauge length, considerably greater than what is possible with UFT [49]. Testing at 390 Hz was shown to have no influence on the fatigue strength of the welded joints in comparison to tests conducted using conventional servo-hydraulic fatigue testing at 20 Hz. The elevated frequency of 390 Hz allowed for tests to be conducted in the VHCF domain, up to a runout value of 5×10^8 cycles. However, the specimens displayed a fatigue limit at 3×10^6 cycles [48]. A fatigue limit at 10^7 cycles was observed by Gao et al. [50] for FCAW cruciform joints of Q355B steel, tested at 110 Hz to a runout value of 2×10^8 cycles.

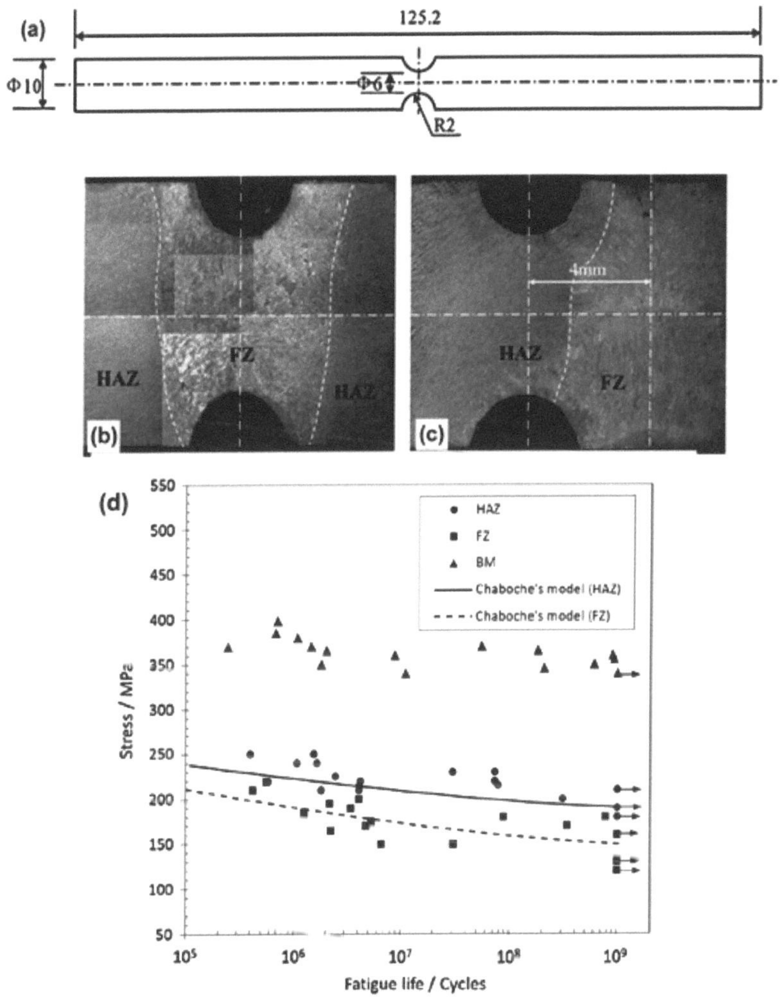

Figure 9. (**a**) Q345 welded fatigue specimen dimensions (mm); (**b**,**c**) etched specimens with the FZ and HAZ in the region of highest stress, respectively; (**d**) *S–N* curve for base metal, HAZ, and FZ specimens [44] (with permission from Elsevier, 2023).

A summary of the research carried out on the VHCF behaviour of low-carbon steel welded joints is provided in Table 1. Generally, it is found that welds of this class of metal show no fatigue limit before 10^7 cycles. Low-carbon steel welds have a reduced VHCF strength when compared to the corresponding base metals, regardless of the welding method used [12,41,43,44]. Almost all this research is conducted using structural steels with a yield strength of approx. 350 MPa (e.g., Q345, S355, EH36), and consequently future research should focus on other, preferably higher-strength grades. The effect of the loading frequency should be considered in subsequent investigations where UFT is employed, as the fatigue strength of low-carbon steels has been identified as showing a strong dependence on this aspect [31].

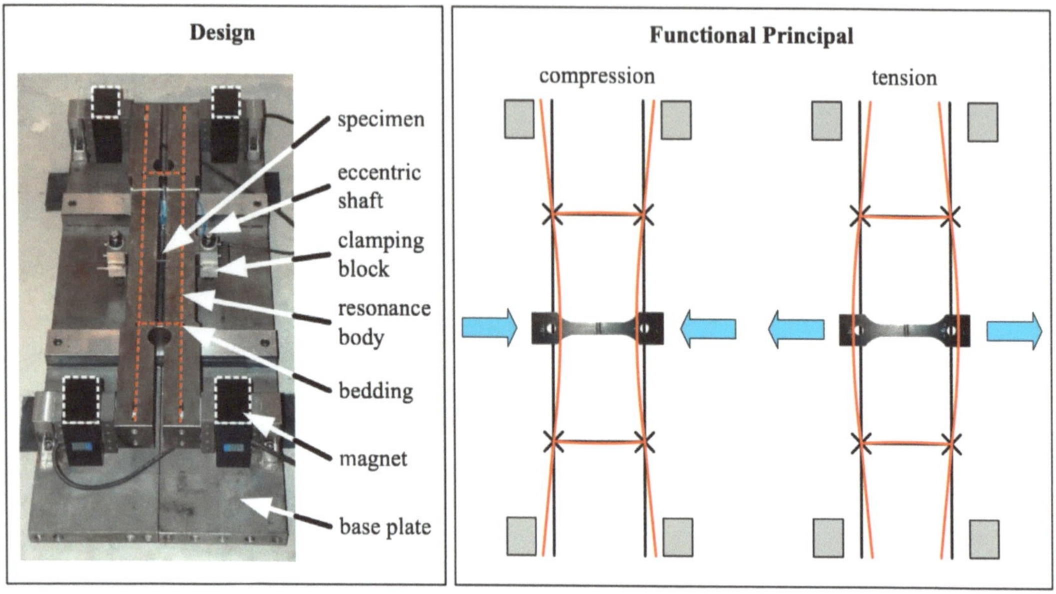

Figure 10. Design and fatigue loading schematic of resonance-based fatigue testing machine [48] (used under Creative Commons CC-BY license).

Table 1. Summary of research on VHCF of low-carbon steel welds.

Base Metal	Welding Method	Specimen Type	Research Focus	Reference
EH36	FCAW	Cylindrical dog-bone	Comparison of fatigue strength to base metal and effect of inclusions	[41]
Q235 and Q345	Arc welding (not specified)	Cruciform joint	Effect of UPT	[15]
Q345	Arc welding (not specified)	Cylindrical hourglass	Comparison of fatigue strength to base metal	[12]
16Mn	GTAW	As-welded cylindrical hourglass and cruciform	Specimen geometry	[43]
Q345	SMAW	Notched cylindrical hourglass	Comparison of fatigue strength of different weld regions	[44]
0.15% carbon steel	-	Cylindrical hourglass	Fatigue strength of HAZ simulated by heat treatment	[47]
S355J2+N	GMAW	Rectangular dog-bone	390 Hz resonance fatigue testing machine	[48]
Q355B	FCAW	Cruciform	Corrosion fatigue testing at 110 Hz	[50]

3.2. *Alloy Steels*

Alloy steels include alloying elements in appreciable amounts, in addition to carbon, that serve to improve mechanical properties [51]. They are generally used in applications where increased strength, hardness, and corrosion resistance are required compared to unalloyed low-carbon steels [51]. Alloy steels often feature inclusions or secondary phases, characteristics of Type II materials in the VHCF regime [8]. The additional alloying elements result in a higher potential for solidification cracking or HAZ softening post-welding; thus, preheating or post-weld heat treatments may be required [52]. Alloy steel welds are used in a range of applications that are subjected to VHCF, e.g., automotive struts and suspension components [13] and impeller blades [53].

As detailed in Section 2.1, the frequency effect is a significant issue with UFT and, consequently, characterising material behaviour in the VHCF regime. Despite this, the only existing research on the frequency effect with UFT of welded joints was conducted by Zhu et al. [53], who tested submerged arc welding (SAW) joints of 25Cr2Ni2MoV alloy steel at 110 Hz and 20 kHz. Cylindrical, dog-bone-shaped specimens were used so that the base metal, HAZ, and FZ regions all experienced the same stress during cyclic loading. As shown in Figure 11, the fatigue strength and fracture initiation sites of the welded joints showed sensitivity to the loading frequency, especially as the number of cycles to failure increased. The percentage increase in fatigue strength from 110 Hz to 20 kHz was 30% at 10^5 cycles, rising to 75% at 10^8 cycles. Additionally, fatigue cracks tended to initiate in the base metal at 110 Hz, only transitioning to the FZ at lower stress amplitudes above 2×10^6 cycles. In contrast, the majority of specimens tested at 20 kHz failed from the FZ, regardless of the stress amplitude applied. For both test frequencies, fatigue cracks initiated at the specimen surface at higher stress amplitudes, but from interior pores and inclusions at lower stress amplitudes. The volume of material subjected to >90% of maximum applied stress, known as the risk volume [54], differed between the specimens tested at each frequency, as well as the proportion comprised by each weld region [53]. These factors limit the certainty that any difference in fatigue behaviour can be solely attributed to the loading frequency.

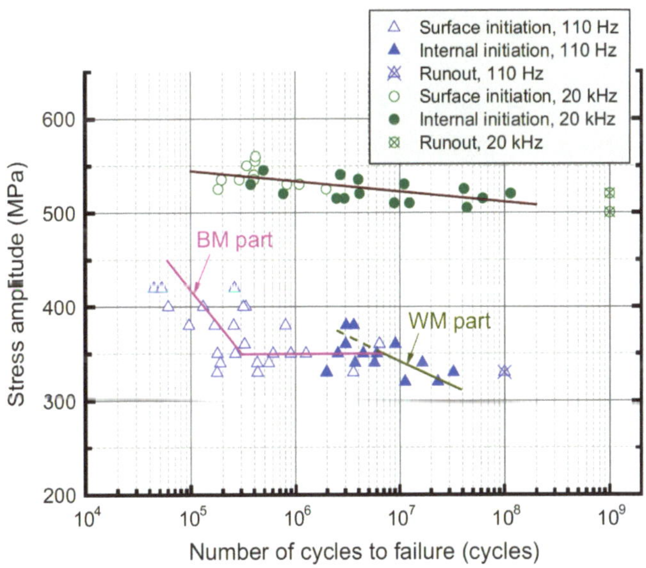

Figure 11. *S–N* curves for 25Cr2Ni2MoV welds at 110 Hz and 20 kHz [53] (with permission from Elsevier, 2023).

Cylindrical, dog-bone-shaped specimens with diameters of 5 mm and 3 mm were used to assess the size effect in UFT of welded joints by Zhu et al. [55]. For the same 25Cr2Ni2MoV alloy steel welds as detailed above [53], an increase of approx. 20% was observed in the fatigue strength at 10^7 cycles for the 3 mm diameter specimens compared to the 5 mm diameter specimens. This phenomenon was attributed to the increased probability of the presence of a larger defect, which a crack can initiate from at a lower stress [55]. SAW joints of the same alloy steel, 25Cr2Ni2MoV, were used to study the heat dissipation during UFT of welds [56]. The heat dissipation behaviour was similar for base metal and welded specimens during ultrasonic loading. The maximum surface temperature of the specimens followed the same trend for continuous and intermittent loading; an initial temperature increase was followed by a stable period. As crack propagation occurred, an increase in maximum surface temperature was observed, and its location correlated with the fracture site. The position of maximum temperature varied throughout tests in the welded specimen, whereas it remained fixed at the centre of base metal specimens [56].

The determination of a potential fatigue limit is the focus of a range of research on alloy steel welds. Yeom et al. [13] investigated the VHCF behaviour of GMAW joints of ferritic-bainitic alloy steels 590 FB and 780 FB. Rectangular cross-section dog-bone specimens with the weld bead removed were tested using UFT in addition to base metal specimens. Neither the base metal nor the respective welded joints displayed a fatigue limit below 10^9 cycles, as shown in Figure 12a. Only a minimal decrease in the VHCF strength (approx. 8%) was measured for both 590 FB and 780 FB due to welding, contrary to other research comparing welded joints to base metals in the VHCF domain [41,43,44]. As depicted in Figure 12b, dissimilar welds of 590 FB and 780 FB showed comparable VHCF strength to 780 FB welds [13]. Moreover, it was found that all specimens that fractured above 10^7 cycles exhibited crack initiation at a single point on the specimen surface [13]. A limited number of specimens were tested for each base metal and welded joint, restricting any conclusions that could be established.

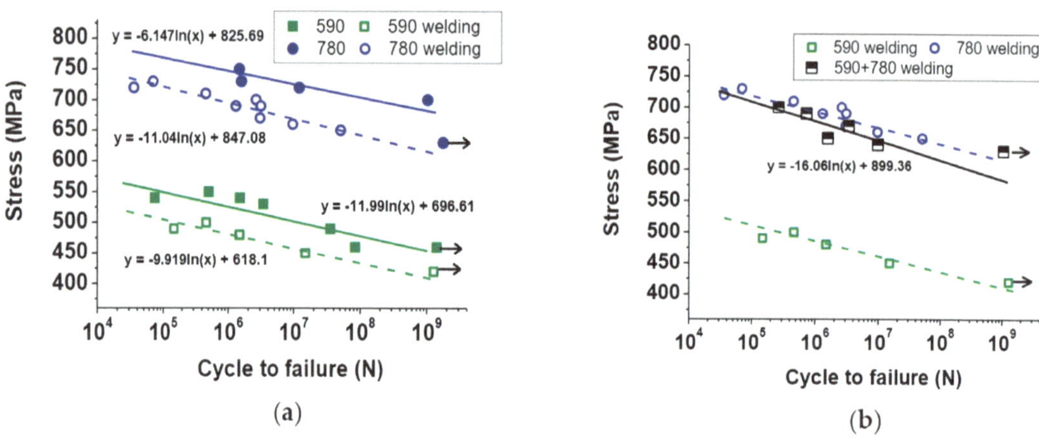

Figure 12. *S–N* curves for (**a**) 590 FB and 780 FB base metals and welded joint; (**b**) 590 FB and 780 FB similar and dissimilar welds [13] (used under Creative Commons CC-BY license).

He et al. [57] studied the VHCF behaviour of GMAW butt joints of SMA490BW, a weathering steel alloy, before and after UPT. A specially designed specimen was employed with the weld toe in the region of highest stress (Figure 13). Both as-welded joints and joints subjected to UPT each showed multiple failures in the VHCF regime, with all fractures originating from the weld toe. UPT resulted in a 38% increase in fatigue strength at 10^9 cycles [57].

The VHCF properties of SAW joints made of KMN-I, a low-alloy steel used for compressor blades, were investigated by Wang et al. [58]. The standard cylindrical hourglass shape was used for both base metal specimens and specimens with the FZ located in the centre. The fatigue strength of welds was similar to the base metal at lower fatigue lives (10^4–10^5 cycles), but the *S–N* curve for welds decreased at a greater rate (Figure 14). All fatigue failures of welds occurred below 3×10^7 cycles, indicating a possible fatigue limit early in the VHCF regime [58].

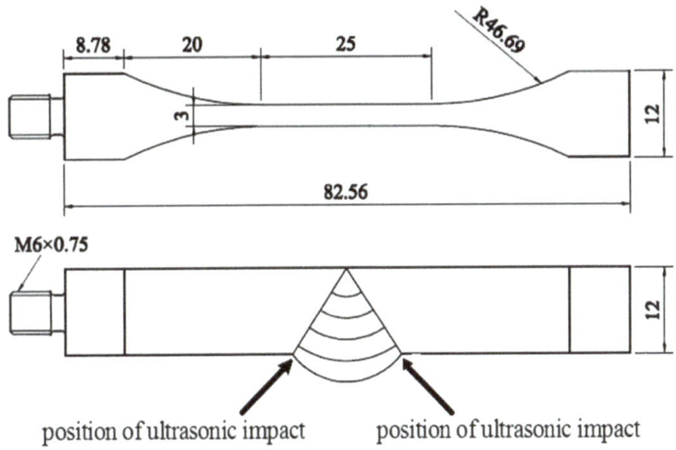

Figure 13. Geometry of SMA490BW butt welded specimens showing ultrasonic impact at weld toe (mm) [57] (with permission from Springer Nature, 2023).

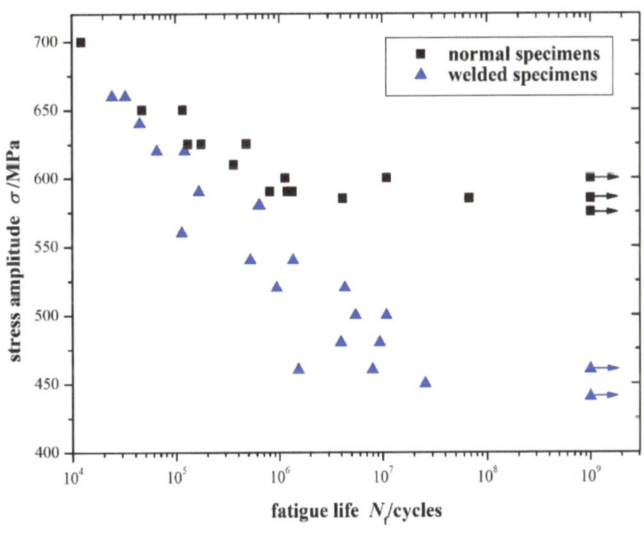

Figure 14. *S–N* diagram of base metal (denoted as normal specimens) and welded specimens of KMN-I [58] (used under Creative Commons CC-BY license).

In a series of papers [59–61], the fatigue properties of alloy steel welds at elevated temperatures were investigated up to 5×10^7 cycles using a loading frequency of 100 Hz. Failures occurred from internal inclusions above 10^7 cycles for SAW joints of a CrMoV steel tested at 370 °C [59]. No fatigue limit was observed for dissimilar SAW joints of CrMoV and 9% Cr steels at 500 °C [60]. Specimens predominantly failed at the HAZ of the weaker CrMoV steel, initiating from inclusions or microstructural inhomogeneities [60]. Li et al. [61] tested two different specimen types of SAW joints of 25Cr2Ni2MoV steel at 300 °C. Specimens with the HAZ in the centre had an approx. 7% greater fatigue strength than specimens with the FZ in the centre, an effect attributed to increased dynamic strain ageing [61].

Table 2 displays a summary of the research on the VHCF properties of alloy steel welds. Failures in the VHCF regime were observed for all welded joints of alloy steels; therefore, the assumption of a fatigue limit before 10^7 cycles is non-conservative. In addition, this alloy class has been used to investigate the frequency and size effects of UFT.

Table 2. Summary of research on VHCF of alloy steel welds.

Base Metal	Welding Method	Specimen Type	Research Focus	Reference
25Cr2Ni2MoV	SAW	Cylindrical dog-bone	Assessment of frequency effect	[53]
			Assessment of size effect	[55]
			Heat dissipation during UFT	[56]
			Elevated temperature testing at 100 Hz	[61]
590 FB and 780 FB	GMAW	Rectangular dog-bone	Comparison of fatigue strength to base metal and dissimilar welds	[13]
SMA490BW	GMAW	As-welded rectangular dog-bone	Effect of UPT	[57]
KMN-I	SMAW	Cylindrical hourglass	Comparison of fatigue strength to base metal	[58]
CrMoV Steel	SAW	Cylindrical hourglass	Elevated temperature testing at 100 Hz	[59]
9% Cr Steel and CrMoV Steel	SAW	Cylindrical dog-bone	Comparison of base metals and dissimilar weld at elevated temperature	[60]

3.3. Stainless Steels

The use of welded joints of stainless steel alloys in applications such as impeller blades [62] and nuclear reactor coolant pipes [63] means that they are subjected to VHCF during service; however, there is only limited research for this class of materials. The VHCF behaviour of SMAW joints of FV520B-I martensitic stainless steel was investigated by Zhang et al. [62]. Their study represents the only published research on the effect of surface roughness on the VHCF of welds, with "smooth" ($R_a \approx 0.05$) and "rough" ($R_a \approx 0.6$) cylindrical hourglass-shaped specimens tested at 20 kHz. The smooth specimens displayed the typical VHCF behaviour of high-strength steels [64], with no fatigue limit and some fractures originating from subsurface non-metallic inclusions or pores (Figure 15a–c). Energy-dispersive spectroscopy revealed the inclusions originated from the flux coating of the welding electrode. Additionally, two smooth specimens fractured from inclusions located at the specimen surface (Figure 15d,e). The rough specimens exhibited a tendency to fail in the HCF regime from surface initiation sites, while only two internally initiated failures in the VHCF were recorded [62]. In a subsequent paper using the same base metal and weld method, the size effect was assessed using smooth specimens [65]. Cylindrical dog-bone-shaped specimens with an approx. 3.3x greater risk volume showed a slight decrease in VHCF strength but no change in crack initiation behaviour. Compared to base metal, both specimen shapes showed a lower fatigue life and greater scatter in fatigue testing results [65].

Xiong et al. [63,66] investigated the VHCF behaviour of GTAW and laser beam welding (LBW) joints of 316L stainless steel using cylindrical dog-bone-shaped specimens. GTAW joints presented no fatigue limit in the VHCF regime, and specimens tended to fail from surface-initiated cracks. The GTAW joints showed comparable fatigue strength with the base metal when stress amplitudes were normalised to the ultimate tensile strength of each region [66]. Despite the presence of large pores, the refined microstructure due to the high heat input of the LBW process resulted in comparable VHCF strength to GTAW joints [63]. A greater scatter was displayed in the fatigue lives of LBW specimens due to the random size distribution of pores [63].

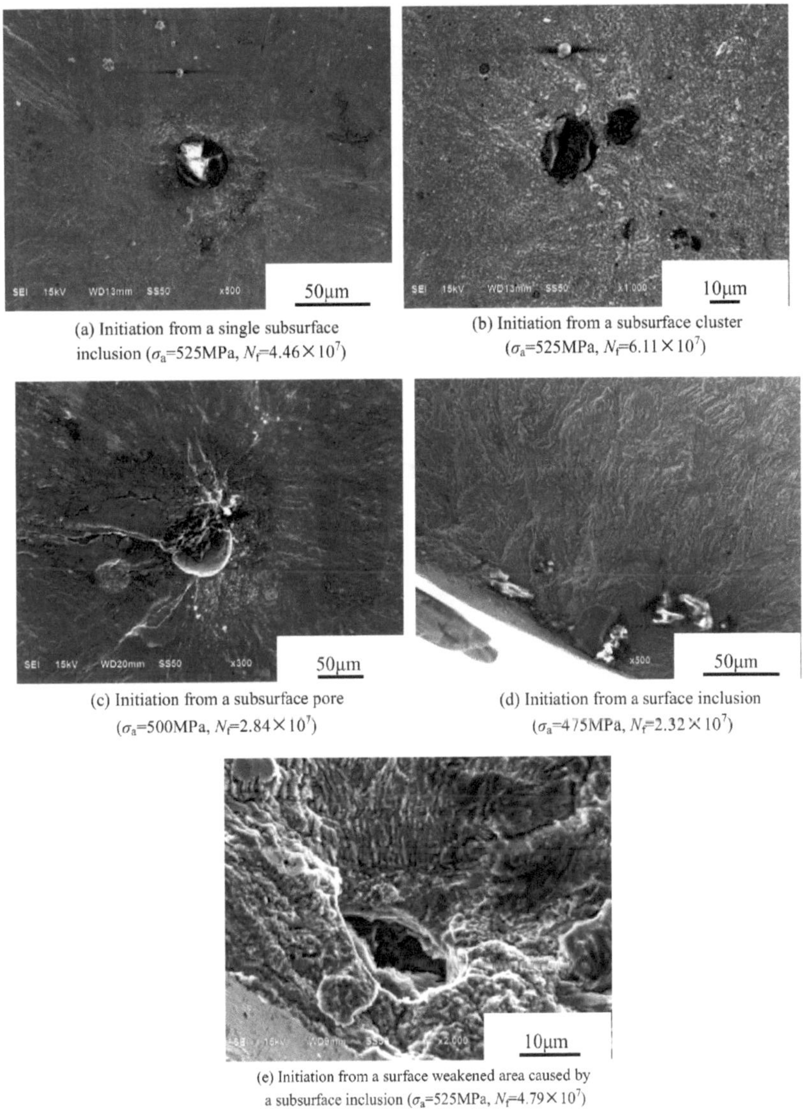

(a) Initiation from a single subsurface inclusion (σ_a=525MPa, N_f=4.46×10^7)

(b) Initiation from a subsurface cluster (σ_a=525MPa, N_f=6.11×10^7)

(c) Initiation from a subsurface pore (σ_a=500MPa, N_f=2.84×10^7)

(d) Initiation from a surface inclusion (σ_a=475MPa, N_f=2.32×10^7)

(e) Initiation from a surface weakened area caused by a subsurface inclusion (σ_a=525MPa, N_f=4.79×10^7)

Figure 15. Fracture initiation sites of smooth specimens of FV520B-I welds in VHCF regime [62] (with permission from Elsevier, 2023).

No definitive conclusions can be drawn from the limited research on the VHCF properties of stainless steel welds (Table 3). The comparison of welding methods, as investigated by Xiong et al. [63], is not seen in research on other metals and could be important for the design of components and structures subjected to VHCF. Future research should focus on expanding the data for this class of materials, with attention to other base metals and welding methods.

Table 3. Summary of research on VHCF of stainless steel welds.

Base Metal	Welding Method	Specimen Type	Research Focus	Reference
FV520B-I	SMAW	Cylindrical hourglass	Effect of surface roughness	[62]
		Cylindrical dog-bone	Size effect	[65]
316L	GTAW	Cylindrical dog-bone	Comparison to base metal	[66]
	GTAW and LBW	Cylindrical dog-bone	Comparison of weld methods	[63]

3.4. Non-Ferrous Alloys

There is limited fatigue data within the engineering literature for welded joints of non-ferrous alloys in the VHCF domain, with only a small number of investigations conducted focusing on aluminium [67–69], magnesium [70], and titanium alloys [71–74]. The VHCF behaviour of GMAW joints of AA6082 was characterised by Cremer et al. [67] using rectangular dog-bone-shaped specimens with the weld reinforcement removed. Three different specimens were used, each featuring either the base metal, HAZ, or FZ at the position of maximum stress. Neither of the weld zones showed a fatigue limit below 2×10^9 cycles. As expected, the base metal exhibited the greatest fatigue strength, while the HAZ presented a slightly inferior strength (reduced by approx. 10%). Due to incomplete fusion and gas pore defects, the lowest fatigue strength was identified in the FZ specimens, with significant scatter in the results. Notched specimens were also tested in different heat-treated conditions, to achieve hardness values representative of the FZ (85 HV) and HAZ (75 HV). The notch factor was chosen to represent the stress concentration at the weld toe. It was reported that the VHCF strength of the notched specimens decreased as the material hardness decreased [67]. One limitation of this research is that although the heat-treated notched specimens allowed for a simulated fatigue test of the weld toe, they did not include defects, such as micro-ripples, spatter droplets, small undercuts, or microscopic slag inclusions [75,76], that can be detrimental to fatigue strength.

Deng et al. [68] examined the VHCF behaviour of AA7050 base metal and friction stir-welded (FSW) butt joints using cylindrical dog-bone-shaped specimens. No fatigue limit was observed for either specimen type up to 10^9 cycles. The welded specimens showed slightly greater fatigue strength than the base metal [68]. The findings were consistent with the literature on the HCF of FSW joints, with the highly refined microstructures caused by the FSW process considerably enhancing fatigue performance [77]. No fatigue limit was found for FSW joints of AA6061 below 10^9 cycles by He et al. [69]. Rectangular dog-bone-shaped specimens were extracted from different regions through the thickness of the weld; it was found that specimens extracted from the weld root displayed the lowest fatigue life due to the softening that occurred in this region during the FSW process [69]. The VHCF behaviour of FSW joints of ZK60 magnesium alloy was investigated by Chen et al. [70] using rectangular dog-bone-shaped specimens. The fatigue strength of the joints was found to continually decrease in the VHCF regime, with no discernible fatigue limit. All failures above 10^6 cycles were identified to originate from porosity defects in the thermo-mechanically affected zone [70].

Electron beam welding (EBW) joints of titanium alloys have received some research attention [71–74] due to their use in aerospace engines, which are subjected to VHCF in service. Liu et al. [71] investigated the VHCF behaviour of TC17 titanium alloy EBW joints. Cylindrical dog-bone-shaped specimens were employed, and welds were tested after PWHT. Both EBW and base metal specimens showed no fatigue limit prior to 10^9 cycles at room temperature. The fatigue strength of EBW joints was pronouncedly decreased compared to the base metal; a reduction of approx. 45% was found in the VHCF regime (Figure 16). The large scatter in the fatigue lives of welded joints was ascribed to the random size distribution of pores and the location of pores relative to the specimen surface [71]. In two subsequent studies [72,73], identical EBW TC17 joints were tested at 400 °C. The *S–N* curve at this elevated temperature revealed a reduced slope in the VHCF regime, although no fatigue limit was found. At 400 °C, the fatigue strength showed a reduction of approx. 20% at 10^9 cycles as compared to room temperature. Pore-initiated cracks were still the predominant mode of failure [72,73]. Zheng et al. [74] found no fatigue limit for TC21 titanium alloy EBW joints and a similar dependence of the VHCF behaviour on pores.

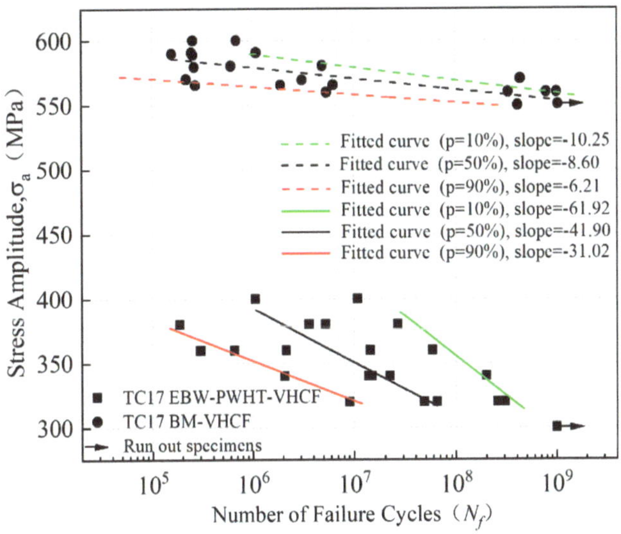

Figure 16. *S–N* curves for TC-17 alloy base metal and EBW joint [71] (used under Creative Commons CC-BY license).

A synopsis of the research carried out to date on the VHCF properties of welded joints of non-ferrous alloys is shown in Table 4. In general, non-ferrous alloy welds presented a similar trend to ferrous alloys, with no fatigue limit observed in the VHCF regime. The VHCF properties developed by more advanced welding methods (FSW and EBW) are investigated more commonly for this class of metals compared to ferrous alloys. Titanium alloys can be joined by a range of conventional welding methods, e.g., GMAW and GTAW [78]; therefore, there is significant merit in considering the VHCF of welds produced by these processes in future research endeavours.

Table 4. Summary of research on VHCF of non-ferrous alloy welds.

Base Metal	Welding Method	Specimen Type	Research Focus	Reference
AA6082	GMAW	Rectangular dog bone and notched	Comparison of fatigue strength of different weld regions	[67]
AA7050	FSW	Cylindrical dog-bone	Comparison of fatigue strength to base metal	[68]
AA6061	FSW	Rectangular dog-bone	Comparison of fatigue strength through-thickness	[69]
ZK60	FSW	Rectangular dog-bone	Effect of welding defects	[70]
TC17	EBW	Cylindrical dog-bone	Effect of welding pores	[71]
			Fatigue strength at elevated temperature	[72]
			Comparison of fatigue strength to base metal at elevated temperature	[73]
TC21	EBW	Cylindrical hourglass	Effect of welding pores	[74]

4. Design Standards and Industrial Case Studies

4.1. Design Standards for Very High Cycle Fatigue of Welds

The guidance for the design of welded joints subjected to VHCF varies between international standards. Within these standards, S–N curves are generally specified in terms of the nominal stress. This refers to the stress remote of the weld bead, with no consideration of the stress concentration caused by the weld toe geometry [79,80]. Diverse welded joint geometries are then assigned to a certain fatigue class corresponding to their fatigue strength. For example, a full penetration butt weld ground flush is assigned a higher fatigue class than a partial penetration butt weld.

One of the main differences is whether a fatigue limit is incorporated into the design S–N curves. BS 7608:2014 [80] assumes a fatigue limit of 1×10^7 cycles for normal stresses in the absence of corrosion. The fatigue limit in AWS D1.1/D1.1M:2020 [81] is defined in the range of 2×10^6 and 2×10^7 cycles, depending on the fatigue class of the welded joint. A decrease in the slope of the S–N curves in the VHCF regime is considered in a number of standards [79,82,83]. This is similar to the observations reported by some researchers in laboratory-based VHCF tests [41,43]. Due to the emergence of VHCF as a separate research field, the IIW removed the consideration of a fatigue limit in the 2006 update of their design recommendations [79,84]. In line with this update, the slope of the S–N curves is reduced at 1×10^7 cycles, so that the rate of decrease in fatigue strength is 10% per decade of loading cycles [79]. BS EN 1993-1-9:2005 [82] specifies a reduction in the slope of the S–N curve between 5×10^6 and 1×10^8 cycles, with a fatigue limit beyond this. For offshore applications in clean air (i.e., no influence of corrosion), DNV-RP-C203 [83] defines a reduction in the slope at 1×10^7 cycles with no fatigue limit. It is evident that there is no consensus in the relevant standards on the fatigue assessment of welded joints in the VHCF domain. The consideration of a fatigue limit could potentially lead to the non-conservative design of structures and the standards that consider this should be updated.

4.2. Industrial Case Studies

Studying failures of welded components subjected to VHCF in an industrial setting allows for further insight into the topic, in the absence of the limitations of the UFT method. Fry [14] gathered data on fatigue failures of welded components of vibrating screens used in the mineral processing industry. Vibrating screens typically operate continually between 14 and 17 Hz; therefore, the VHCF domain is reached within a month of usage [14]. Due to the common occurrence of fatigue failures beyond 10^7 cycles, an extension to the design S–N curves used within BS7608:2014 [80] was proposed (Figure 17). The extension of the S–N curves involves a change in slope at 10^7 cycles and a fatigue limit at 4×10^8 cycles for welds protected from corrosion [14].

More recently, the failure of a welded joint after 6.2×10^7 cycles on a drive beam of a vibrating screen was investigated by Rumiche et al. [85]. Comparing the fatigue life of the component to an appropriate fatigue class from Fry's modified S–N curves [14] returns an approximate stress range of 20 MPa, which is within typical operating stress ranges for vibrating screen components [85]. Other industrial applications where VHCF failures of welded components have been observed include conveyor pulleys [86] and railway bridges [87].

The VHCF behaviour of welds is an important consideration in the context of the service life extension of existing components. As an illustration, welded wind turbine monopiles approaching 20 years in service are being considered for further use after their initial design life [88]. Service life extension of existing welded components would result in both environmental and economic benefits [88].

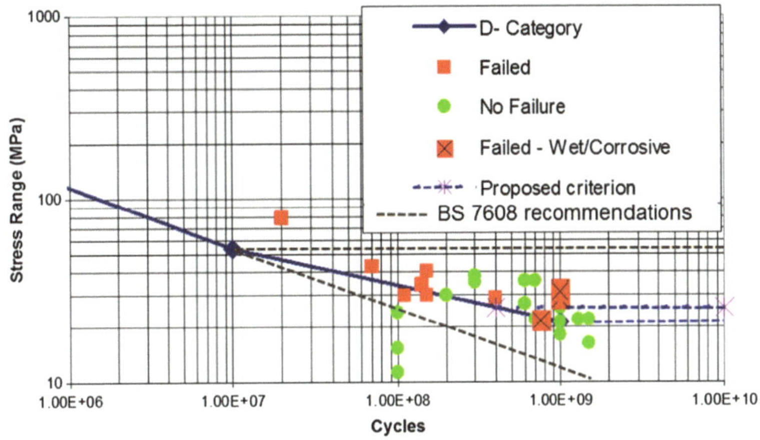

Figure 17. Data from failures of vibrating screen components and proposed modification to BS7608 *S–N* curves [14] (with permission from Elsevier, 2023).

5. Future Directions

As highlighted in this paper, research on the VHCF behaviour of various metals and weld methods has been conducted, but there are a number of knowledge gaps to be addressed in future research. Investigation of the fundamental challenges of the UFT method is required to promote a better understanding of the VHCF performance of welds. The frequency effect of welded joints has only been studied in one investigation, despite being identified as one of the primary limitations of UFT [27]. As the frequency effect is heavily dependent on crystalline structure [27], research is required for diverse metals, e.g., aluminium and titanium alloys. Unalloyed low-carbon steels show a significant frequency effect [31], yet there is no research comparing the results of conventional fatigue testing and UFT for welds of this class of material. The control of specimen surface temperature generally requires the combination of air or water cooling, and intermittent driving so the effect of these should be considered more often in relevant studies. A wide range of specimen geometries for UFT of welded joints have been exhibited in the literature, limiting comparisons that can be made between different investigations. The standardisation of specimen design for UFT of welded joints would allow for the VHCF behaviour of various base metals and welding methods to be suitably compared.

UFT machines require the use of specimens that are significantly smaller than welded joints in real components, and the limitation of this is often not considered. The local geometry of welded components causes stress concentration regions [5], and although some as-welded UFT specimen designs have been exhibited, this area needs further investigation. All studies described herein used uniaxial tensile-compressive loading in the VHCF regime. Therefore, the effect of various loading conditions, namely bending, torsional, and multiaxial loading, should be explored for welded joints using UFT. Applying mean stress is recommended when fatigue testing specimens of welded joints to represent the effect of residual stresses [79], but this has not been investigated to date. Additionally, the effect of corrosion on VHCF strength has not been thoroughly examined for welded joints. Fundamental studies on the effects of different weld methods and PWHT would also enhance the scientific understanding of this subject area.

6. Conclusions

UFT has been proven to be a viable method to study the VHCF of welded joints in a reasonable timeframe. The elevated loading frequency significantly accelerates testing; hence, the VHCF of welds can be investigated in a controlled laboratory environment rather than relying on reported failures of real-world components. Several conclusions can be drawn from the present review of the engineering literature on the VHCF behaviour of welds:

- The VHCF strength of welded joints is generally lower than that of base metals, even when the stress concentration of the weld toe is removed by machining.
- A continuous decrease in the *S–N* curve is observed beyond 10^7 loading cycles for all metallic welds studied to date (steel, aluminium, titanium, and magnesium alloys), challenging the classical assumption of a fatigue limit. This is further demonstrated by examples of in-service failures of welded components in the VHCF regime in various applications.
- A change in the design *S–N* curves of standards that include a fatigue limit, e.g., BS7608:2014, should be implemented to guarantee the structural integrity of welded components operating in the VHCF regime.
- The primary factor for the reduction in VHCF strength of welds is welding defects, principally inclusions and gas pores. Failures in the VHCF domain typically originate from these defects.
- Limited research on UPT has shown the process to significantly increase the VHCF strength of welded joints due to the grain refinement and compressive residual stresses caused by the process.
- Testing programmes that cover the HCF and VHCF regimes tend to show that there is a reduction in the slope of the *S–N* curve in the VHCF regime, an observation that has been reflected in some design standards.

Although the scientific understanding of the VHCF behaviour of welds has been greatly expanded in the past 20 years, the research field remains relatively unexplored; thus, more fundamental studies are required.

Author Contributions: Conceptualization, A.E., A.T. and Y.G.; investigation, A.E.; writing—original draft preparation, A.E.; writing—review and editing, A.T. and Y.G.; supervision, A.T. and Y.G. All authors have read and agreed to the published version of the manuscript.

Funding: This research received no external funding.

Data Availability Statement: No new data were created or analysed in this study. Data sharing is not applicable to this article.

Conflicts of Interest: The authors declare no conflict of interest.

References

1. Campbell, F.C. *Fatigue and Fracture—Understanding the Basics*; ASM International: Materials Park, OH, USA, 2012; ISBN 978-1-61503-976-0.
2. Masoudi Nejad, R.; Shariati, M.; Farhangdoost, K. Prediction of Fatigue Crack Propagation and Fractography of Rail Steel. *Theor. Appl. Fract. Mech.* **2019**, *101*, 320–331. [CrossRef]
3. Moore, P.; Booth, G. *The Welding Engineer's Guide to Fracture and Fatigue*; Elsevier: Amsterdam, The Netherlands, 2015; ISBN 9781782423706.
4. Tsutsumi, S.; Fincato, R.; Luo, P.; Sano, M.; Umeda, T.; Kinoshita, T.; Tagawa, T. Effects of Weld Geometry and HAZ Property on Low-Cycle Fatigue Behavior of Welded Joint. *Int. J. Fatigue* **2022**, *156*, 106683. [CrossRef]
5. Ma, M.; Lai, R.; Qin, J.; Wang, B.; Liu, H.; Yi, D. Effect of Weld Reinforcement on Tensile and Fatigue Properties of 5083 Aluminum Metal Inert Gas (MIG) Welded Joint: Experiments and Numerical Simulations. *Int. J. Fatigue* **2021**, *144*, 106046. [CrossRef]
6. Zhan, R.; Wang, D.; Ren, Z.; Deng, C.; Xu, X.; Liang, H. Evolution of Welding Residual Stresses Involving the Cutting Process and Its Effect on Fatigue Performance. *Int. J. Press. Vessel. Pip.* **2022**, *197*, 104636. [CrossRef]
7. Zerbst, U.; Ainsworth, R.A.; Beier, H.T.; Pisarski, H.; Zhang, Z.L.; Nikbin, K.; Nitschke-Pagel, T.; Münstermann, S.; Kucharczyk, P.; Klingbeil, D. Review on Fracture and Crack Propagation in Weldments—A Fracture Mechanics Perspective. *Eng. Fract. Mech.* **2014**, *132*, 200–276. [CrossRef]
8. Sharma, A.; Oh, M.C.; Ahn, B. Recent Advances in Very High Cycle Fatigue Behavior of Metals and Alloys—A Review. *Metals* **2020**, *10*, 1200. [CrossRef]
9. Pineau, A. Low-Cycle Fatigue. In *Fatigue of Materials and Structures—Fundamentals*; John Wiley & Sons: Hoboken, NJ, USA, 2010; pp. 113–177. ISBN 9781118623435.
10. Bathias, C.; Drouillac, L.; Le François, P. How and Why the Fatigue S-N Curve Does Not Approach a Horizontal Asymptote. *Int. J. Fatigue* **2001**, *23*, 143–151. [CrossRef]

11. Bathias, C. There Is No Infinite Fatigue Life in Metallic Materials. *Fatigue Fract. Eng. Mater. Struct.* **1999**, *22*, 559–565. [CrossRef]
12. He, C.; Liu, Y.; Fang, D.; Wang, Q. Very High Cycle Fatigue Behavior of Bridge Steel Welded Joint. *Theor. Appl. Mech. Lett.* **2012**, *2*, 31010. [CrossRef]
13. Yeom, H.; Choi, B.; Seol, T.; Lee, M.; Jeon, Y. Very High Cycle Fatigue of Butt-Welded High-Strength Steel Plate. *Metals* **2017**, *7*, 103. [CrossRef]
14. Fry, P.R. High Cycle Fatigue of Welded Structures: Design Guidelines Validated by Case Studies. *Eng. Fail. Anal.* **2014**, *46*, 179–187. [CrossRef]
15. Yin, D.; Wang, D.; Jing, H.; Huo, L. The Effects of Ultrasonic Peening Treatment on the Ultra-Long Life Fatigue Behavior of Welded Joints. *Mater. Des.* **2010**, *31*, 3299–3307. [CrossRef]
16. Mayer, H. Recent Developments in Ultrasonic Fatigue. *Fatigue Fract. Eng. Mater. Struct.* **2016**, *39*, 3–29. [CrossRef]
17. Kuhn, H.; Medlin, D. Ultrasonic Fatigue Testing. In *Mechanical Testing and Evaluation*; ASM International: Materials Park, OH, USA, 2000; Volume 8, pp. 717–729. ISBN 978-1-62708-176-4.
18. Trško, L.; Nový, F.; Bokůvka, O.; Jambor, M. Ultrasonic Fatigue Testing in the Tension-Compression Mode. *J. Vis. Exp.* **2018**, *2018*, 57007. [CrossRef]
19. Wan, J.; Zhao, J.; Zhou, L.; Zhang, S.; Zhao, H. A Novel Methodology for Bending Ultrasonic Fatigue Testing in the VHCF Regime. *Int. J. Fatigue* **2023**, *170*, 107562. [CrossRef]
20. Marines-Garcia, I.; Doucet, J.P.; Bathias, C. Development of a New Device to Perform Torsional Ultrasonic Fatigue Testing. *Int. J. Fatigue* **2007**, *29*, 2094–2101. [CrossRef]
21. da Costa, P.R.; Soares, H.; Reis, L.; Freitas, M. Ultrasonic Fatigue Testing under Multiaxial Loading on a Railway Steel. *Int. J. Fatigue* **2020**, *136*, 105581. [CrossRef]
22. Shimadzu Corporation. *USF-2000A Hardware Instruction Manual*. Shimadzu Corporation, 2017. Available online: https://www.shimadzu.com/an/sites/shimadzu.com.an/files/pim/pim_document_file/manuals/13951/349-04408.pdf (accessed on 20 September 2023).
23. Feng, N.; Wang, X.; Guo, J.; Li, Q.; Yu, J.; Zhang, X. Design Theory and Experimental Research of Ultrasonic Fatigue Test. *Machines* **2022**, *10*, 635. [CrossRef]
24. The Japan Welding Engineering Society. *Method for Ultrasonic Fatigue Testing in Metallic Material*; WES 1112: Tokyo, Japan, 2022.
25. Furuya, Y.; Shimamura, Y.; Takanashi, M.; Ogawa, T. Standardization of an Ultrasonic Fatigue Testing Method in Japan. *Fatigue Fract. Eng. Mater. Struct.* **2022**, *45*, 2415–2420. [CrossRef]
26. Bach, J.; Göken, M.; Höppel, H.-W. Fatigue of Low Alloyed Carbon Steels in the HCF/VHCF-Regimes. In *Fatigue of Materials at Very High Numbers of Loading Cycles*; Chris, H.J., Ed.; Springer Spektrum: Berlin/Heidelberg, Germany, 2018; pp. 1–23. ISBN 978-3-658-24531-3.
27. Hong, Y.; Hu, Y.; Zhao, A. Effects of Loading Frequency on Fatigue Behavior of Metallic Materials—A Literature Review. *Fatigue Fract. Eng. Mater. Struct.* **2023**, *46*, 3077–3098. [CrossRef]
28. Guennec, B.; Ueno, A.; Sakai, T.; Takanashi, M.; Itabashi, Y.; Ota, M. Dislocation-Based Interpretation on the Effect of the Loading Frequency on the Fatigue Properties of JIS S15C Low Carbon Steel. *Int. J. Fatigue* **2015**, *70*, 328–341. [CrossRef]
29. Milne, L.; Gorash, Y.; Comlekci, T.; MacKenzie, D. Frequency Effects in Ultrasonic Fatigue Testing (UFT) of Q355B Structural Steel. *Procedia Struct. Integr.* **2022**, *42*, 623–630. [CrossRef]
30. Gorash, Y.; Comlekci, T.; Styger, G.; Kelly, J.; Brownlie, F.; Milne, L. Ultrasonic Fatigue Testing of Structural Steel S275JR+AR with Insights into Corrosion, Mean Stress and Frequency Effects. *Materials* **2023**, *16*, 1799. [CrossRef]
31. Guennec, B.; Kinoshita, T.; Horikawa, N.; Oguma, N.; Sakai, T. Loading Frequency Effect on the Fatigue Endurance of Structural Carbon Steels: Estimation Based on Dislocation Motion Theory and Experimental Verification of the Model. *Int. J. Fatigue* **2023**, *172*, 107634. [CrossRef]
32. Zhao, A.; Xie, J.; Sun, C.; Lei, Z.; Hong, Y. Effects of Strength Level and Loading Frequency on Very-High-Cycle Fatigue Behavior for a Bearing Steel. *Int. J. Fatigue* **2012**, *38*, 46–56. [CrossRef]
33. Mughrabi, H. Specific Features and Mechanisms of Fatigue in the Ultrahigh-Cycle Regime. *Int. J. Fatigue* **2006**, *28*, 1501–1508. [CrossRef]
34. Pyttel, B.; Schwerdt, D.; Berger, C. Very High Cycle Fatigue—Is There a Fatigue Limit? *Int. J. Fatigue* **2011**, *33*, 49–58. [CrossRef]
35. Lee, J.; Park, S.Y.; Choi, B.H. Evaluation of Fatigue Characteristics of Aluminum Alloys and Mechanical Components Using Extreme Value Statistics and C-Specimens. *Metals* **2021**, *11*, 1915. [CrossRef]
36. Murakami, Y.; Nomoto, T.; Ueda, T. Factors Influencing the Mechanism of Superlong Fatigue Failure in Steels. *Fatigue Fract. Eng. Mater. Struct.* **1999**, *22*, 581–590. [CrossRef]
37. Sakai, T.; Sato, Y.; Oguma, N. Characteristics S-N Properties of High-Carbon-Chromium-Bearing Steel under Axial Loading in Long-Life Fatigue. *Fatigue Fract. Eng. Mater. Struct.* **2002**, *25*, 765–773. [CrossRef]
38. De Li, Y.; Zhang, L.L.; Fei, Y.H.; Liu, X.Y.; Li, M.X. On the Formation Mechanisms of Fine Granular Area (FGA) on the Fracture Surface for High Strength Steels in the VHCF Regime. *Int. J. Fatigue* **2016**, *82*, 402–410. [CrossRef]
39. Sippel, J.P.; Kerscher, E. Properties of the Fine Granular Area and Postulated Models for Its Formation during Very High Cycle Fatigue—A Review. *Appl. Sci.* **2020**, *10*, 8475. [CrossRef]
40. Chen, Z.; Dong, Z.; Liu, C.; Dai, Y.; He, C. Characterization on Crack Initiation and Early Propagation Region of Nickel-Based Alloys in Very High Cycle Fatigue. *Materials* **2022**, *15*, 5806. [CrossRef] [PubMed]

41. Zhao, X.; Dongpo, W.; Deng, C.; Liu, Y.; Zongxian, S. The Fatigue Behaviors of Butt Welds Ground Flush in the Super-Long Life Regime. *Int. J. Fatigue* **2012**, *36*, 1–8. [CrossRef]
42. Tridello, A.; Boursier Niutta, C.; Berto, F.; Tedesco, M.M.; Plano, S.; Gabellone, D.; Paolino, D.S. Design against Fatigue Failures: Lower Bound P-S-N Curves Estimation and Influence of Runout Data. *Int. J. Fatigue* **2022**, *162*, 106934. [CrossRef]
43. Liu, Y.; He, C.; Huang, C.; Khan, M.K.; Wang, Q. Very Long Life Fatigue Behaviors of 16Mn Steel and Welded Joint. *Struct. Eng. Mech.* **2014**, *52*, 889–901. [CrossRef]
44. He, C.; Huang, C.; Liu, Y.; Wang, Q. Fatigue Damage Evaluation of Low-Alloy Steel Welded Joints in Fusion Zone and Heat Affected Zone Based on Frequency Response Changes in Gigacycle Fatigue. *Int. J. Fatigue* **2014**, *61*, 297–303. [CrossRef]
45. Shang, D.G. Measurement of Fatigue Damage Based on the Natural Frequency for Spot-Welded Joints. *Mater. Des.* **2009**, *30*, 1008–1013. [CrossRef]
46. Jonsson, B.; Dobmann, G.; Hobbacher, A.F.; Kassner, M.; Marquis, G. *IIW Guidelines on Weld Quality in Relationship to Fatigue Strength*; Springer International Publishing: Cham, Switzerland, 2016; ISBN 978-3-319-19197-3.
47. Nishikawa, H.; Furuya, Y. Gigacycle Fatigue Fracture of Low Strength Carbon Steel, Tested Using a Simulated Heat Affected Zone Microstructure. *ISIJ Int.* **2019**, *59*, 1926–1928. [CrossRef]
48. Schaumann, P.; Steppeler, S. Fatigue Tests of Axially Loaded Butt Welds up to Very High Cycles. *Procedia Eng.* **2013**, *66*, 88–97. [CrossRef]
49. Invernizzi, S.; Montagnoli, F.; Carpinteri, A. Experimental Evidence of Specimen-Size Effects on EN-AW6082 Aluminum Alloy in VHCF Regime. *Appl. Sci.* **2021**, *11*, 4272. [CrossRef]
50. Gao, Z.; Wang, D.; Gong, B.; Deng, C.; Wu, S.; Zhang, H. VHCF Behavior of Welded Joints with HFMI Treatment under Moisture Conditions. *Weld. J.* **2022**, *101*, 27–42. [CrossRef]
51. Oyyaravelu, R.; Kuppan, P.; Arivazhagan, N. Metallurgical and Mechanical Properties of Laser Welded High Strength Low Alloy Steel. *J. Adv. Res.* **2016**, *7*, 463–472. [CrossRef] [PubMed]
52. Kou, S. *Welding Metallurgy*, 2nd ed.; John Wiley & Sons: Hoboken, NJ, USA, 2003; ISBN 3175723993.
53. Zhu, M.L.; Liu, L.L.; Xuan, F.Z. Effect of Frequency on Very High Cycle Fatigue Behavior of a Low Strength Cr-Ni-Mo-V Steel Welded Joint. *Int. J. Fatigue* **2015**, *77*, 166–173. [CrossRef]
54. Murakami, Y. *Metal Fatigue: Effects of Small Defects and Nonmetallic Inclusions*, 2nd ed.; Elsevier: New York, NY, USA, 2019; ISBN 9780128138779.
55. Zhu, M.L.; Xuan, F.Z. Failure Mechanisms and Fatigue Strength Assessment of a Low Strength Cr−Ni−Mo−V Steel Welded Joint: Coupled Frequency and Size Effects. *Mech. Mater.* **2016**, *100*, 198–208. [CrossRef]
56. Hu, J.-T.; Chen, R.; Zhu, G.; Wang, C.; Zhu, M.-L.; Xuan, F.-Z. Heat Dissipation Behavior of a Low-Strength-Steel Welded Joint in Ultrasonic Fatigue. *Metals* **2022**, *12*, 1857. [CrossRef]
57. He, B.; Deng, H.; Jiang, M.; Wei, K.; Li, L. Effect of Ultrasonic Impact Treatment on the Ultra High Cycle Fatigue Properties of SMA490BW Steel Welded Joints. *Int. J. Adv. Manuf. Technol.* **2018**, *96*, 1571–1577. [CrossRef]
58. Wang, P.; Wang, W.; Zhang, M.; Zhou, Q.; Gao, Z. Effects of Specimen Size and Welded Joints on the Very High Cycle Fatigue Properties of Compressor Blade Steel KMN-I. *Coatings* **2021**, *11*, 1244. [CrossRef]
59. Zhu, M.L.; Xuan, F.Z.; Du, Y.N.; Tu, S.T. Very High Cycle Fatigue Behavior of a Low Strength Welded Joint at Moderate Temperature. *Int. J. Fatigue* **2012**, *40*, 74–83. [CrossRef]
60. Zhang, W.C.; Zhu, M.L.; Wang, K.; Xuan, F.Z. Failure Mechanisms and Design of Dissimilar Welds of 9%Cr and CrMoV Steels up to Very High Cycle Fatigue Regime. *Int. J. Fatigue* **2018**, *113*, 367–376. [CrossRef]
61. Li, S.C.; Zhang, W.C.; Zhu, M.L.; Xuan, F.Z. On Specimen Design for High Cycle Fatigue Testing of Welded Joint. *Int. J. Fatigue* **2020**, *136*, 105597. [CrossRef]
62. Zhang, M.; Wang, W.; Wang, P.; Liu, Y.; Li, J. Fatigue Behavior and Mechanism of FV520B-I Welding Seams in a Very High Cycle Regime. *Int. J. Fatigue* **2016**, *87*, 22–37. [CrossRef]
63. Xiong, Z.; Peng, E.; Zeng, L.; Xu, Q. Giga-Cycle Fatigue Behavior of the Nuclear Structure of 316L Weldments. *Front. Energy Res.* **2021**, *9*, 696608. [CrossRef]
64. Himmelbauer, F.; Winter, G.; Grün, F.; Kiesling, C. VHCF Properties and Assessment of Specimens and Thin-Walled Component-like Structures Made of High-Strength Steel X5CrNiCuNb16-4. *Int. J. Fatigue* **2022**, *156*, 106645. [CrossRef]
65. Zhang, M.; Zhang, H.; Li, M.; Liu, L.; Wang, P. Fatigue Behavior and Mechanism of Dog-Bone-Shaped Specimens of FV520B-I in a Very High Cycle Regime. *Fatigue Fract. Eng. Mater. Struct.* **2022**, *45*, 3658–3676. [CrossRef]
66. Xiong, Z.; Wei, D.; Wang, H.; Shi, H.J.; Ma, X. Fatigue Behavior of 316 L Stainless Steel Weldment up to Very-High-Cycle Fatigue Regime. *Mater. Res. Express* **2019**, *6*, 076514. [CrossRef]
67. Cremer, M.; Zimmermann, M.; Christ, H.J. High-Frequency Cyclic Testing of Welded Aluminium Alloy Joints in the Region of Very High Cycle Fatigue (VHCF). *Int. J. Fatigue* **2013**, *57*, 120–130. [CrossRef]
68. Deng, C.; Wang, H.; Gong, B.; Li, X.; Lei, Z. Effects of Microstructural Heterogeneity on Very High Cycle Fatigue Properties of 7050-T7451 Aluminum Alloy Friction Stir Butt Welds. *Int. J. Fatigue* **2016**, *83*, 100–108. [CrossRef]
69. He, C.; Liu, Y.; Dong, J.; Wang, Q.; Wagner, D.; Bathias, C. Through Thickness Property Variations in Friction Stir Welded AA6061 Joint Fatigued in Very High Cycle Fatigue Regime. *Int. J. Fatigue* **2016**, *82*, 379–386. [CrossRef]

70. Chen, Y.; He, C.; Yang, K.; Zhang, H.; Wang, C.; Wang, Q.; Liu, Y. Effects of Microstructural Inhomogeneities and Micro-Defects on Tensile and Very High Cycle Fatigue Behaviors of the Friction Stir Welded ZK60 Magnesium Alloy Joint. *Int. J. Fatigue* **2019**, *122*, 218–227. [CrossRef]
71. Liu, F.; Zhang, H.; Liu, H.; Chen, Y.; Kashif, K.M.; Wang, Q.; Liu, Y. Influence of Welded Pores on Very Long-Life Fatigue Failure of the Electron Beam Welding Joint of TC17 Titanium Alloy. *Materials* **2019**, *12*, 1825. [CrossRef] [PubMed]
72. Liu, F.; Chen, Y.; He, C.; Wang, C.; Li, L.; Liu, Y.; Wang, Q. Very Long Life Fatigue Failure Mechanism of Electron Beam Welded Joint for Titanium Alloy at Elevated Temperature. *Int. J. Fatigue* **2021**, *152*, 106446. [CrossRef]
73. Liu, F.; Chen, Y.; Li, L.; Wang, C.; Wang, Q.; Liu, Y. Influence of Welded Pores on Fatigue Behavior of TC17 Titanium Alloy Welded Joints Subjected to Gigacycle Regime at Room and High Temperatures. *J. Mater. Sci. Technol.* **2023**. [CrossRef]
74. Zheng, Y.; Zhao, Z.; Zhang, Z.; Zong, W.; Dong, C. Internal Crack Initiation Characteristics and Early Growth Behaviors for Very-High-Cycle Fatigue of a Titanium Alloy Electron Beam Welded Joints. *Mater. Sci. Eng. A* **2017**, *706*, 311–318. [CrossRef]
75. Matsuda, K.; Kodama, S. Observation of Fatigue Microcracks and Estimation of Fatigue Strength of a Thin Sheet Arc Welded Part Considering Micro-Ripples. *Int. J. Fatigue* **2021**, *145*, 106087. [CrossRef]
76. Schork, B.; Kucharczyk, P.; Madia, M.; Zerbst, U.; Hensel, J.; Bernhard, J.; Tchuindjang, D.; Kaffenberger, M.; Oechsner, M. The Effect of the Local and Global Weld Geometry as Well as Material Defects on Crack Initiation and Fatigue Strength. *Eng. Fract. Mech.* **2018**, *198*, 103–122. [CrossRef]
77. Toumpis, A.; Galloway, A.; Molter, L.; Polezhayeva, H. Systematic Investigation of the Fatigue Performance of a Friction Stir Welded Low Alloy Steel. *Mater. Des.* **2015**, *80*, 116–128. [CrossRef]
78. Xing, S.; Dong, P. Fatigue of Titanium Weldments: S-N Testing and Analysis for Data Transferability among Different Joint Types. *Mar. Struct.* **2017**, *53*, 1–19. [CrossRef]
79. Hobbacher, A.F. *Recommendations for Fatigue Design of Welded Joints and Components*; Springer International Publishing: Cham, Switzerland, 2016; ISBN 978-3-319-23756-5.
80. *BS 7608:2014*; Guide to Fatigue Design and Assessment of Steel Products. British Standards Institution: London, UK, 2014.
81. *AWS D1.1/D1.1M:2020*; Structural Welding Code—Steel. American Welding Society: Doral, FL, USA, 2020.
82. *BS EN 1993-1-9:2005*; Eurocode 3: Design of Steel Structures—Fatigue. British Standards Institution: London, UK, 2005.
83. *DNV-RP-C203*; Fatigue Design of Offshore Steel Structures. DNV: Høvik, Norway, 2019.
84. Hobbacher, A.F. The New IIW Recommendations for Fatigue Assessment of Welded Joints and Components—A Comprehensive Code Recently Updated. *Int. J. Fatigue* **2009**, *31*, 50–58. [CrossRef]
85. Rumiche, F.; Noriega, A.; Lean, P.; Fosca, C. Metallurgical Failure Analysis of a Welded Drive Beam of a Vibrating Screen. *Eng. Fail. Anal.* **2020**, *118*, 104936. [CrossRef]
86. Wolf, T. Application of Weldment Fatigue Methods to Conveyor Pulley Design. In *Proceedings of the Bulk Material Handling by Conveyor Belt II*; Precision Pulley, Inc.: Orlando, FL, USA, 1998.
87. Alencar, G.; de Jesus, A.; da Silva, J.G.S.; Calçada, R. Fatigue Cracking of Welded Railway Bridges: A Review. *Eng. Fail. Anal.* **2019**, *104*, 154–176. [CrossRef]
88. Pakenham, B.; Ermakova, A.; Mehmanparast, A. A Review of Life Extension Strategies for Offshore Wind Farms Using Techno-Economic Assessments. *Energies* **2021**, *14*, 1936. [CrossRef]

metals

Review

Fracture Behaviour of Aluminium Alloys under Coastal Environmental Conditions: A Review

Ibrahim Alqahtani *, Andrew Starr and Muhammad Khan

Centre for Life-Cycle Engineering and Management, School of Aerospace, Transport and Manufacturing, Cranfield University, College Road, Cranfield MK43 0AL, UK
* Correspondence: i.alqahtani@cranfield.ac.uk

Abstract: Aluminium alloys have been integral to numerous engineering applications due to their favourable strength, weight, and corrosion resistance combination. However, the performance of these alloys in coastal environments is a critical concern, as the interplay between fracture toughness and fatigue crack growth rate under such conditions remains relatively unexplored. This comprehensive review addresses this research gap by analysing the intricate relationship between fatigue crack propagation, fracture toughness, and challenging coastal environmental conditions. In view of the increasing utilisation of aluminium alloys in coastal infrastructure and maritime industries, understanding their behaviour under the joint influences of cyclic loading and corrosive coastal atmospheres is imperative. The primary objective of this review is to synthesise the existing knowledge on the subject, identify research gaps, and propose directions for future investigations. The methodology involves an in-depth examination of peer-reviewed literature and experimental studies. The mechanisms driving fatigue crack initiation and propagation in aluminium alloys exposed to saltwater, humidity, and temperature variations are elucidated. Additionally, this review critically evaluates the impact of coastal conditions on fracture toughness, shedding light on the vulnerability of aluminium alloys to sudden fractures in such environments. The variability of fatigue crack growth rates and fracture toughness values across different aluminium alloy compositions and environmental exposures was discussed. Corrosion–fatigue interactions emerge as a key contributor to accelerated crack propagation, underscoring the need for comprehensive mitigation strategies. This review paper highlights the pressing need to understand the behaviour of aluminium alloys under coastal conditions comprehensively. By revealing the existing research gaps and presenting an integrated overview of the intricate mechanisms at play, this study aims to guide further research and engineering efforts towards enhancing the durability and safety of aluminium alloy components in coastal environments.

Keywords: fatigue crack growth; fracture toughness; aluminium alloys; coastal environments; temperature; humidity; corrosion

Citation: Alqahtani, I.; Starr, A.; Khan, M. Fracture Behaviour of Aluminium Alloys under Coastal Environmental Conditions: A Review. *Metals* **2024**, *14*, 336. https://doi.org/10.3390/met14030336

Academic Editors: Umberto Prisco, Tomasz Tański, Francesca Borgioli and Denis Benasciutti

Received: 8 February 2024
Revised: 26 February 2024
Accepted: 13 March 2024
Published: 15 March 2024

1. Introduction

In the engineering world, the initiation and propagation of cracks in the material components of aircraft structures are difficult challenges that demand attention. If failure remains undetected in its early stages, it can result in severe consequences, ultimately leading to catastrophic damage. The catastrophic damage to the structures will be more expensive to replace. For this reason, researchers have performed extensive research to find solutions and minimise crack growth.

Engineering materials, particularly metals, are prone to developing cracks and fractures during service. Numerous application structures, including those in automotive, aircraft, and power plants, operate under diverse loading conditions [1]. These structures, particularly those made of aluminium alloys, become increasingly susceptible to fatigue and failure under the combined influence of mechanical loads [2]. Aluminium alloys are essential in engineering because they provide a balance of strength and light weight [3–6]. Further technological advances enhanced the aluminium alloy, allowing it to be used in the aerospace industry [7]. Aluminium alloys in aircraft structures are advantageous due to their corrosion resistance, durability, and low maintenance requirements [8]. These properties collectively contribute to structures' effectiveness and sustainability in challenging corrosive conditions.

In corrosive conditions, environmental factors such as temperature and humidity impact the fatigue growth and fracture behaviour of aluminium alloys [9]. Researchers can evaluate a material's ability to withstand cracks or other defects caused by the corrosive environment or other loading conditions by analysing its fracture toughness [10]. High fracture toughness values indicate that a material is less likely to experience sudden failures [11], making it more reliable for corrosive environments. The combined influence of the temperature and humidity creates unique challenges to the material properties. These conditions can accelerate material degradation and cause structural failures [12]. Therefore, addressing the interaction of materials and structures with such harsh conditions is essential.

This literature review presents the significant efforts of the past and reviews the experimental and modelling challenges on fracture and fatigue crack growth under a corrosive environment. Focusing on qualitative descriptions of fracture behaviour in aluminium alloys is crucial due to their widespread use across industries. Qualitative analysis is instrumental in ensuring the safety and reliability of structures by identifying potential failure modes such as fatigue cracks and stress corrosion cracking. By studying fracture surfaces, researchers can detect manufacturing defects and impurities influencing the material's fracture behaviour. This analysis aids in understanding corrosion-related fracture behaviour, facilitating the development of alloys with enhanced corrosion resistance for applications in corrosive environments. Moreover, qualitative analysis allows tailoring alloys for specific purposes by comprehending the impact of different compositions and heat treatments on fracture behaviour. Overall, this approach contributes to a deeper understanding of the relationships between microstructure, alloy composition, and fracture properties, providing a foundation for further advancements in the field of aluminium alloys.

The literature review provides an overview of the research complexities in the coastal environment. This review organises the previous research under four significant approaches: the fundamentals of fracture and fatigue crack growth behaviour, coastal environmental conditions and their effects, fracture mechanisms in coastal environments, and modelling and predictive methods. The conclusion highlights the need for comprehensive efforts to address existing gaps in the current research field.

2. Fundamentals of Fracture and Fatigue Crack Growth Behaviour

In the context of materials engineering and science application, a crack refers to a discontinuity in the structure of a material, which can be initiated by various factors such as manufacturing defects, stress, fatigue, or corrosion. Crack formation in aluminium is influenced by various manufacturing-related factors that must be carefully considered to ensure the structural integrity of the material. Inclusions, impurities, and porosity arising from incomplete refinement or casting processes can act as stress concentration points, initiating cracks. Welding issues, such as incomplete fusion or a lack of penetration, can create weak zones susceptible to cracking. Heat treatment effects, including improper processes or quenching, may lead to variations in hardness and microstructural defects. Alloy composition and exposure to hydrogen during manufacturing can also affect aluminium's susceptibility to cracking [13]. The appearance of a crack is influenced by the material's microstructure, loading conditions, and environmental factors. Within engineering applications, the frequent use of fracture toughness and fatigue crack growth values attends to characterizing and evaluating the material's performance and integrity [14]. Fracture toughness represents a material's resistance to crack propagation under stress, ensuring its ability to withstand flaws and defects. On the other hand, the fatigue crack growth rate examines how cracks initiate and propagate over time under cyclic loading conditions, offering essential insights into the fatigue behaviour of materials [15]. Researchers and engineers rely on these fundamental principles to ensure the durability and safety of structures in different applications.

Crack initiation is a complex process integral to understanding the failure of materials and structures, particularly in the realm of fatigue and fracture mechanics. The initiation of cracks often occurs under cyclic loading conditions, where repeated or fluctuating stress is applied to a material. Microscopic defects, impurities, or inclusions within the material act as stress concentration points, creating localised areas of increased stress. These stress raisers promote the nucleation of cracks, marking the initiation phase. Moreover, material flaws such as inhomogeneities, microstructural defects, or manufacturing imperfections can serve as natural sites for crack initiation. Environmental factors, such as corrosion or exposure to aggressive chemicals, further accelerate this process by degrading material properties and promoting crack formation. Overloading, exceeding a material's capacity, can induce localized plastic deformation, creating conditions conducive to crack initiation, particularly in brittle materials.

The stress intensity factor (K) plays a pivotal role in fracture mechanics, offering a quantitative measure of stress concentration near the tip of a crack. It is expressed by the Equation (1):

$$\mathrm{K} = \sigma\sqrt{\pi a} \tag{1}$$

where K is the stress intensity factor, σ is the applied stress, and a is the crack length. This factor is essential for evaluating the severity of stress near the crack tip and understanding crack growth behaviour. As the stress intensity factor increases, it approaches a critical value known as fracture toughness (K_{Ic}). This critical point signifies the threshold beyond which crack propagation becomes unstable, leading to catastrophic failure. The stress intensity factor aids in predicting under what conditions cracks will propagate, guiding engineers in designing structures to avoid failure due to crack growth. In essence, it provides a critical parameter for assessing the susceptibility of materials to fracture, ensuring the reliability and safety of engineering components subjected to various loading conditions [16].

There will be yielding (σ_{ys}) and a plastic zone due to the postulated infinite elastic stresses at the fracture tip. Irwin determined the size of the plastic zone (r_y) [17]; Equation (2):

$$r_y = \frac{1}{2\pi}\left(\frac{K_{Ic}}{\sigma_{ys}}\right)^2 \tag{2}$$

A more precise estimate of the size of the plastic zone (r_p) considers the redistribution of stresses when the zone begins to yield; Equation (3):

$$r_p = 2r_y = \frac{1}{\pi}\left(\frac{K_{Ic}}{\sigma_{ys}}\right)^2 \tag{3}$$

Figure 1a illustrates this redistribution effect, r indicates the radial distance from the crack tip, and σ_{yy} represents the stress in the radial direction. The plastic zone's geometry under plane stress and plane strain conditions is illustrated in Figure 1b.

2.1. Fracture Toughness

Fracture toughness measures a material's capacity to resist the crack growth [18]. The quantification and establishment of standardised fracture toughness are achieved through the application of fracture mechanic theories such as linear elastic fracture mechanics (LEFM) and elastic–plastic fracture mechanics (EPFM) [19]. These theories find widespread use in the evaluation of structural integrity, residual strength analysis, fitness-for-service assessments, and the design of damage-tolerant structures [11]. Consequently, the assessment and testing of fracture toughness have become crucial for advancing the practical engineering applications of the fracture mechanics approach [16]. After recognising this importance, the American Society for Testing and Materials (ASTM) stipulated standardised terminology and procedures for fracture toughness testing and assessment. These guidelines are outlined in fatigue and fracture testing E1820 [20] and linear–elastic plane strain fracture toughness of metallic materials E399 [21].

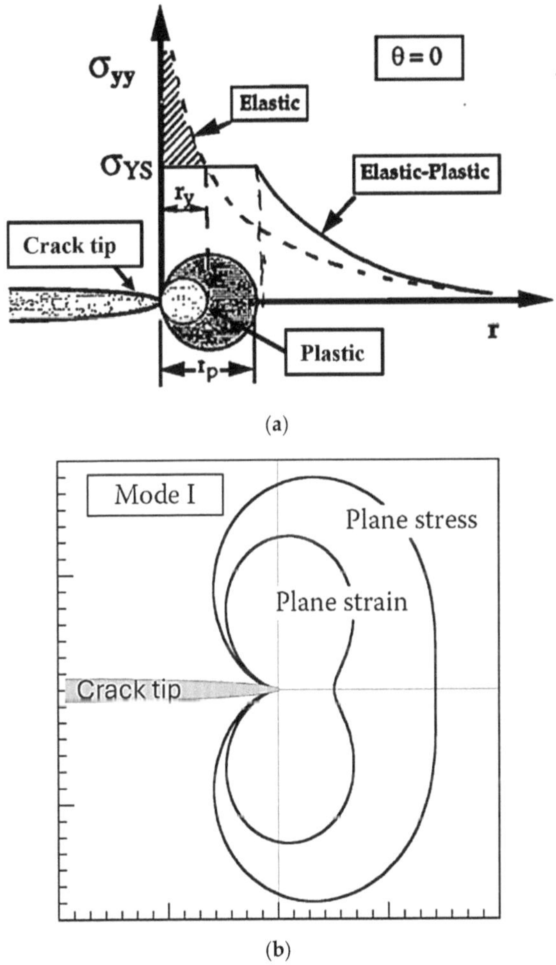

Figure 1. (**a**): Crack tip plasticity. (**b**): Crack tip plastic zone shapes and the von Mises yield criterion.

Figure 2 illustrates the three distinct modes of loading, Mode I, Mode II, and Mode III, that can be applied to a crack. Mode I involves forces pulling the crack surfaces apart, often encountered in tensile scenarios. Mode II features sliding forces along the crack plane, essential for plane shear or sliding stress analysis. Mode III involves forces that make one side of the crack move out-of-plane shear, leading to tearing. Many standard geometries of specimens in materials testing, such as compact tension (CT), disk-shaped compact specimen (DCT), arc-shape (AS) specimen, and single-edge notch bend (SENB) specimens, are designed for opening loading Mode I during testing, as shown in Figure 3A [16]. These specimens are specifically engineered to apply controlled tensile forces to study various material properties, including fracture toughness and fatigue crack growth [16]. Doddamani et al. [15,22–24] studied how various geometric parameters affect fracture toughness in Mode I. Also, they utilised different specimen types, including CT, SENB, and circumferential notched tensile (CNT) specimens, and their findings indicate that all these specimen types produced consistent results.

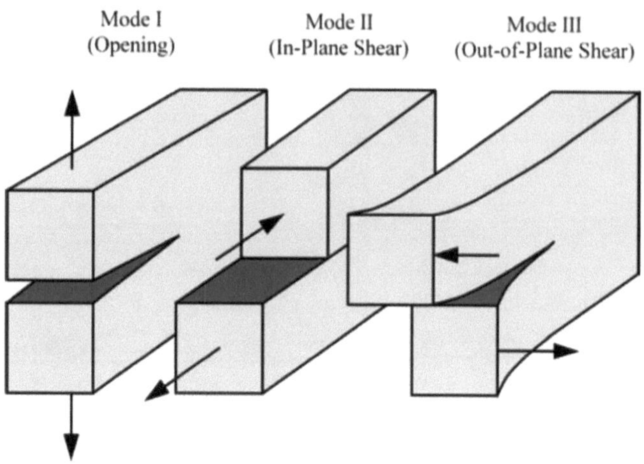

Figure 2. Three loading modes imposed on a crack. Adapted from ref. [16].

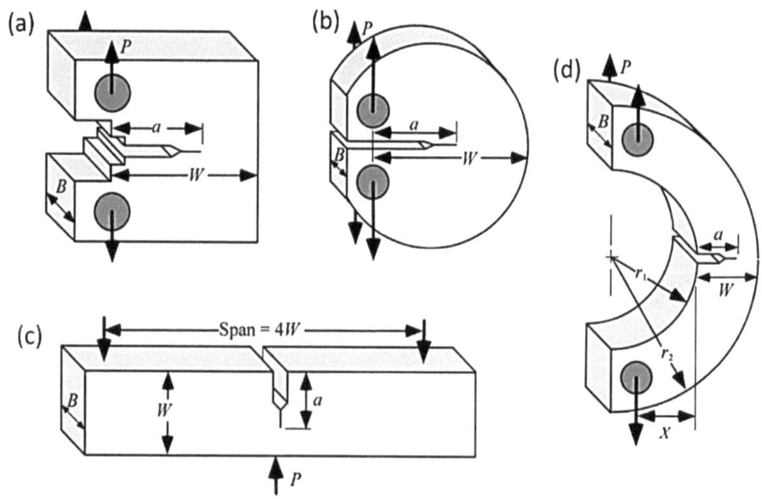

(**A**)

Figure 3. *Cont.*

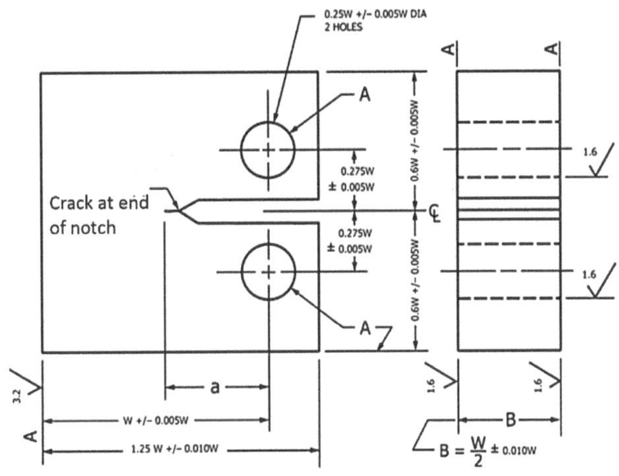

Figure 3. (**A**): Standardised test specimens: (**a**) CT, (**b**) DCT, (**c**) SENB, and (**d**) AS specimen adapted from ref. [16]. (**B**): Geometry of CT specimen. Adapted from ref. [21].

Researchers [8,12,25–28] generally prefer CT specimens for fracture toughness testing due to their standardised nature, widespread acceptance, and the assurance of consistent and comparable outcomes. CT specimens are well-suited for Mode I crack propagation, which is common in engineering applications, and they create near-plane strain conditions, simplifying crack growth analysis. Figure 3B illustrates the geometry of the CT specimen employed in fracture toughness testing, as per ASTM E399 standard.

The fracture toughness testing machine records the load and the crack mouth opening displacement data with three distinct curves, as shown in Figure 4. The three different types are Type I, II, and III, as classified by ASTM E399 [21].

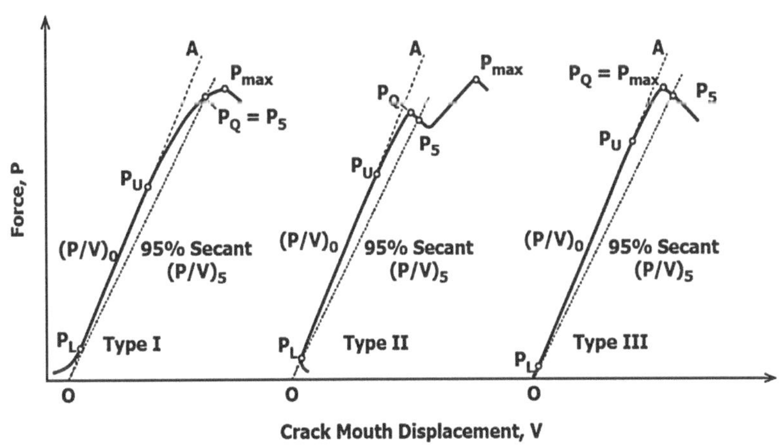

Figure 4. Load vs. CMOD curves. Adapted from ref. [21].

The ASTM E399 standard introduced a method to define the conditional stress intensity factor (K_Q), which depends on the value of conditional load P_Q and maximum load P_{max}. Conditional load value P_Q is determined by the secant line (OP_5) with a slope $(P/V)_5$ equal to 0.95 times the slope of the tangent OA between (PL) and (PU), denoted as (P/V)o. In practical terms, when determining the secant line OP_5, the pivotal point for adjusting the slope rotation should align with the intersection of the line OA and the displacement axis. The force P_Q is subsequently defined as follows: if the maximum force occurs after P_5 (Types I and II), then P_5 is designated as P_Q. In cases where a preceding maximum force before P_5 surpasses it (Type III), this preceding maximum force is acknowledged as P_Q [21].

Determining the critical load P_Q at a 95% secant offset contributes to achieving the valid plane strain fracture toughness value, K_{Ic}, which depends on calculating the P_{max}/P_Q ratio. This ratio serves as validation criteria, ensuring the fracture toughness measurement complies with the specified testing standards and maintains reliability [29]. The fracture toughness of metallic material was determined using the Equation (4) [21].

$$K_Q = \frac{P_Q}{\sqrt{BB_N}\sqrt{W}} f\left(\frac{a}{W}\right) \tag{4}$$

where, for CT Specimens,

$$f\left(\frac{a}{W}\right) = \frac{\left(2+\frac{a}{W}\right)}{\left(1-\frac{a}{W}\right)^{\frac{3}{2}}}\left[0.886+4.64\left(\frac{a}{W}\right)-13.32\left(\frac{a}{W}\right)^2+14.72\left(\frac{a}{W}\right)^3-5.6\left(\frac{a}{W}\right)^4\right]$$

The condition for the plane strain fracture toughness is given in Equations (5) and (6) below [30]:

$$B \geq 2.5\left(\frac{K_Q}{\sigma_y}\right)^2 \text{ and } a \geq 2.5\left(\frac{K_Q}{\sigma_y}\right)^2 \tag{5}$$

$$W \geq 5.0\left(\frac{K_Q}{\sigma_y}\right)^2 \tag{6}$$

where B_N is the thickness of the specimen at the notch, B is the thickness of the specimen, a is the crack length, W is the specimen width, P_{max} is the load-carrying capacity, σ_y is the yield strength of the material.

2.2. Fatigue Crack Growth Rate (FCGR)

FCGR is a measure used to quantify the rate at which a crack in a material grows over time under cyclic loading conditions. Consequently, ASTM E647 is a standard test method that provides guidelines for measuring the FCGR in metallic materials. CT specimens are frequently employed in FCGR testing due to their standardised geometry, which ensures precise and replicable testing conditions. Also, it requires the least amount of test material to evaluate crack growth behaviour. Their suitability for investigating crack propagation is attributed to a simplified analysis of crack growth rate [31].

Figure 5 illustrates a comprehensive schematic fatigue crack growth curve, explaining distinct regimes in the progression of cracks under cyclic loading conditions. In the initial regime, short/small fatigue cracks initiate at discontinuities, showcasing exponential growth. This transitions into a second regime with non-inspectable slow fatigue crack growth (FCG), marking a critical juncture. Short/small cracks evolve into 0.25–0.5 mm long/large cracks in aluminium alloys, representing in-service inspectable FCG. Subsequently, increased fatigue loads expedite crack growth, ultimately leading to component failure [32].

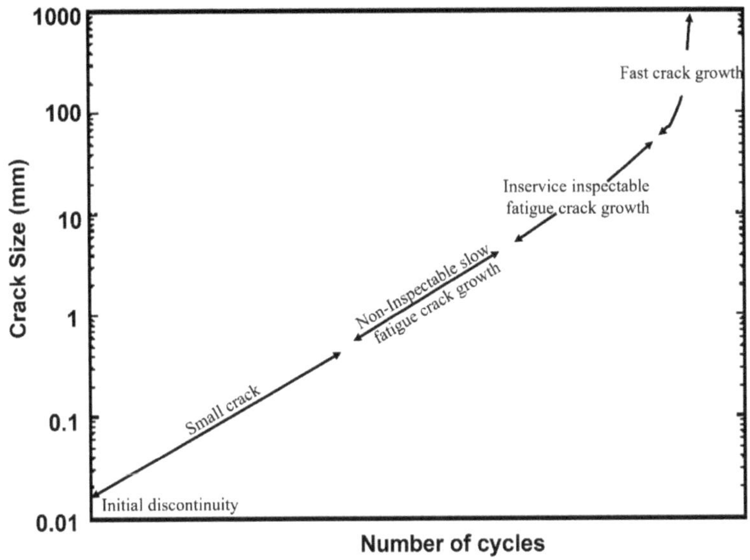

Figure 5. Schematic fatigue crack growth curve showing different regimes of crack growth. Adapted from ref. [32].

All the following authors used the CT specimens and established the standard procedure for fracture and fatigue testing of aluminium alloys and their composites. Lee [33] focused on assessing the fatigue crack growth performance of aluminium metal matrix composites (MMCs). R. Yuan et al. [34] performed tests to evaluate fracture toughness and cyclic fatigue properties of aluminium alloy and its composites. Abdul Budan et al. [35] studied the fatigue and mechanical characteristics of Al6061 aluminium alloy and composites. Y. Uematsu et al. [36] conducted fatigue tests at elevated temperatures, employing smooth specimens of aluminium alloy and its composites with diverse particle sizes while keeping a consistent weight percentage of silicon carbide (SiC) particles. D. P. Myriounis et al. [37] investigated the fatigue and fracture toughness characteristics of aluminium reinforced with SiC particles. J. Huang et al. [38] studied the impact of microstructural variability on the very high-cycle fatigue behaviour of discontinuously reinforced aluminium metal matrix composites (MMCs). Sharma et al. [39] investigated the fatigue characteristics of the aluminium alloy, and their experimental results showcased improvements in the fatigue and fracture properties of the material. Bikash Joadder et al. [40] performed experimental and finite element analyses to predict failure cycles. Shahani et al. [41] employed CT specimens to estimate methods for fatigue life prediction. Furthermore, numerous other researchers [42,43] conducted similar experiments on CT specimens to predict fatigue life.

A crack employed at the end of the notch, shown in Figure 3, can be introduced using a servo-hydraulic testing machine [8]. The prepared CT specimen is securely mounted in the grips of the testing machine. Crack initiation is then started by applying controlled loading conditions, such as load ratio, frequency, and strain rate [44]. Even though employing digital image correlation (DIC) and/or high-speed camera [45] techniques, monitoring crack propagation carefully throughout the fatigue testing process is challenging.

Fatigue in materials exposed to repetitive cyclic loading can be described as a developing process involving three stages: regime I, which is crack initiation; regime II, which is stable crack propagation as per Paris law; and regime III, unstable crack propagation or fracture failure, shown in Figure 6 [46]. The Paris law is an empirical equation used to model materials' fatigue crack growth rate. Using a power-law equation plotted in logarithmic coordinates, Paris' law (Equation (7)) establishes a connection between the stress intensity factor range (ΔK) and the crack growth rate (da/dN). The alterations made to Paris' law for determining crack growth rates proved highly effective in enhancing the accuracy of fatigue life predictions in engineering design [47].

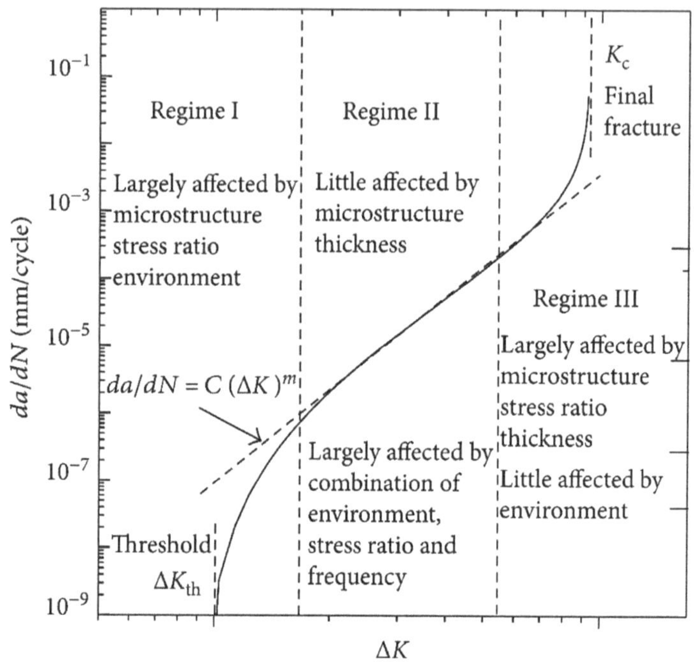

Figure 6. The stress intensity factor curve for metallic materials typically exhibits three distinct regimes when it comes to fatigue crack growth rate. Reprinted from ref. [47].

2.3. Environmental Fracture

Figure 6 depicts the curve correlating the stress intensity factor with the fatigue crack growth rate. Moreover, the figure visually represents the regimes of fatigue crack growth rate that are notably impacted by environmental factors. Environmental factors such as temperature, humidity, and exposure to corrosive agents are commonly encountered in coastal regions [48]. In a corrosive medium, regime I still begins with small crack initiation, but it happens faster due to corrosion weakening the material at potential stress points [49]. This can lead to the formation of tiny corrosion-induced cracks, often detectable only under a microscope. During regime II, when small cracks expand, the corrosive agents attack the crack tips, promoting rapid material degradation and creating a favourable environment for crack growth [50]. Cracks can spread more quickly and deeply compared to non-corrosive conditions, posing a significant threat to material integrity [51]. In regime III, the corrosive medium intensifies rapid crack advancement, potentially resulting in unstable crack propagation. Cyclic loading, combined with corrosion-induced weakening, leads to a sudden failure and severe loss of structural strength [52]. The cross-sectional area of the material reduces further due to corrosion, thus increasing the risk of unexpected failure.

Several deterministic FCGR functions have been proposed and extensively used in the determination of fatigue crack development rate under corrosive conditions [53]. These functions include the Trantina–Johnson Equation (8), Walker Equation (9), and Forman Equation (10). Although adjustments were made to account for varying stress ratios, the Paris model is only capable of explaining linear or stable fracture growth rates at a given stress ratio. Nonetheless, Equation (7) is still utilised to describe the behaviour of corrosion fatigue crack development (FCG) [53].

$$\frac{da}{dN} = C(\Delta K)^m \quad (7)$$

Trantina–Johnson Model:

$$\frac{da}{dN} = C(\Delta K - \Delta K_{th})^m \quad (8)$$

Walker Model:

$$\frac{da}{dN} = C(\Delta K)^{m_1}(1-R)^{m_2} \quad (9)$$

Forman model:

$$\frac{da}{dN} = C\frac{(\Delta K)^m}{(1-R)K_c - \Delta K} \quad (10)$$

where $\frac{da}{dN}$ is the crack growth per stress cycle, ΔK is the stress intensity factor (SIF) range; R is the stress ratio; K_c is the fracture toughness depending on the thickness of the specimen; a is the crack length or size; C is the material-specific Paris constant; $m,\ m_1,\ m_2$ are the material constant; and ΔK_{th} is the threshold stress intensity factor.

One of the primary environmental factors affecting fatigue crack growth is corrosion. Materials can be specifically engineered to resist corrosion by selecting corrosion-resistant alloys or incorporating corrosion inhibitors. Applying protective coatings is a common strategy to shield materials from environmental factors. These coatings act as barriers, preventing direct contact between the material and corrosive agents or environmental moisture [54]. Various coating materials, such as polymers, ceramics, or corrosion-resistant paints, can be tailored to provide an additional layer of defence against environmental degradation, thus slowing down fatigue crack initiation and growth [55].

The composition and microstructure of materials play a vital role in their response to environmental factors. By carefully selecting alloying elements and controlling the microstructural features, engineers can enhance the material's resistance to fatigue crack growth. Advances in material science and technology offer opportunities to develop innovative materials with superior fatigue resistance. Nanotechnology, for instance, allows for the creation of nanostructured materials with unique mechanical and environmental resistance properties [56]. Designing materials for enhanced durability involves considering the entire lifecycle of the material, from manufacturing to usage and eventual disposal. Sustainable materials and manufacturing processes that minimize environmental impact contribute to a more holistic approach to material design.

2.4. Threshold Stress Intensity Factor

The threshold stress intensity factor (ΔK_{th}) indicates the minimum stress intensity required to initiate stable crack growth in a material [57]. The crack growth testing data was utilised to determine ΔK_{th} values. The linear Paris curve was studied in the evaluation process to extrapolate the values corresponding to extremely low crack growth rates, such as 10^{-10} m/cycle or 10^{-6} mm/cycle [58]. This extrapolation aids in estimating the threshold stress-intensity range, shown in Figure 6.

Perez N [59] studied the linear Paris growth–rate curve to estimate the threshold stress-intensity range, making the critical limiting variable the threshold ΔK_{th} value. Equation (11) states that small-crack fatigue thresholds can be calculated from the Paris law constants C and m [58,59].

$$\Delta K_{th,smallcrack} = \left(\frac{10^{-6}}{C}\right)^{\frac{1}{m}} \quad (11)$$

Corrosion-induced small cracks in materials often have a lower ΔK_{th} than large cracks. Environmental factors influence this phenomenon, emphasising the importance of assessing the threshold stress intensity range when evaluating the structural integrity of materials exposed to coastal conditions.

Table 1 provides a comparative overview of the fatigue strength and fracture toughness of various aluminium alloys, presenting key mechanical properties essential for assessing their performance in structural applications.

Table 1. Fatigue strength and fracture toughness of different aluminium alloys [45].

Al Alloy	Elastic Modulus (GPa)	Yield Strength (MPa)	Fatigue Strength (MPa)	Fracture Toughness (MPa.$\sqrt{m}$)	Reference
AA6061	68–74	193–290	207	18.21	[60]
AA6082	67.1	276	-	19–25	[60]
AA7075	71	482	-	27.5	[61]
AA7050	70–80	455	240	27.5	[62]
AA2024	72–75.7	345–381	138	18.5	[63]
AA5083	70–73.6	269–297	-	28.3	[64]
AA8090	77	370	100	28	[65]

3. Coastal Environmental Conditions and Their Effects

The coastal region experiences intense, dry summers and high humidity [66]. These climatic particulars underscore the challenges of coastal region environmental conditions, particularly for materials, infrastructure, and equipment exposed to these harsh coastal environments. The corrosive nature of the coastal environment, characterised by exposure to saltwater, high humidity, and temperature, challenges the durability and integrity of various engineering materials [67]. You Tang et al. [68] conducted a study investigating how the coastal environment and concentration of salt ions affect the materials and structures. These conditions have a saltwater potential to advance material corrosion due to the corrosive nature of salt ions. Figure 7 illustrates the unique challenges and environmental factors that structures in coastal areas are exposed to, which can substantially impact their integrity and durability.

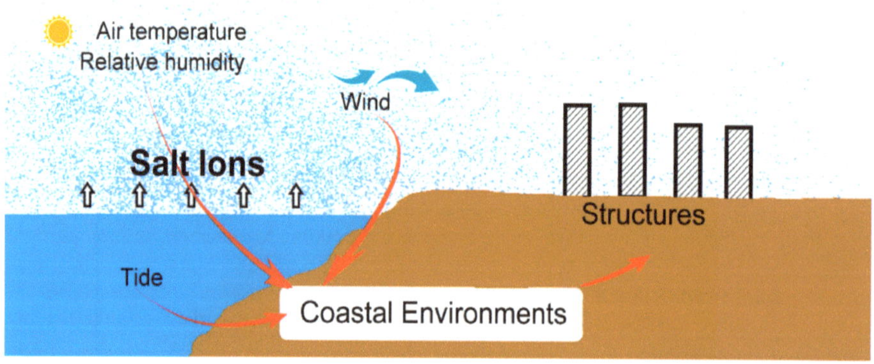

Figure 7. Typical coastal environment conditions on nearby structures.

Additionally, coastal areas are marked by higher humidity levels, mainly due to their proximity to vast bodies of water. This higher humidity can induce moisture absorption by materials, resulting in possible corrosion. It was observed that temperature variations are common in coastal areas, leading to thermal damage that influences fatigue performance and long-term durability [69]. Furthermore, the occurrence of marine vaporisers in coastal regions, comprised of airborne seawater droplets and particles, amplifies corrosion processes by depositing corrosive salts on structural surfaces.

3.1. Simulation of Coastal Conditions

Simulation of coastal conditions involves creating controlled laboratory environments that mimic the environmental factors found in coastal areas. This simulation is critical for studying how materials, such as aluminium alloys, react to the unique challenges of coastal environments.

3.1.1. Corrosion Simulation

Ramesh et al. [12] studied the fracture toughness of the aluminium alloy under a corrosive environment using CT specimens. In this work, the corrosion simulation of aluminium alloys was accomplished by immersing prepared CT specimens in a 3.5% sodium chloride (NaCl) solution for a specified number of days. Cavalcante et al. [70] and Y Zheng et al. [71] chose a 3.5% NaCl solution for the immersion, as it closely replicates the salinity of seawater. Similarly, Zakaria [72] also mentioned in his research that the specimens are subjected to immersion in the 3.5% NaCl solution for a specified number of days, with the duration based on the research objectives and the desired level of corrosion simulation.

Specimens can be immersed in actual seawater from a coastal area to simulate natural conditions [73]. Alternatively, synthetic seawater solutions can be prepared to mimic the composition of natural seawater. B J Little et al. [74] mentioned the composition of synthetic seawater solutions, which often include salts like NaCl, $MgCl_2$, $CaCl_2$, and other ions found in seawater. Researchers can control each component's concentration for precise corrosion studies in this approach.

Over time, the corrosion process on the surface of the aluminium alloy specimens is initiated and accelerated by the NaCl solution, replicating the corrosion observed in coastal regions due to salt ions in the atmosphere and water. After the specified immersion period, the specimens are carefully removed from the NaCl solution, cleaned, and thoroughly examined [75].

3.1.2. Environmental Chamber

Sarah et al. [76] used the environmental chamber to simulate the induced crack on the specimens. An environmental chamber replicates the typical conditions available in coastal regions [77]. Sahand et al. [78] and Faridah et al. [79] mentioned the environmental chamber, also called the climate chamber, that precisely controlled temperature and humidity levels. Coastal environments are characterised by notable temperature fluctuations, ranging from elevated daytime temperatures to cooler nighttime conditions [80]. Temperature cycling can be implemented within the environmental chamber to replicate these daytime variations accurately. Furthermore, maintaining high humidity levels is necessary for accurately simulating coastal conditions. The climate chamber offers the capability to finely control humidity levels, creating a humid atmosphere similar to that experienced in coastal regions. This controlled setup is essential for scientific investigations aiming to replicate and study the effects of temperature variations and high humidity available in coastal environments on material behaviour.

The specimens are positioned within the environmental chamber for a predetermined duration [76]. Since utilising the environmental chamber represents a novel technique, standardising its procedural application becomes essential. This standardisation ensures consistency and repeatability in the experimental process, enabling reliable and comparable results. Once this exposure period is completed, the samples are retrieved from the environmental chamber. Subsequently, they were immediately taken to the testing machine through the desiccator [81] and fixed in the machine for mechanical testing. The desiccator maintains the specified temperature and humidity conditions until the specimens are ready to be taken to the mechanical testing machine [82]. This step is essential to prevent changes in the specimens' condition and properties between the exposure phase and the subsequent testing. This instant transition ensures that the specimens are assessed quickly after exposure to the simulated coastal conditions, enabling the accurate evaluation of their material properties and performance.

3.2. *Effect of Corrosive Solution*

Aluminium alloys are susceptible to corrosion when they are exposed to a NaCl solution. Corrosion can lead to pits and cracks on the material's surface. These defects can act as stress concentrators and reduce the material's fracture toughness [12]. The cyclic loading and exposure to a corrosive environment, such as a NaCl solution, could result in corrosion fatigue. This phenomenon occurs when the cyclic stresses applied to the material work in conjunction with the corrosive environment to accelerate crack growth [83]. Ramesh et al. [12] mentioned in their research that aluminium typically develops a protective oxide layer on its surface in atmospheric conditions. However, the presence of corrosion pits can significantly disrupt the uniformity of this oxide layer [84]. This effect is particularly pronounced with a longer duration, resulting in a defective oxide layer.

Notably, the intermetallic compounds commonly found in Al6000 alloys, the Mg_2Si phase, and AlFeSi particles [85] are noteworthy. AlFeSi particles can cause localised corrosion in the surrounding matrix because they function as cathodic phases [86]. Conversely, the Mg_2Si phase is less noble than the adjacent Al matrix [87], creating a potential difference promoting localised corrosion at the interface. Additionally, corroded samples often accumulate corrosive ions on their surfaces, such as Cl^-, which inevitably interact with the oxide film, compromising its corrosion resistance.

Figure 8 provides a schematic representation of the localised corrosion process observed in the Al6061 alloy. In the initial stages of corrosion, chloride ions (Cl^-) penetrate the matrix through imperfections originating from either the AlFeSi or Mg_2Si particles. This penetration leads to the dissolution of the oxide film that typically protects the metal surfaces, consequently initiating pitting corrosion, as mentioned by Can Peng et al. [88]. An oxide layer, in the context of corrosion, refers to a protective film that forms on the surface of a metal exposed to environmental conditions. This layer consists of metal oxides and can act as a barrier, preventing further corrosion by isolating the metal from the surrounding environment.

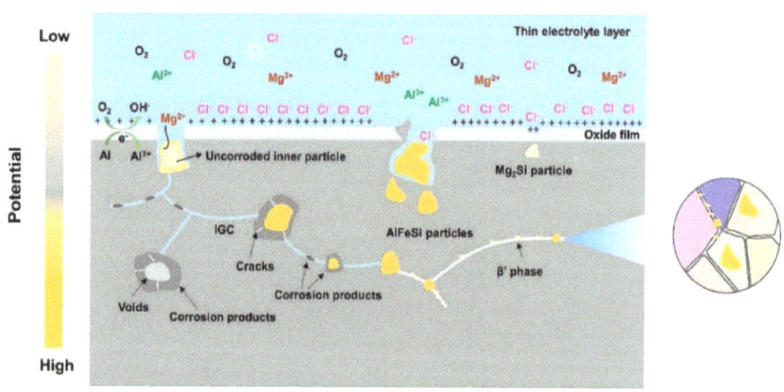

Figure 8. Diagrammatic representation of the Al6061 alloy's corrosion process in a marine environment. Reprinted with permission from ref. [88] Copyright 2022 Elsevier.

Pits are small, localised cavities that form on the surface of a material due to exposure to a NaCl solution [89]. They are typically deeper than they are wide and can penetrate the material, compromising its integrity. This leads to the formation of cracks and corrosion products, which decreases corrosion resistance, as shown in Figure 9. Corrosion products are compounds or substances that result from the chemical reactions between a metal and its environment during the corrosion process. Peng et al. [88] mentioned that corrosion products, such as metal oxides, hydroxides, and salts, often contribute to the visible changes in the appearance of a corroded surface.

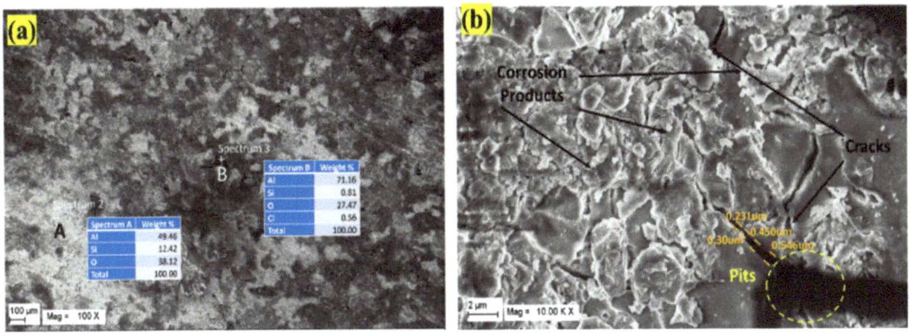

Figure 9. Corroded samples showing (**a**) oxide layer, (**b**) corrosion products, cracks, and pits. Reprinted with permission from ref. [12] Copyright 2023 Elsevier.

3.3. *Effect of Temperature*

The effect of temperature in coastal regions on the fracture toughness and FCGR of aluminium alloys is influenced by saltwater, humidity, and exposure to marine atmospheres. Coastal temperatures generally exhibit moderate effects due to the influence of large bodies of water. The combination of temperature variations and humidity can contribute to the degradation of material properties [90].

According to the researchers [91–93], environmental variables were responsible for the failures of aircraft components made of aluminium alloys while in service. The corrosive environment significantly influenced the operation of the aircraft components in coastal areas. Temperatures exceeding 70 °C may result from the heat generated by aircraft equipment during its operation, potentially impacting component performance [94].

Otieno et al. [95] and Zheng et al. [71] worked on aluminium alloys and a chloride medium in marine exposure environments. From the outcomes, it can be reported that temperature can increase the salinity of seawater, making it more aggressive [71]. Structures can be impacted by the evaporation of hazardous airborne chemicals such as chlorides [95]. On the other hand, faster and deeper chloride penetration causes the material to develop microscopic cracks and corrosion pits. The rate of chloride penetration and its depth into the material is influenced by factors like temperature and the diffusion coefficient.

Vargel et al. [96] collectively identified pitting corrosion in an aluminium alloy subjected to corrosion within 20 to 80 °C. However, aluminium is susceptible to pitting corrosion at room temperature and in a higher pH range (8.1 for seawater) [94]. In marine environments, the oxide layers, FeO, Al_2O_3, CuO, and ZnO, were formed at 30–60 °C [90]. The tendency for pitting corrosion gradually reduces at temperatures of 70 °C and up to 150 °C when an aluminium oxide layer forms [96].

Cao et al. [90] conducted a thorough investigation to enhance the comprehension of the corrosion characteristics exhibited by the 2A02 aluminium alloy. They subjected samples of the 2A02 alloy to 200 h of corrosion exposure at 30 °C, 60 °C, and 80 °C. As the temperature increased from 30 °C to 60 °C and further to 80 °C, it became evident that the number of pits decreased accordingly, as shown in Figure 10a.

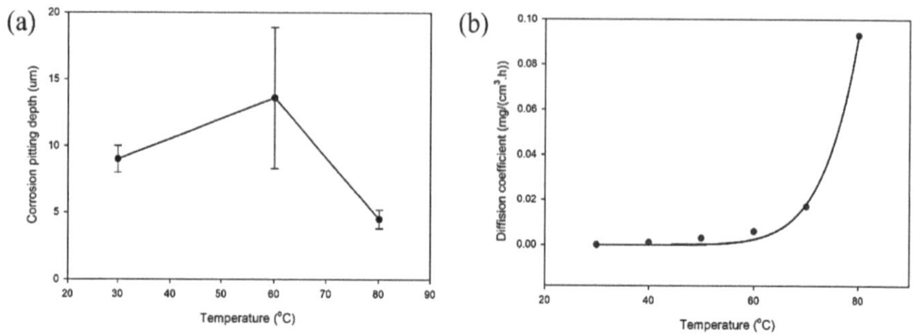

Figure 10. Effect of temperature and corrosion on aluminium alloy: (**a**) corrosion depth and (**b**) diffusion coefficient relationship trend chart [90].

Figure 10b, illustrating the correlation between temperature and the diffusion coefficient, offers valuable insights into how particle or molecule diffusion varies with temperature. The diffusion coefficient is linked to the rate at which reactants or ions can migrate through the material or the oxide layer. Generally, temperature increases the diffusion coefficient [80], facilitating the more rapid diffusion of reactants, such as oxygen or water molecules [90]. This accelerated diffusion can enhance reactions at the metal–oxide interface, potentially fostering the growth of protective oxide layers.

3.4. Effect of Humidity

The corrosion behaviour and mechanical performance of 7085 aluminium alloy in a hot, humid marine environment are studied by Tao J et al. [97]. Humid conditions can significantly influence the material properties of aluminium alloys, often increasing the challenges posed by the marine atmosphere. Daming et al. [98] studied the corrosion resistance of aluminium alloy under high humidity conditions. High humidity levels prevalent in coastal areas enhance moisture absorption by aluminium alloys. This moisture accelerates the corrosion process when combined with the corrosive nature of salt-laden coastal air. The chloride ions in the air can penetrate the alloy's surface, initiating and accelerating corrosion, leading to surface degradation, pitting, and the formation of corrosion products, as mentioned by Ramesh et al. [12].

Sarah Dorman et al. [99] worked on corrosion fatigue under moist air atmospheric conditions in samples with surface salt deposits. The result concludes that the structure and component are often subjected to cyclic loading under moisture conditions, which can lead to fatigue failure. In humid conditions, combining moisture and salt particles can further reduce the fatigue life of aluminium alloys by promoting faster crack growth rates. This effect is particularly detrimental, as it can compromise the long-term durability of structures. Intergranular and pitting corrosion are common localised types of corrosion observed in humid coastal environments, as mentioned by Yingchang et al. [100]. These types of corrosion can develop cracks, voids, and localised material degradation, affecting the material's structural integrity. Bray et al. [101] work on moist air and state that humid conditions can alter the mechanical properties of aluminium alloys. This may include reductions in tensile strength, flexibility, and toughness. These changes can affect the load-bearing capacity and overall structural performance of components and structures.

Ahmad [102] and Bradshaw et al. [103] worked on gaseous environments and the corrosion of aluminium alloys. When the relative humidity in the gaseous environment exceeds the equilibrium relative humidity over any saturated solution on the metal's surface, atmospheric corrosion occurs. This is particularly important for aluminium alloys, highlighting that the ambient air acts as a corrosive medium, causing chemical reactions on fracture surfaces through atmospheric moisture [103].

Davidson et al. [104] studied the effect of water vapour and ambient air on fatigue crack tip mechanics in Al7075 alloy. When aluminium is subjected to cyclic loading in ambient air, its fracture growth rates rise, and its threshold stress intensities drop in comparison to vacuum circumstances. Moisture causes a fracture tip's ability to tolerate cyclic plastic strain to decrease, which leads to this phenomenon. According to Holper et al. [105], diffusion carries water vapour from the ambient air to the tip of the crack, where it initiates chemical interactions with exposed fracture surfaces to produce hydroxide, hydrated oxides, and hydrogen absorption.

Young et al. [106] and Mahdieh Safyari et al. [107] worked on the hydrogen embrittlement mechanisms and mentioned that cracking in humid air is more accurately described as hydrogen-environment-assisted cracking. According to Equation (12) provided below, the reaction of water vapour at the alloy surface can result in hydrogen atoms that have the potential to be absorbed into the alloy and then recombine to generate H_2 gas [106]:

$$2Al(s) + (3 + X)\ H_2O(g) \rightarrow 3H_2\ (g) + Al_2O_3 \cdot X\ (H_2O)\ (s) \quad (12)$$

where X is the degree of hydration, g is gas (vapour), and s is solid.

These detailed studies regarding coastal environments help cultivate an understanding of the combined effect of various factors in corrosive conditions. The combined effects of these conditions also need to be understood by the multiple mechanisms explained in further sections.

4. Fracture Mechanisms in Coastal Environments

Fracture and fatigue crack growth involve the initiation, expansion, and eventual instability of cracks within materials. In coastal conditions, factors like corrosion, temperature fluctuations, and humidity play critical roles in these processes, as mentioned by Sarah et al. [99,108]. In the fracture process, small cracks usually originate at stress concentration points or material flaws and slowly develop under cyclic loading. In the case of fatigue crack growth, Khodor et al. [109] mentioned the microscopic defects within the material, which can evolve into small cracks due to repeated cyclic stresses. González et al. [110] thought that, in humid environments, corrosive agents can speed up crack initiation. As these cracks initiate, they may gradually propagate through the material, with the rate of growth influenced by factors like material properties, stress levels, and environmental conditions.

The mechanisms associated with fracture and fatigue crack growth in metals involve various elements and occurrences essential for understanding the initiation and propagation of cracks and their interactions with the corrosive environment. The following are the fracture mechanisms identified to occur in corrosive environmental conditions.

4.1. Oxide Layer

Studying the effect of the oxide layer in crack initiation and propagation is essential, particularly in coastal regions where the presence of salt and moisture is common. Understanding the beginning and spread of cracks, especially in corrosion conditions like those seen in coastal locations, depends on an oxide layer on the surface of materials like aluminium alloys [111].

Oxide layers frequently and naturally occur on metals exposed to air conditions [112]. As seen in Figure 11, these layers serve as protective barriers that prevent the metal surface from touching corrosive substances like moisture and salts available in corrosive areas. This protection is vital because it helps delay or prevent crack initiation [113]. The oxide layer can enhance the material's corrosion resistance by reducing the rate at which the underlying metal corrodes [114]. This is essential in coastal regions where the corrosive nature of salt ions, NaCl, in the atmosphere can accelerate material degradation. By impeding corrosion, the oxide layer plays a key role in preserving the material's integrity and structural strength [115]. However, the oxide layer is not always uniformly perfect [116] and can have defects or imperfections, which can become potential sites for crack initiation [117].

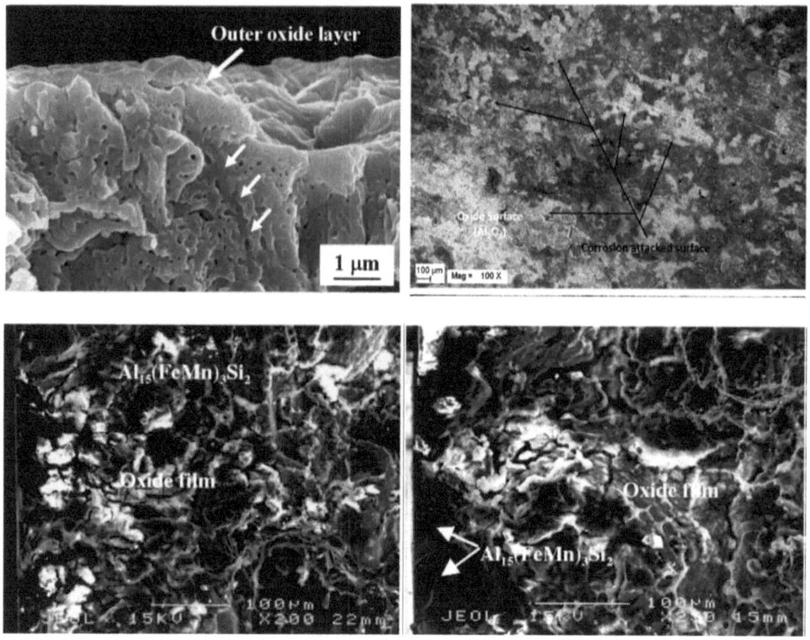

Figure 11. Oxide layers formed on aluminium alloys. Reprinted with permission from ref. [12] Copyright 2023 Elsevier, [118], reprinted with permission from ref. [119] Copyright 2013 Taylor & Francis.

The presence of oxide layers on metal surfaces, particularly in coastal regions, is crucial for preventing corrosive substances like moisture and salts from initiating cracks. While these layers enhance corrosion resistance, defects in the oxide layer can become potential sites for crack initiation, highlighting the importance of studying their effects on material integrity.

4.2. Crack Closure

At the fracture tip, the phenomenon of crack closure acts as a mechanism to lessen the crack's driving force [120]. Yamada et al. [121] highlighted, in their research, that crack closure plays a significant role in governing crack growth behaviour. Pokorny et al. [120] mentioned the residual plastic deformations, the roughness on the crack surfaces, and debris formation along these surfaces partially drive the crack closure phenomenon.

As illustrated in Figure 12, several elements, such as an oxide layer, phase particles, and plastic deformation in front of the crack tip, impact crack closure. This can temporarily reduce the effective stress intensity factor (K_{eff}), which drives crack propagation, helping to extend the life of aluminium structures exposed to corrosive conditions. Cracks can initiate and propagate due to the corrosive attack on the material's surface in the presence of corrosion, which is common in coastal environments.

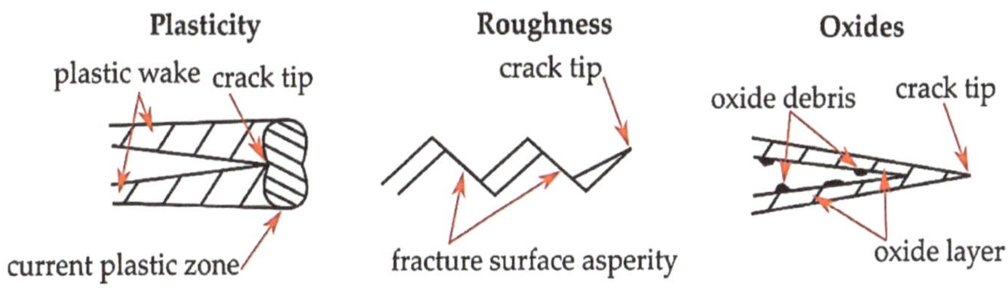

Figure 12. Crack closure mechanisms. Reprinted from ref. [120].

Moreover, coastal regions often experience temperature variations and high humidity levels [95]. The wide temperature range in coastal regions results in the expansion and contraction of materials, including aluminium structures. This thermal expansion and contraction can cause thermal fatigue [80], which leads to the narrowing of existing cracks, a form of crack closure [122]. The elevated humidity levels in coastal areas can introduce moisture to the cracks in aluminium structures. This moisture can initiate corrosion or oxidation processes, potentially causing the formation of corrosion products on the crack surfaces [12]. The formation of corrosion products causes the surface's roughness, partially closing the crack, and contributes to the crack closure mechanism [123].

Crack closure at the fracture tip, influenced by factors like an oxide layer and plastic deformations, temporarily reduces the effective stress intensity factor, extending the life of aluminium structures in corrosive coastal conditions. Temperature variations and high humidity in coastal regions lead to thermal fatigue and moisture-induced corrosion, contributing to crack closure mechanisms and influencing crack propagation in aluminium structures.

4.3. Phase Particles

The formation of phase particles in aluminium alloys exposed to coastal environmental conditions, including corrosion, temperatures up to 50 °C, and humidity levels reaching 95%, is a multifaceted process influenced by various interacting factors. Alloy composition is pivotal, as elements like magnesium, silicon, and manganese can precipitate out of the aluminium matrix under specific conditions, as mentioned by Culliton et al. [124].

Figure 13a depicts the formation of intermetallic phase particles in a corrosion medium. It also illustrates different types of phase particles formed on various aluminium alloys, as seen in Figure 13b,c. Dessi et al. [125] and Shrivastava et al. [126] studied the chloride-induced corrosion on aluminium alloys. They concluded that corrosion, driven by chloride ions from saltwater, can alter the microstructure and contribute to phase particle formation.

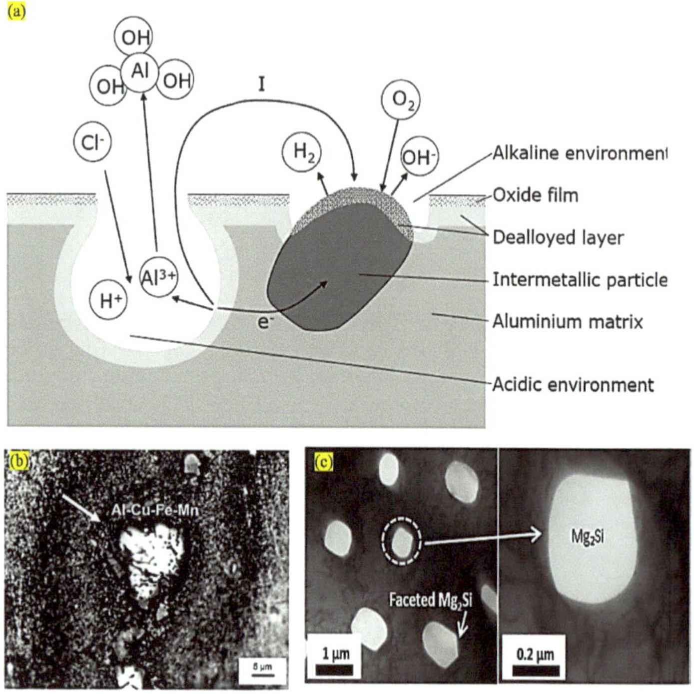

Figure 13. (**a**) Formation of intermetallic particles in corrosion medium. Reprinted from ref. [127]; (**b**,**c**) different phase particles formed on other aluminium alloys. Reprinted from ref. [128,129].

Blau P J [130] studied the mobility of alloying elements at elevated temperatures and concluded that temperature enhances alloying elements' mobility, promoting their interaction and subsequent particle formation. Gain et al. [131] and Arrabal et al. [132] worked on the Sn–Ag–Cu and Mg/Al alloys, respectively, in high-humidity environments, and they concluded that high humidity levels accelerate corrosion, further influencing the material's microstructure. Cavalcante et al. [70] studied different aeronautic aluminium alloys in air and saline environments, emphasising the critical consideration of exposure duration. The cumulative effects of corrosion and humidity gradually lead to microstructural changes, including the development of phase particles. Thus, particles can form through various precipitation mechanisms depending on alloy composition, environmental variables, and other factors.

In coastal environments, alloy composition, chloride-induced corrosion, and high humidity contribute to complex processes leading to the formation of phase particles in aluminium alloys. Elevated temperatures enhance alloying elements' mobility, influencing particle formation and emphasizing the cumulative effects of corrosion and humidity over time.

4.4. Striations Spaces

Fatigue striations refer to fine, repetitive patterns or lines observed on the fracture surface of a material placed to fatigue loading [133]. Monitoring and analysing fatigue striations are essential in understanding the mechanisms and behaviour of materials subjected to cyclic loading conditions, particularly in fatigue failure [134]. It was observed that a striation mechanism of failure in materials becomes prominent within the power law region or Paris law regime [135]. These striations indicate a characteristic pattern of crack growth within this regime. They occur as the crack tip repeatedly advances and changes during each loading cycle. The spacing between fatigue striations is equivalent to the crack propagation rate in one cycle, particularly during stage II, stable fatigue crack growth [136]. However, in some instances, the distance between fatigue striations is significantly wider than the rate at which the fatigue crack propagates in a single cycle. This suggests that the striations develop over multiple cycles of cyclic damage accumulation. It is important to note that environmental factors can also influence the formation of fatigue striations [133].

The features of fatigue striations at the stable crack growth are depicted in Figure 14a–d. As seen in Figure 14d, some second-phase particles are chopped, and secondary cracks are frequently obstructed when they come into contact with the secondary-phase particles. Additionally, the dislocation will accumulate around the second-phase particles due to the dislocation reciprocating through them, which leads to stress concentration and cracks on both sides of the particles [137].

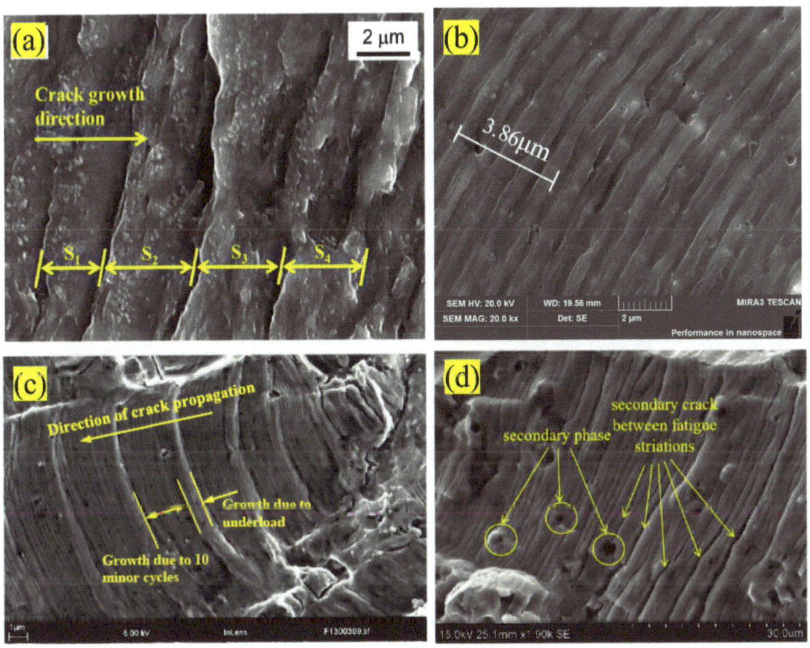

Figure 14. Striation spaces observed in aluminium alloys due to fatigue loading: (**a**) Reprinted with permission from ref. [138] Copyright 2019 Elsevier, (**b**) [139], (**c**) Reprinted from ref. [140], (**d**) Reprinted from ref. [137].

Due to the combined effects of fatigue and surface concentration, plastic deformation becomes confined to specific grains with favourable orientations, minimizing constraints from neighbouring grains. As the number of cycles increases, the damage progresses, giving rise to persistent slip bands (PSB) induced by shear stresses, resulting in intrusions and extrusions (stage 1). The localized damage areas give rise to microcracks, reaching sizes similar to the microstructure (e.g., Figure 15, mode II crack arrested at a grain boundary, stage 2). The early propagation of microcracks ultimately leads to the formation of macrocracks (stage 3), and subsequent engineering crack propagation leads to the final failure or fracture (stages 4 and 5). Sharma et al. [141] investigated the fatigue crack growth striation and threshold behaviour of the Al2219 alloy. Their work utilised the micrograph mapping technique, explicitly mapping the striations observed at regimes I, II, and III of the fatigue crack growth rate curve.

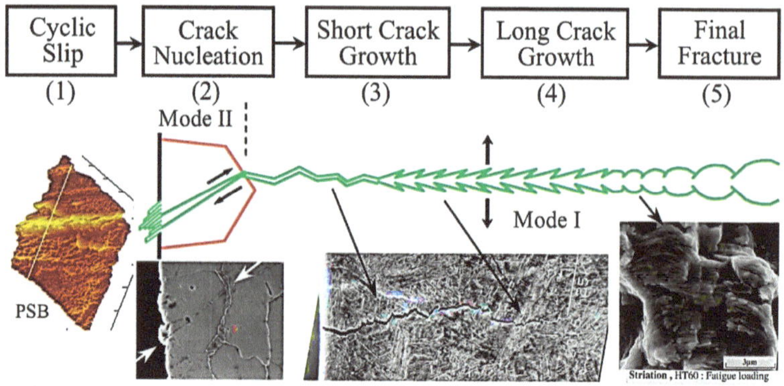

Figure 15. Stages of fatigue damage in metals. Reprinted from ref. [142].

In stable crack growth, the spacing between striations reflects the crack propagation rate, influenced by environmental factors. The progression from persistent slip bands to macrocracks delineates stages of fatigue damage leading to final failure, as investigated in studies on aluminium alloy.

4.5. Crack Propagation Path

Capturing the crack propagation path is critical to understanding material behaviour and structural integrity. Researchers primarily rely on valuable techniques such as digital image correlation (DIC) [143] and optical microscopy for this purpose [144]. Fatigue loading repeatedly stresses the material, leading tiny cracks or imperfections on the surface to grow into small cracks over time [145]. These cracks can be challenging to detect with the naked eye. Therefore, researchers utilise digital image correlation (DIC) to monitor crack propagation and growth speed [146]. DIC can generate detailed crack maps illustrating the propagation path [147]. These maps can be analysed to understand how environmental conditions impact the direction and rate of fatigue crack growth in aluminium alloys.

Anna et al. [148] utilised a high-resolution camera positioned in the front of their experiment to continuously monitor crack propagation in the specimen. They also measured crack lengths on the surface of the CT specimens using this camera to validate data from the compliance method, as depicted in Figure 16.

The morphological characteristics of fatigue fracture of the alloy under different conditions are shown in the optical micrographs in Figure 17. Figure 17a shows the most complex expansion path with a "Z" pattern in this example. In contrast, the sample in Figure 17b exhibits a relatively straight fatigue fracture propagation path with only a slight curvature in the early stages. The fatigue fracture of the sample is shown in Figure 17c, starting gradually and developing into a curved path as it expands. Based on previous studies, complex fatigue crack growth paths indicate slower crack growth rates, supported by experimental fatigue crack growth data [136].

Understanding crack propagation paths is crucial for assessing material behaviour and structural integrity. Techniques like digital image correlation (DIC) and optical microscopy help monitor and analyze fatigue crack growth in aluminium alloys. Researchers utilize high-resolution cameras and DIC to capture and validate crack propagation data, revealing diverse patterns in fatigue fracture paths, with complex paths indicating slower crack growth rates.

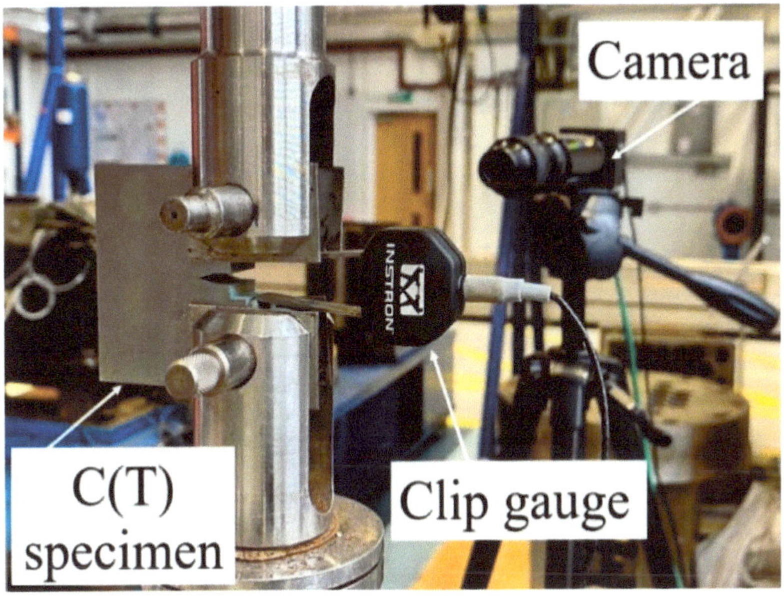

Figure 16. Experimentation setup with high-speed camera. Reprinted from ref. [148].

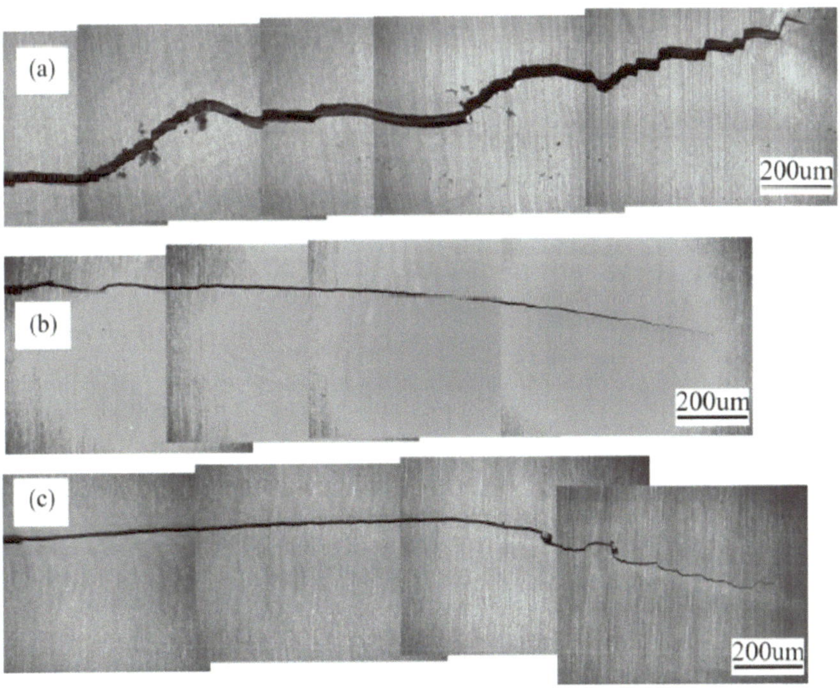

Figure 17. (**a–c**) Crack propagation paths showing different patterns. Reprinted with permission from ref. [136] Copyright 2022 Elsevier.

4.6. Corrosion–Fatigue Interaction

Corrosion fatigue, which is viewed as a combination of two mechanical behaviours in components, is one of the most significant factors in fatigue analysis [149]. Understanding the fatigue and corrosion behaviour of these materials is crucial in specialised industries like submarines [150,151], aeroplanes [152,153], aviation engine combustion systems [154], and other applications [155,156] because of the development of new manufacturing systems in the modern world.

When a material is subjected to cyclic mechanical loading in a corrosive environment, it can degrade and experience corrosion fatigue. It is a particularly challenging and damaging form of degradation because it combines the effects of mechanical stress and chemical corrosion, leading to a more rapid deterioration of materials [157].

Corrosion fatigue often begins with microscopic defects or stress concentrators in the material [158]. In a corrosive environment, the material surface may be attacked by chemical agents, forming corrosion pits or localised corrosion [159]. When the material is subjected to cyclic mechanical loading, it introduces alternating stress on the material, which leads to the initiation and propagation of cracks from the stress concentrators or corrosion pits [160]. The cracks tend to grow incrementally during fatigue loading. The corrosive environment can accelerate crack propagation by facilitating the nucleation and growth of corrosion-assisted fatigue cracks, leading to the localised weakening of the material around the cracks [52]. Corrosion fatigue increases crack growth rate compared to purely mechanical fatigue or corrosion alone [161].

Corrosion fatigue, a significant factor in fatigue analysis, combines mechanical stress and chemical corrosion in materials subjected to cyclic loading in corrosive environments. This damaging degradation accelerates crack initiation and propagation, leading to localized material weakening and a faster deterioration rate compared to purely mechanical fatigue or corrosion alone. Understanding corrosion fatigue is crucial in specialized industries such as submarines, aeroplanes, and aviation engine combustion systems.

4.7. Moisture-Assisted Crack Propagation

Moisture-assisted crack growth occurs when materials are exposed to higher humidity. This causes moisture to be absorbed and diffused into the material and promotes the formation of a thin electrolyte layer (TEL), a thin water layer on the material's surface [162]. In the presence of chloride ions, this microenvironment creates localised areas conducive to corrosion, accelerating localised corrosion processes and causing material degradation near the crack tip. Moisture promotes localised corrosion and accelerates crack growth with corrosion products (oxides, hydroxides, or chlorides) [163,164]. Moisture distribution in materials plays a crucial role in the movement of chloride ions at different depths, which was supported by Huague et al. [165].

Chloride ions are known for their corrosive nature, and their presence near the crack tip can initiate and accelerate corrosion processes. The synergy between moisture and these corrosion products significantly amplifies the localised corrosion process. As a result, the material's integrity becomes compromised and more susceptible to crack propagation [166]. This combined effect of moisture and chloride ions weakens the material's resistance to crack growth. Consequently, cracks can advance rapidly, potentially reducing the material's fatigue life [161].

Moisture-assisted crack growth in higher humidity creates a thin electrolyte layer, promoting localized corrosion and accelerating cracks with chloride ions, compromising material integrity and potentially reducing fatigue life.

4.8. Hydrogen Embrittlement

The phenomenon known as "hydrogen embrittlement" occurs when hydrogen atoms seep into a material's structure, impairing its mechanical qualities and increasing its brittleness and susceptibility to failure and cracking [167]. This process can occur in various materials, including aluminium alloys. The schematic diagram of the hydrogen-enhanced localised plasticity (HELP) mechanism is shown in Figure 18. In the HELP region, hydrogen is trapped inside the aluminium alloy through the weak areas (in the absence of an oxide layer). The plastic zone without hydrogen characterises localised deformation, crack blunting, micro-void coalescence, and deep dimples [168].

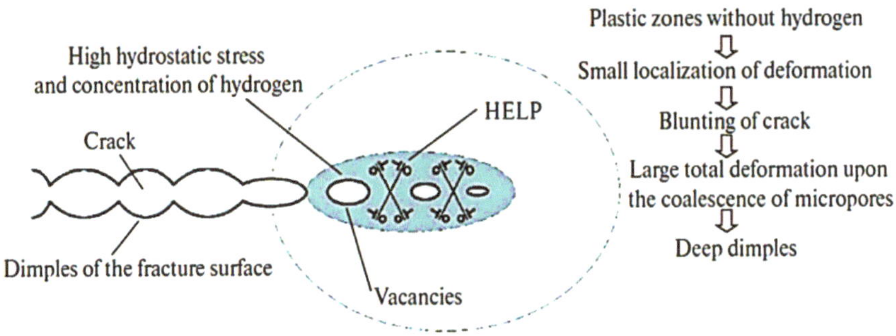

Figure 18. Schematic diagram of the HELP mechanism. Adapted from ref. [168].

When aluminium alloys are exposed to corrosive environments [169,170], they react with water vapour and cause hydrogen embrittlement (HE) [171], as shown in Figure 19. Hydrogen atoms can be introduced into aluminium alloys through various sources, such as exposure to moisture or corrosive environments [107]. Inside aluminium alloys, hydrogen atoms accumulate at defects, grain boundaries, dislocations, and other imperfections. These regions act as trapping sites for the accumulation of hydrogen atoms [172].

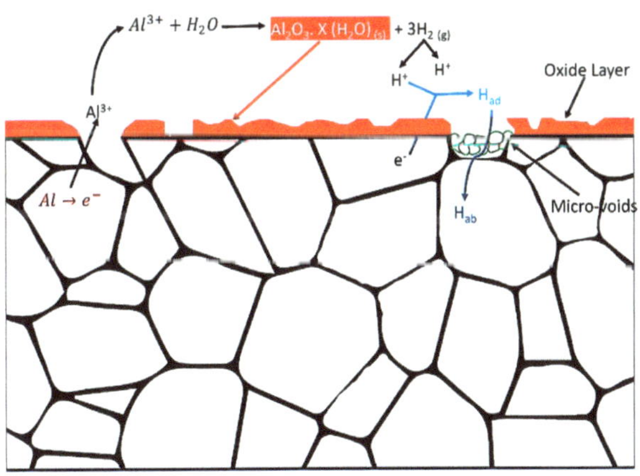

Figure 19. Hydrogen absorption mechanism in aluminium under humid conditions [10].

The presence of hydrogen in these trapping sites can cause various detrimental effects. It can create internal pressure within the material, which leads to cracking. Thus, hydrogen can also weaken the atomic bonds within the material's crystal lattice, reducing its mechanical strength and ductility [173]. The weakening of the material's structure can make it more brittle and susceptible to cracking, even under relatively low-stress conditions. This embrittlement phenomenon can initiate and propagate cracks within the aluminium alloy, reducing fatigue strength and life [174].

The identified mechanisms highlight material responses in coastal environments, including oxide layers, crack closure, phase particle edge spaces, crack propagation paths, corrosion fatigue interactions, moisture-assisted crack propagation, and hydrogen embrittlement. These mechanisms serve as key parameters and fill a gap in the existing knowledge on the influences of temperature and humidity on aluminium structures under coastal conditions. A systematic study of these failure mechanisms can provide a scientific understanding of the complex interactions between temperature and humidity and their impact on the structural integrity of aluminium materials under coastal conditions.

Hydrogen embrittlement occurs when hydrogen atoms seep into a material, increasing brittleness and susceptibility to failure. In aluminium alloys exposed to corrosive environments, hydrogen can accumulate at defects, weakening atomic bonds, reducing mechanical strength, and initiating cracks, ultimately diminishing fatigue strength and life.

5. Modelling and Predictive Approaches

Modelling and prediction methods for estimating fracture toughness and fatigue crack growth of aluminium alloy typically often use a variety of techniques, including empirical modelling [175] and simulations by using finite element analysis (FEA) [176,177]. These methods are needed to characterise the fracture behaviour of aluminium alloys and ensure their safe and effective use in various applications [178,179]. It is difficult to predict how materials would fracture when exposed to various coastal environmental conditions like temperature, humidity, and corrosion. Conventional modelling tools do not help much with this process.

However, under stable crack growth, the Paris equation can become a valuable tool for predicting fatigue crack growth rate (FCGR) even in corrosive environments [53,180,181]. The relationship between temperature humidity and its effect on fracture toughness, FCGR, and fatigue life estimation has yet to be fully understood. Empirical modelling will be one method to build models for predicting fracture toughness, fatigue crack growth parameters, and fatigue life cycles of materials in coastal environments.

Curve fitting is used in material fracture behaviour analysis to predict fracture toughness and fatigue life. The effect of curve fitting on the failure probability, crack size, and fracture toughness of a failed component depends strongly on the type of distribution assumed and the quality of the fit [182]. In this method, the parameters of the curve are adjusted until they substantially match the experimental data. Once a suitable curve is created, it can predict material behaviour under various conditions. Curve fitting is often used when a simplified mathematical representation of a complex relationship is required.

The empirical modelling approach through curve fitting involves conducting experiments under controlled conditions. During these experiments, data were collected on fracture toughness, fatigue crack growth parameters, and fatigue life cycle. These experimental data are then analysed to develop empirical models describing the relationships between material properties, specific factors, and their levels [176]. Many authors [23,24,183] developed equations for predicting fracture behaviour using regression methods, a type of empirical modelling technique that involves fitting mathematical models to the experimental data to describe the relationships between corrosive environment variables.

Hassan et al. [184] used optimized neural networks to predict the fatigue crack growth rate (FCG) guided by the mutation-guided algorithm (MLA). Daniel Kovalov et al. [185] studied the corrosion fatigue of aluminium alloy 2024-T351 and found the effects of electrochemical potential, NaCl concentration, loading frequency, and temperature. Zhang T et al. [186] developed a model considering the corrosion and fatigue cycle effects of aluminium alloy 2024-T4. They used curve fitting techniques to relate the corrosion time and depth of aluminium alloys. The result of the relationship between the corrosion depth of the fatigue crack tip (d, mm) and the obtained corrosion time (t, h) is an exponential fit given by Equation (13), with a correlation coefficient of $R^2 = 0.949$.

$$d = 0.00981 t^{0.39159} \tag{13}$$

Yichen et al. [136] used the curve fitting technique to correlate the stress range (s, Mpa) and life cycles (N), mentioned in Equation (14), and the linear fitting was obtained with $R^2 = 0.797$.

$$s = 233.4906 - 12.7622\log(N) \tag{14}$$

Bergner et al. [187] used the curve fitting technique to correlate Paris constants C and m and obtained the linear fit with $R^2 = 0.96$. Alqahtani et al. [188] utilised the curve fitting method to find the Paris constants C and m from the Paris equation for different temperature and humidity conditions. The regression equations obtained by curve fitting were linear for fatigue life cycles and C.

Mohammad Khan et al. [189] proposed the Khan–He model, which is more suitable for experimental crack depth and frequency data at different temperatures than the established Ostachowicz model. Yang Yali et al. [190] used a curve-fitting method to predict fatigue life from surface roughness. The results demonstrated a correlation between surface roughness and fatigue life (N_i) and found that a cubic polynomial equation provided the best fit shown in Equations (15) and (16).

$$K_t = 0.93 + 0.37b - 0.04b^2 - 0.07b\left(\frac{a}{b}\right) + 0.02\left(\frac{a}{b}\right)^2 + 0.002b^3 + 0.004b^2\left(\frac{a}{b}\right) + 0.006b\left(\frac{a}{b}\right)^2 - 0.003\left(\frac{a}{b}\right)^3 \tag{15}$$

where a and b are single micro-notch parameters, and K_t is the stress concentration factor.

$$N_i = 3.53 \times 10^{13}\left[(207.25K_t)^{1.84} - 127^{1.84}\right]^{-2} \tag{16}$$

The calculated fatigue life values were compared to experimental results, revealing a maximum error of 15.65%. This suggests that the formulated equation obtained from curve fitting is reasonable and effective in predicting fatigue life.

These findings show the flexibility and efficiency of curve fitting techniques in developing the relationship between various parameters, aiding in understanding materials' behaviour under any conditions.

Although various models [185,186,191] are used to analyse the obtained data and predict the life cycle in non-corrosive and corrosive conditions, the current understanding still needs to consider the specific effects of temperature and humidity. Investigating how temperature and humidity interact to affect fracture toughness and FCGR behaviour is critical to ensure aluminium alloy components' reliable and safe performance in coastal environments. The effects of humidity [192,193] and temperature [48,194–196] on the fatigue properties of materials [197–200] were independently studied in earlier research. For the purpose of forecasting fatigue life and creating empirical equations for predicting and associating FCGR with fracture toughness, a thorough understanding of the combined effects of these components is essential.

Simulation techniques involve using computer-based models to replicate the behaviour of materials under various coastal environmental scenarios [201]. Finite element simulations are a standard tool used for this purpose [202,203]. In FEA, the material and environmental conditions are mathematically modelled and the simulation predicts how the material will behave. By altering the input parameters, such as temperature and humidity levels, researchers can assess their impact on fracture toughness, fatigue crack growth, and fatigue life.

Figure 20a depicts an FE simulation model of the CT specimen used in fracture toughness tests. In the contour plot, Figure 20b, the distribution of Von Mises stress in the deformed CT specimen can be observed as simulated by the FEM analysis [204]. Along with the stress distribution, fracture toughness, and fatigue life cycles, FCGR can also be predicted using commercially available simulation software.

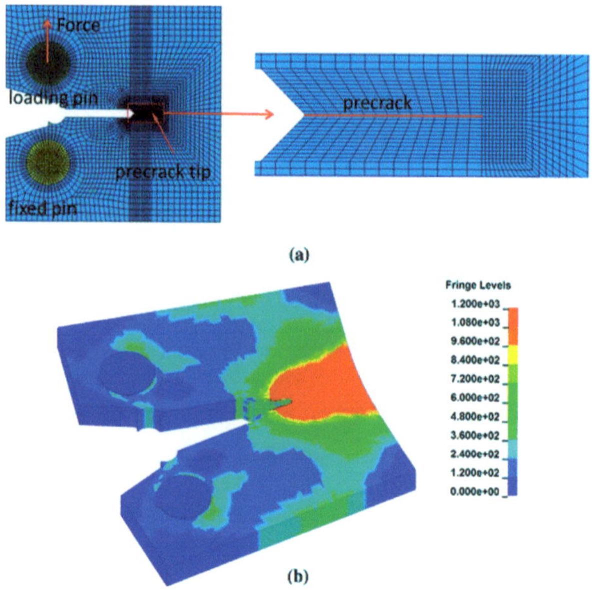

Figure 20. (**a**) Simulation model of the CT specimen for fracture toughness tests. (**b**) Stress distribution Adapted from ref. [204].

Table 2 provides an overview of the various models used to predict the fracture behaviour of aluminium alloys, along with their respective methods, applications, and limitations. However, some models focus on predicting fracture behaviour under conditions including only corrosion, temperature, or humidity. Still, the combined effects of temperature and humidity in coastal environments remain unexplored. Structures in coastal areas are frequently exposed to temperature, humidity, and corrosion.

Table 2. Models used to predict the fracture behaviour of the aluminium alloys.

Author	Model Used	Methodology	Application	Limitation
Zhiying et al. (2016) [205]	Corrosion–fatigue	Paris model and Trantina–Johnson model	Accurate results in corrosion conditions	Does not consider the temperature and humidity
C.Q. Wang et al. (2023) [53]	Corrosion–fatigue	Trantina–Johnson model		
Huang et al. (2016) [206]	Pre-corrosion fatigue	Equivalent crack size (ECS) models and experiment	It focuses on single and multi-crack initiations	Pre-corrosion; does not consider the temperature and humidity
Ping et al. (2016) [207]	Theoretical model and numerical simulation	Pitting corrosion model	Fatigue damage was evaluated for the model of pit growth	Does not consider the temperature and humidity
Safyari et al. (2021) [107] and (2023) [208]	Humidity model	Hydrogen embrittlement mechanism	Hydrogen sensitivity index	Does not consider the temperature and corrosion
Delshad et al. (2020) [209]	Temperature model	Ductility and Yielding	Mechanical properties	Does not consider the humidity and corrosion
Mouritz et al. (2012) [210]	Temperature model	Fracture toughness	Aerospace materials	Does not consider the humidity and corrosion

Table 2. *Cont.*

Author	Model Used	Methodology	Application	Limitation
Kimberly et al. (2020) [108]	Temperature and corrosion model	Pitting corrosion	Mg_2Si intermetallic formation	Does not consider the humidity conditions
Sarah et al. (2023) [99]	Atmospheric corrosion	Frequency and salinity, atmospheric conditions	Atmospheric corrosion FCGR	Does not consider the Temperature and humidity

6. Conclusions

Coastal areas present unique challenges for materials and structures. A comprehensive understanding of the material's fracture behaviour under such conditions is essential. This literature study investigates various fracture mechanisms under corrosive conditions and the interaction between corrosion, temperature, humidity, and material characteristics. Temperature and humidity influence the fatigue and fracture behaviour of materials and structures, a significant concern in engineering applications. The exposure of structures to high temperature and humidity conditions can change material properties and affect crack propagation.

Coastal environments face unique challenges due to saltwater exposure, humidity, temperature fluctuations, and corrosive agents. Saltwater accelerates material corrosion, while humidity can cause moisture absorption and deterioration. Temperature and humidity variations can affect the corrosion process. Many elements and processes can contribute to fracture mechanisms, such as corrosion fatigue, hydrogen embrittlement, and moisture-assisted crack propagation. These factors play an important role in material degradation in corrosive environments.

The existing literature has not yet provided sufficient explanations about the fatigue and fracture behaviour of the materials under temperature and humidity conditions across the structure. The following conclusions were drawn from the existing literature:

- Aluminium alloys, especially the Al6000 series, are prone to corrosion in NaCl solutions, leading to reduced fracture toughness and corrosion fatigue under cyclic loading. Elevated temperatures exceeding 70 °C impact aircraft component performance, causing microscopic cracks and corrosion pits. Pitting corrosion is observed between 20 to 80 °C, decreasing above 70 °C with the formation of an aluminium oxide layer. Research on the aluminium alloy shows a reduction in corrosion pits with rising temperatures, emphasizing the correlation between temperature and the diffusion coefficient for protective oxide layer growth.
- Humidity significantly accelerates corrosion and affects mechanical properties in coastal areas, causing surface degradation and fatigue failure. Common corrosion types include intergranular and pitting corrosion. Cyclic loading in humid air increases fracture growth rates, and hydrogen embrittlement mechanisms involve water vapor reactions leading to hydrogen absorption.

Understanding the combined effects of temperature, humidity, and corrosive marine atmospheres is crucial for addressing challenges to aluminium alloy performance in coastal environments. Addressing coastal challenges for aluminium alloys requires a deeper understanding of corrosion mechanisms, comprehensive field studies, and accurate predictive models. Exploring emerging materials and coatings tailored to coastal conditions holds promise for effective mitigation in the future.

Author Contributions: Writing—original draft, writing—review and editing, information collection, modifications, validation, resources, I.A.; review, modifications, supervision, A.S.; conceptualization, methodology, review, supervision, M.K. All authors have read and agreed to the published version of the manuscript.

Funding: This research received no external funding.

Conflicts of Interest: The authors declare no conflict of interest.

References

1. Mohammed, Y. Abdellah Ductile Fracture and S–N Curve Simulation of a 7075-T6 Aluminum Alloy under Static and Constant Low-Cycle Fatigue. *J. Fail. Anal. Prev.* **2021**, *21*, 1476–1488. [CrossRef]
2. Salam, I.; Muhammad, W.; Ejaz, N. Fatigue Crack Growth in an Aluminum Alloy-Fractographic Study. In Proceedings of the IOP Conference Series: Materials Science and Engineering, Volume 146, 14th International Symposium on Advanced Materials, Islamabad, Pakistan, 12–16 October 2015; National Centre for Physics: Islamabad, Pakistan, 2016.
3. Menan, F.; Henaff, G. Influence of Frequency and Exposure to a Saline Solution on the Corrosion Fatigue Crack Growth Behavior of the Aluminum Alloy 2024. *Int. J. Fatigue* **2009**, *31*, 1684–1695. [CrossRef]
4. Abdellah, M.Y.; Alharthi, H. Fracture Toughness and Fatigue Crack Growth Analyses on a Biomedical Ti-27Nb Alloy under Constant Amplitude Loading Using Extended Finite Element Modelling. *Materials* **2023**, *16*, 4467. [CrossRef]
5. Kamal, A.-S.H.; Ghazaly, N.M.; Abdellah, M.Y.; Seleem, A.-E.H.A.; Abdel-Jaber, G. Influence Parameters on the Essential Work of Fracture of 5754-H111 Aluminum Alloy Plate: Comparative Study. *SVU-Int. J. Eng. Sci. Appl.* **2023**, *4*, 243–259. [CrossRef]
6. Abdellah, M.Y.; Ghazaly, N.M.; Kamal, A.-S.H.; Hagag, A.-E.; Seleem, A.; Abdel-Jaber, G. Ductile Fracture Toughness of Al 5754-H111 Alloy Using Essential Work of Fracture Method. *AIMS Mater. Sci.* **2023**, *10*, 370–389. [CrossRef]
7. Abdellah, M.Y.; Fadhl, B.M.; Abu El-Ainin, H.M.; Hassan, M.K.; Backar, A.H.; Mohamed, A.F. Experimental Evaluation of Mechanical and Tribological Properties of Segregated Al-Mg-Si Alloy Filled with Alumina and Silicon Carbide through Different Types of Casting Molds. *Metals* **2023**, *13*, 316. [CrossRef]
8. Doddamani, S.; Kaleemulla, M. Effect of Graphite on Fracture Toughness of 6061Al-Graphite. *Strength Fract. Complex.* **2018**, *11*, 295–308. [CrossRef]
9. Igwemezie, V.; Mehmanparast, A. Waveform and Frequency Effects on Corrosion-Fatigue Crack Growth Behaviour in Modern Marine Steels. *Int. J. Fatigue* **2020**, *134*, 105484. [CrossRef]
10. Alqahtani, I.; Starr, A.; Khan, M. Coupled Effects of Temperature and Humidity on Fracture Toughness of Al–Mg–Si–Mn Alloy. *Materials* **2023**, *16*, 4066. [CrossRef]
11. Murphy, F.A.A.; Murphy, T.P. Influence of Material Toughness on Fracture Reliability in Steel Bridges. *Natl. Acad. Sci. Transp. Res. Board* **2021**, 1–13. [CrossRef]
12. Ramesh, R.S.; Santhosh Kumar, M.V.; Yasmin, B.; Doddamani, S.; Mohamed Kaleemulla, K. Fracture Toughness Investigations of AA6061-SiC Composites: Effect of Corrosion Parameters. *Mater. Chem. Phys.* **2023**, *308*, 128224. [CrossRef]
13. Tenkamp, J.; Awd, M.; Siddique, S.; Starke, P.; Walther, F. Fracture–Mechanical Assessment of the Effect of Defects on the Fatigue Lifetime and Limit in Cast and Additively Manufactured Aluminum–Silicon Alloys from HCF to VHCF Regime. *Metals* **2020**, *10*, 943. [CrossRef]
14. Abdellah, M.Y. Essential Work of Fracture Assessment for Thin Aluminium Strips Using Finite Element Analysis. *Eng. Fract. Mech.* **2017**, *179*, 190–202. [CrossRef]
15. Saleemsab, D.; Kiran, J.O.; Kaleemulla, M.; Bakkappa, B. Fracture Toughness Testing of 6061Al-Graphite Composites Using SENB Specimens. *J. Inst. Eng. (India)-Ser. D Springer* **2019**, *100*, 195–201. [CrossRef]
16. Anderson, T.L. *Fracture Mechanics—Fundamentals and Applications*, 3rd ed.; CRC Press: Boca Raton, FL, USA, 2005.
17. Dowling, N.E. *Mechanical Behavior of Materials—Engineering Methods for Deformation, Fracture and Fatigue*, 4th ed.; Pearson Education Limited: Essex, UK, 2013.
18. Golewski, G.L. The Phenomenon of Cracking in Cement Concretes and Reinforced Concrete Structures: The Mechanism of Cracks Formation, Causes of Their Initiation, Types and Places of Occurrence, and Methods of Detection—A Review. *Buildings* **2023**, *13*, 765. [CrossRef]
19. Golewski, G.L. Fracture Performance of Cementitious Composites Based on Quaternary Blended Cements. *Materials* **2022**, *15*, 6023. [CrossRef]
20. *ASTME1823-10a*; Standard Terminology Relating to Fatigue and Fracture Testing. ASTM: West Conshohocken, PA, USA, 2011.
21. *ASTME399-22*; Standard Test Method for Linear-Elastic Plane Strain Fracture Toughness KIc of Metallic Materials. ASTM: West Conshohocken, PA, USA, 2022.
22. Bharath, K.N.; Saleemsab Doddamani, A.M.; Rajesh, K.M.K. Simulation of Various Fracture Models by Varying Geometrical Parameters Using Taguchi's DOE. *Struct. Integr. Life* **2023**, *23*, 191–195.
23. Dhummansure, V.; Kalyanrao, A.A.; Doddamani, S. Optimization of Process Parameters for Fracture Toughness of Al6061-Graphite Composites. *Struct. Integr. Life* **2020**, *20*, 51–55.
24. Doddamani, S.; Kaleemulla, M. Comparisons of Experimental Fracture Toughness Testing Methods of Al6061–Graphite Particulate Composites. *J. Fail. Anal. Prev.* **2019**, *19*, 730–737. [CrossRef]
25. Hegde, R.; Sivaram, N.M.; Ajaykumar, B.S.; Kirthan, L.J. Evaluation of Heat Treatment Effect on Fracture Behavior of Aluminum Silicon Carbide Graphite Hybrid Composite. *Int. J. Appl. Eng. Res.* **2017**, *12*, 605–610.
26. Bergant, Z.; Trdan, U.; Grum, J. Effects of Laser Shock Processing on High Cycle Fatigue Crack Growth Rate and Fracture Toughness of Aluminium Alloy 6082-T651. *Int. J. Fatigue* **2016**, *87*, 444–455. [CrossRef]
27. Raviraj, M.S.; Sharanaprabhu, C.M.; Mohankumar, G.C. Experimental Investigation of Effect of Specimen Thickness on Fracture Toughness of Al-TiC Composites. *Frattura ed Integrita Strutturale* **2016**, *10*, 360–368. [CrossRef]

28. Chen, C.J.; Su, M.N.; Wang, Y.H.; Deng, X.W. Experimental Research on the Fatigue Crack Growth Behaviour of Q420C. *J. Constr. Steel Res.* **2022**, *192*, 107241. [CrossRef]
29. Doddamani, S. Fracture Toughness Investigations of Al6061-Graphite Particulate Composite Using Compact Specimens. *Frattura ed Integrita Strutturale* **2017**, *41*, 484–490. [CrossRef]
30. Zhu, X.-K.; Joyce, J.A. Review of Fracture Toughness (G, K, J, CTOD, CTOA) Testing and Standardization. *Eng. Fract. Mech.* **2012**, *85*, 1–46. [CrossRef]
31. *ASTM E647−13*; Standard Test Method for Measurement of Fatigue Crack Growth Rates. ASTM: West Conshohocken, PA, USA, 2014; Volume 3, pp. 1–50.
32. Wanhill, R.J.H.; Bray, G.H. Fatigue Crack Growth Behavior of Aluminum–Lithium Alloys. In *Aluminum-Lithium Alloys*; Prasad, N.E., Gokhale, A.A., Wanhill, R.J.H., Eds.; Butterworth-Heinemann: Oxford, UK, 2014; pp. 381–413.
33. Lee, E.E. *Fatigue Behavior of Silicon Carbide Whisker/Aluminum Composite*; Naval Air Development Center: Warminster, PA, USA, 1988.
34. Yuan, R.; Kruzic, J.; Zhang, X.; De Jonghe, L.; Ritchie, R. Ambient to High-Temperature Fracture Toughness and Cyclic Fatigue Behavior in Al-Containing Silicon Carbide Ceramics. *Acta Mater.* **2003**, *51*, 6477–6491. [CrossRef]
35. Achutha, M.V.; Sridhara, B.K.; Budan, A. Fatigue Life Estimation of Hybrid Aluminium Matrix Composites. *Int. J. Des. Manuf. Technol.* **2008**, *2*, 14–21.
36. Uematsu, Y.; Tokaji, K.; Kawamura, M. Fatigue Behavior of SiC-Particulate-Reinforced Aluminium Alloy Composites with Different Particle Sizes at Elevated Temperatures. *Compos. Sci. Technol.* **2008**, *68*, 2785–2791. [CrossRef]
37. Myriounis, D.P.; Hasan, S.T. Fatigue and Fracture Behaviour of Al-SiCp MMC NDE by IR Detection. In Proceedings of the 18TH International Conference on Composite Materials, Jeju Island, Republic of Korea, 21–26 August 2011.
38. Huang, J.; Spowart, J.; Jones, J. The Role of Microstructural Variability on the Very High-Cycle Fatigue Behavior of Discontinuously-Reinforced Aluminum Metal Matrix Composites Using Ultrasonic Fatigue. *Int. J. Fatigue* **2010**, *32*, 1243–1254. [CrossRef]
39. Sharma, M.; Ziemian, C.; Eden, T. Fatigue Behavior of SiC Particulate Reinforced Spray-Formed 7XXX Series Al-Alloys. *Mater. Des.* **2011**, *32*, 4304–4309. [CrossRef]
40. Joadder, B.; Shit, J.; Acharyya, S.; Dhar, S. Fatigue Failure of Notched Specimen—A Strain-Life Approach. *Mater. Sci. Appl.* **2011**, *2*, 1730–1740. [CrossRef]
41. Shahani, A.; Kashani, H.M. Assessment of Equivalent Initial Flaw Size Estimation Methods in Fatigue Life Prediction Using Compact Tension Specimen Tests. *Eng. Fract. Mech.* **2013**, *99*, 48–61. [CrossRef]
42. Ciavarella, M.; D'antuono, P.; Demelio, G. Generalized Definition of "Crack-like" Notches to Finite Life and SN Curve Transition from "Crack-like" to "Blunt Notch" Behavior. *Eng. Fract. Mech.* **2017**, *179*, 154–164. [CrossRef]
43. Raposo, P.; Correia, J.A.F.O.; De Jesus, A.M.P.; Calçada, R.A.B.; Lesiuk, G.; Hebdon, M.; Fernández-Canteli, A. Probabilistic Fatigue S-N Curves Derivation for Notched Components. *Frattura ed Integrita Strutturale* **2017**, *42*, 105–118. [CrossRef]
44. Saleemsab Doddamani, M.K.K. Effect of Thickness on Fracture Toughness of Al6061-Graphite. *J. Solid Mech.* **2019**, *11*, 635–643.
45. Farheen Kulsum, A.S.D. Review on Corrosion Behavior, Fatigue Behavior and Fracture Toughness of Al Alloy MMCS. *Mater. Today Proc.* 2024; *in press*. [CrossRef]
46. Antunes, F.V.; Serrano, S.; Branco, R.; Prates, P. Fatigue Crack Growth in the 2050-T8 Aluminium Alloy. *Int. J. Fatigue* **2018**, *115*, 79–88. [CrossRef]
47. Li, H.; Li, J.; Yuan, H. Application of a Cohesive Zone Model for Simulating Fatigue Crack Growth from Moderate to High ΔK Levels of Inconel 718. *Int. J. Aerosp. Eng.* **2018**, *2018*, 4048386. [CrossRef]
48. Alqahtani, I.M.; Starr, A.; Khan, M. Experimental and Theoretical Aspects of Crack Assisted Failures of Metallic Alloys in Corrosive Environments—A Review. *Mater. Today Proc.* **2022**, *66*, 2530–2535. [CrossRef]
49. Niazi, H.; Eadie, R.; Chen, W.; Zhang, H. High PH Stress Corrosion Cracking Initiation and Crack Evolution in Buried Steel Pipelines: A Review. *Eng. Fail. Anal.* **2021**, *120*, 105013. [CrossRef]
50. Blanc, C.; Creus, J.; Touzet-Cortina, M. Stress Corrosion Cracking. Between the Corrosion Defect and the Long Crack: The Phase of the Initiation of the Cracks. In *Mechanics—Microstructure—Corrosion Coupling*; Christine Blanc, I.A., Ed.; Elsevier: Amsterdam, The Netherlands, 2019; pp. 287–312.
51. Alqahtani, I.; Starr, A.; Khan, M. Crack Propagation Behaviour under Corrosion and Thermomechanical Loads. In Proceedings of the 7th International Conference on Sustainable Materials and Recent Trends in Mechanical Engineering (SMARTME-2023), Bengaluru, India, 11–12 August 2023; pp. 37–45.
52. Pedeferri, P. Stress Corrosion Cracking and Corrosion-Fatigue. In *Corrosion Science and Engineering*; Engineering Materials; Springer: Cham, Switzerland, 2018; pp. 243–273.
53. Wang, C.Q.; Xiong, J.J.; Shenoi, R.A.; Liu, M.D.; Liu, J.Z. A Modified Model to Depict Corrosion Fatigue Crack Growth Behavior for Evaluating Residual Lives of Aluminum Alloys. *Int. J. Fatigue* **2016**, *83*, 280–287. [CrossRef]
54. Adeboye, S.A.; Adebowale, A.D.; Siyanbola, T.O.; Ajanaku, K.O. Coatings and the Environment: A Review of Problems, Progress and Prospects. In Proceedings of the IOP Conference Series: 6th International Conference on Science and Sustainable Development (ICSSD 2022), Virtual, 9–10 December 2022; Volume 1197.
55. Panda, R.; Fatma, K.; Tripathy, J. Anti-Corrosion and Anti-Wear Ceramic Coatings. In *Advanced Ceramic Coatings*; Gupta, R.K., Motallebzadeh, A., Kakooei, S., Nguyen, T.A., Behera, A., Eds.; Elsevier: Amsterdam, The Netherlands, 2023; pp. 197–217.

56. Shaikh, S. Nanomaterials Applied in Coatings: Synthesis, Structures, Properties, and Applications. *Coatings* **2022**, *12*, 1773. [CrossRef]
57. Putatunda, S.K.; Bajajand, S.; Boileau, J. A Novel Method for Determining the Fatigue Threshold and Fracture Toughness from a Single Test. *Int. J. Metall. Met. Phys.* **2018**, *3*, 1–11. [CrossRef]
58. Robertson, S.W.; Ritchie, R.O. A Fracture-Mechanics-Based Approach to Fracture Control in Biomedical Devices Manufactured from Superelastic Nitinol Tube. *J. Biomed. Mater. Res.—Part B Appl. Biomater.* **2008**, *84*, 26–33. [CrossRef] [PubMed]
59. Perez, N. *Fracture Mechanics*, 2nd ed.; Springer: Berlin/Heidelberg, Germany, 2016; ISBN 978-3-319-24997-1.
60. Saleemsab Doddamani, M.K.K. Review of Experimental Fracture Toughness (K IC) of Aluminium Alloy and Aluminium MMCs Metal Matrix Composites. *Int. J. Fract. Damage Mech.* **2016**, *2*, 38–51.
61. Lai, M.O.; Ferguson, W.G. Fracture Toughness of Aluminium Alloy 7075-T6 in the as-Cast Condition. *Mater. Sci. Eng.* **1985**, *74*, 133–138. [CrossRef]
62. Hill, M.R.; Panontin, T.L. Micromechanical Modeling of Fracture Initiation in 7050 Aluminum. *Eng. Fract. Mech.* **2002**, *69*, 2163–2186. [CrossRef]
63. Urrego, L.F.; García-Beltrán, O.; Arzola, N.; Araque, O. Mechanical Fracture of Aluminium Alloy (AA 2024-T4), Used in the Manufacture of a Bioproducts Plant. *Metals* **2023**, *13*, 1134. [CrossRef]
64. Tsangarakis, N. All Modes Fracture Toughness of Two Aluminium Alloys. *Eng. Fract. Mech.* **1987**, *26*, 313–321. [CrossRef]
65. Prasad, N.E.; Kamat, S.V.; Prasad, K.S.; Malakondaiah, G.; Kutumbarao, V. In-Plane Anisotropy in the Fracture Toughness of an Al-Li 8090 Alloy Plate. *Eng. Fract. Mech.* **1993**, *46*, 209–223. [CrossRef]
66. Alrashed, F.; Asif, M. Climatic Classifications of Saudi Arabia for Building Energy Modelling. *Energy Procedia* **2015**, *75*, 1425–1430. [CrossRef]
67. Alqahtani, I.M.; Starr, A.; Khana, M. Fracture Toughness Investigation of AL6082-T651 Alloy under Corrosive Environmental Conditions. In Proceedings of the 11th International Conference on Fracture Fatigue and Wear (FFW-2023), Ghent, Belgium, 29–31 August 2023.
68. Tang, Y.; Meng, Q.; Ren, P. Spatial Distribution and Concentrations of Salt Fogs in a Coastal Urban Environment: A Case Study in Zhuhai City. *Build. Environ.* **2023**, *234*, 110156. [CrossRef]
69. Loader, P.B.C. *Effect of High Temperature Exposure on the Mechanical Properties of Cold Expanded Open Holes in 7050-T7451 Aluminium Alloy*; Defence Science and Technology Organisation: Edinburgh, Australia, 2008.
70. Cavalcante, T.R.F.; Pereira, G.S.; Koga, G.Y.; Bolfarini, C.; Bose Filho, W.W.; Avila, J.A. Fatigue Crack Propagation of Aeronautic AA7050-T7451 and AA2050-T84 Aluminum Alloys in Air and Saline Environments. *Int. J. Fatigue* **2022**, *154*, 106519. [CrossRef]
71. Zheng, Y.Y.; Luo, B.H.; He, C.; Gao, Y.; Bai, Z.H. Corrosion Evolution and Behaviour of Al–2.1Mg–1.6Si Alloy in Chloride Media. *Rare Met.* **2021**, *40*, 908–919. [CrossRef]
72. Zakaria, H.M. Microstructural and Corrosion Behavior of Al/SiC Metal Matrix Composites. *Ain Shams Eng. J.* **2014**, *5*, 831–838. [CrossRef]
73. Cai, Y.; Xu, Y.; Zhao, Y.; Zhang, W.; Yao, J.; Wei, M.; Zhou, K.; Ma, X. Quantitative Understanding of the Environmental Effect on B10 Copper Alloy Corrosion in Seawater. *Metals* **2021**, *11*, 1080. [CrossRef]
74. Little, B.J.; Ray, R.I.; Lee, J. Understanding Marine Biocorrosion: Experiments with Artificial and Natural Seawater. In *Understanding Biocorrosion*; Liengen, T., Féron, D., Basséguy, R., Beech, I.B., Eds.; Woodhead Publishing: Sawston, UK, 2014; pp. 329–340.
75. Zai, B.A.; Khan, M.A.; Khan, K.A.; Mansoor, A.; Shah, A.; Shahzad, M. The Role of Dynamic Response Parameters in Damage Prediction. *Proc. Inst. Mech. Eng. Part C J. Mech. Eng. Sci.* **2019**, *233*, 4620–4636. [CrossRef]
76. Sarah Galyon Dorman, S.A.F. Examination of the Effects of Specimen Geometry on Single Edge-Cracked Tension Specimens. *Eng. Fract. Mech.* **2019**, *209*, 221–227. [CrossRef]
77. Ramsden, E. Development Tools. In *Hall-Effect Sensors*; Ramsden, E., Ed.; Newnes: Oxford, UK, 2006; pp. 187–202. ISBN 9780750679343.
78. Darehshouri, S.; Michelsen, N.; Schüth, C.; Schulz, S. A Low-Cost Environmental Chamber to Simulate Warm Climatic Conditions. *Vadose Zone J.* **2020**, *19*, e20023. [CrossRef]
79. Kurniawan, I.; Faridah; Utami, S.S. Characterizing of Climate Chamber Thermal Environment Using the CFD Simulation Method Using IES VE. *AIP Conf. Proc.* **2020**, *2223*, 050010. [CrossRef]
80. Czerwinski, F. Thermal Stability of Aluminum Alloys. *Materials* **2020**, *13*, 3441. [CrossRef] [PubMed]
81. Aboura, Y.; Garner, A.; Euesden, R.; Barrett, Z.; Engel, C.; Holroyd, N.; Prangnell, P.; Burnett, T. Understanding the Environmentally Assisted Cracking (EAC) Initiation and Propagation of New Generation 7xxx Alloys Using Slow Strain Rate Testing. *Corros. Sci.* **2022**, *199*, 110161. [CrossRef]
82. Brown, M.D. *Moisture Absorption and Desorption Effects on Mechanical Behavior in Specialty Polyamide Products*; Lehigh University: Bethlehem, PA, USA, 2019.
83. Yadav, V.; Gaur, V.; Singh, I. Corrosion-Fatigue Behavior of Welded Aluminum Alloy 2024-T3. *Int. J. Fatigue* **2023**, *173*, 107675. [CrossRef]
84. Gyarmati, G.; Bubonyi, T.; Fegyverneki, G.; Tokár, M.; Mende, T. Interactions of Primary Intermetallic Compound Particles and Double Oxide Films in Liquid Aluminum Alloys. *Intermetallics* **2022**, *149*, 107681. [CrossRef]
85. Liu, Y.; Kang, S.; Kim, H. The Complex Microstructures in an As-Cast Al–Mg–Si Alloy. *Mater. Lett.* **1999**, *41*, 267–272. [CrossRef]

86. Eckermann, F.; Suter, T.; Uggowitzer, P.J.; Afseth, A.; Schmutz, P. The Influence of MgSi Particle Reactivity and Dissolution Processes on Corrosion in Al-Mg-Si Alloys. *Electrochim. Acta* **2008**, *54*, 844–855. [CrossRef]
87. Wei, Z.; Goehring, T.; Mioduszewski, M.; Luo, L.; Kotrba, A.; Rybarz, M.; Ellinghaus, K.; Pieszkalla, M. Failure Mechanisms and Modes Analysis of Vehicle Exhaust Components and Systems. In *Handbook of Materials Failure Analysis with Case Studies from the Aerospace and Automotive Industries*; Abdel Salam Hamdy Makhlouf, M.A., Ed.; Butterworth-Heinemann: Oxford, UK, 2016; pp. 393–432.
88. Peng, C.; Cao, G.; Gu, T.; Wang, C.; Wang, Z.; Sun, C. The Corrosion Behavior of the 6061 Al Alloy in Simulated Nansha Marine Atmosphere. *J. Mater. Res. Technol.* **2022**, *19*, 709–721. [CrossRef]
89. Noell, P.J.; Karasz, E.; Schindelholz, E.J.; Polonsky, A.T.; Campbell, I.; Katona, R.M.; Melia, M.A. The Evolution of Pit Morphology and Growth Kinetics in Aluminum during Atmospheric Corrosion. *npj Mater. Degrad.* **2023**, *7*, 12. [CrossRef]
90. Cao, M.; Liu, L.; Fan, L.; Yu, Z.; Li, Y.; Oguzie, E.E.; Wang, F. Influence of Temperature on Corrosion Behavior of 2A02 Al Alloy in Marine Atmospheric Environments. *Materials* **2018**, *11*, 235. [CrossRef]
91. Amura, M.; Aiello, L.; Colavita, M.; De Paolis, F.; Bernabei, M. Failure of a Helicopter Main Rotor Blade. *Procedia Mater. Sci.* **2014**, *3*, 726–731. [CrossRef]
92. Romeyn, A. *Main Rotor Blade Failure Analysis Report*; IOP Publishing: Philadelphia, PA, USA, 2005.
93. Sunrise Helicopters Inc., Canada. In-Flight Separation of Main Rotor Blade and Collision with Terrain. 2011. Available online: https://www.bst-tsb.gc.ca/eng/rapports-reports/aviation/2011/a11o0205/a11o0205.pdf (accessed on 1 March 2024).
94. Cicolin, D.; Trueba, M.; Trasatti, S. Effect of Chloride Concentration, PH and Dissolved Oxygen, on the Repassivation of 6082-T6 Al Alloy. *Electrochim. Acta* **2014**, *124*, 27–35. [CrossRef]
95. Otieno, M.; Thomas, M. Marine Exposure Environments and Marine Exposure Sites. In *Marine Concrete Structures*; Alexander, M.G., Ed.; Woodhead Publishing: Sawston, UK, 2016; pp. 171–196.
96. Vargel, C. Chapter D.1—Freshwater. In *Corrosion of Aluminium*; Vargel, C., Ed.; Elsevier: Amsterdam, The Netherlands, 2004; pp. 299–327. ISBN 9780080444956.
97. Tao, J.; Xiang, L.; Zhang, Y.; Zhao, Z.; Su, Y.; Chen, Q.; Sun, J.; Huang, B.; Peng, F. Corrosion Behavior and Mechanical Performance of 7085 Aluminum Alloy in a Humid and Hot Marine Atmosphere. *Materials* **2022**, *15*, 7503. [CrossRef]
98. Luo, D.; Li, F.; Xing, G. Corrosion Resistance of 6061-T6 Aluminium Alloy and Its Feasibility of near-Surface Reinforcements in Concrete Structure. *Rev. Adv. Mater. Sci.* **2022**, *61*, 638–653. [CrossRef]
99. Free, B.; Marino, G.; Schindelholz, E.; Galyon Dorman, S.; (Warner) Locke, J.S. Measurement of Atmospheric Corrosion Fatigue Crack Growth Rates on AA7085-T7451. *Int. J. Fatigue* **2023**, *167*, 107368. [CrossRef]
100. Mao, Y.; Zhu, Y.; Deng, C.-M.; Sun, S.; Xia, D.-H. Analysis of Localized Corrosion Mechanism of 2024 Aluminum Alloy at a Simulated Marine Splash Zone. *Eng. Fail. Anal.* **2022**, *142*, 106759. [CrossRef]
101. Bray, G.H.; Bucci, R.J.; Brazill, R.L. Lessons Neglected: Effects of Moist Air on Fatigue and Fatigue Crack Growth in Aluminum Alloys. *Mater. Sci. Forum* **2000**, *331–337*, 1413–1426.
102. Ahmad, Z. Chapter 10—Atmospheric Corrosion. In *Principles of Corrosion Engineering and Corrosion Control*; Butterworth-Heinemann: Oxford, UK, 2006; pp. 550–575. ISBN 9780750659246.
103. Bradshaw, F.J.; Wheeler, C. The Influence of Gaseous Environment and Fatigue Frequency on the Growth of Fatigue Cracks in Some Aluminium Alloys. *Int. J. Fract. Mech.* **1969**, *5*, 255–268. [CrossRef]
104. Davidson, D.L.; Lankford, J. The Effect of Water Vapor on Fatigue Crack Tip Mechanics in 7075-T651 Aluminium Alloy. *Fatigue Fract. Eng. Mater. Struct.* **1983**, *6*, 241–256. [CrossRef]
105. Holper, B.; Mayer, H.; Vasudevan, A.; Stanzl-Tschegg, S. Near Threshold Fatigue Crack Growth in Aluminium Alloys at Low and Ultrasonic Frequency: Influences of Specimen Thickness, Strain Rate, Slip Behaviour and Air Humidity. *Int. J. Fatigue* **2003**, *25*, 397–411. [CrossRef]
106. Young, G.A.; Scully, J.R. The Effects of Test Temperature, Temper, and Alloyed Copper on the Hydrogen-Controlled Crack Growth Rate of an Al-Zn-Mg-(Cu) Alloy. *Met. Mater. Trans. A* **2002**, *33*, 1167–1181. [CrossRef]
107. Safyari, M.; Hojo, T.; Moshtaghi, M. Effect of Environmental Relative Humidity on Hydrogen-Induced Mechanical Degradation in an Al–Zn–Mg–Cu Alloy. *Vacuum* **2021**, *192*, 110489. [CrossRef]
108. L'haridon-Quaireau, S.; Laot, M.; Colas, K.; Kapusta, B.; Delpech, S.; Gosset, D. Effects of Temperature and PH on Uniform and Pitting Corrosion of Aluminium Alloy 6061-T6 and Characterisation of the Hydroxide Layers. *J. Alloys Compd.* **2020**, *833*, 155146. [CrossRef]
109. Khodor, J.; Özenç, K.; Lin, G.; Kaliske, M. Characterization of Fatigue Crack Growth by Cyclic Material Forces. *Eng. Fract. Mech.* **2021**, *243*, 107514. [CrossRef]
110. González-Velázquez, J. Environmentally-Assisted Fracture. In *Fractography and Failure Analysis. Structural Integrity*; Springer: Cham, Switzerland, 2018.
111. Navidirad, M.; Stepniowski, W.J.; Cartier, E.; Christ, T.; Watanabe, M.; Misiolek, W.Z. Investigation on the Strain Induced Oxide Layer Fracture and Bonding During Cold Rolling of Aluminum Alloys. In *The Minerals, Metals & Materials Series*; Daehn, G., Cao, J., Kinsey, B., Tekkaya, E.V., Eds.; Springer: Cham, Switzerland, 2021.
112. Song, H.; Liu, C.; Zhang, H.; Leen, S.B. A DIC-Based Study on Fatigue Damage Evolution in Pre-Corroded Aluminum Alloy 2024-T4. *Materials* **2018**, *11*, 2243. [CrossRef]

113. Pu, J.; Zhang, Y.; Zhang, X.; Yuan, X.; Ren, P.; Jin, Z. Mapping the Fretting Corrosion Behaviors of 6082 Aluminum Alloy in 3.5% NaCl Solution. *Wear* **2021**, *482–483*, 203975. [CrossRef]
114. Chugh, B.; Taraphdar, P.K.; Biswal, H.J.; Devi, N.R.; Dorothy, R.; Manimaran, N.; Rajendran, S. Corrosion Inhibition by Aluminum Oxide. In *Inorganic Anticorrosive Materials*; Verma, C., Aslam, J., Hussain, C.M.I., Eds.; Elsevier: Amsterdam, The Netherlands, 2022; pp. 231–249.
115. Wang, D.; Yang, D.; Zhang, D.; Li, K.; Gao, L.; Lin, T. Electrochemical and DFT Studies of Quinoline Derivatives on Corrosion Inhibition of AA5052 Aluminium Alloy in NaCl Solution. *Appl. Surf. Sci.* **2015**, *357*, 2176–2183. [CrossRef]
116. Vargel, C. Oxides and Peroxides. In *Corrosion of Aluminium*; Vargel, C., Ed.; Elsevier: Amsterdam, The Netherlands, 2004; pp. 357–365.
117. Dias, C.; Ventura, J. Understanding the Influence of Metal Oxide Layer Thickness and Defects on Resistive Switching Behavior Through Numerical Modeling. *Phys. Status Solidi A-Appl. Mater. Sci.* **2022**, *220*, 202200730. [CrossRef]
118. Cao, Y.; Xu, H.; Zhan, J.; Zhang, H.; Cui, S.; Tang, W. Microstructure, Growth Kinetics and Formation Mechanism of Oxide Layers on AlN Ceramic Substrates. *J. Ceram. Sci. Technol.* **2018**, *9*, 263–270. [CrossRef]
119. Cao, X.; Campbell, J. Oxide Inclusion Defects in Al-Si-Mg Cast Alloys. *Can. Metall. Q.* **2005**, *44*, 435–448. [CrossRef]
120. Pokorný, P.; Vojtek, T.; Jambor, M.; Náhlík, L.; Hutař, P. Effect of Underload Cycles on Oxide-Induced Crack Closure Development in Cr-Mo Low-Alloy Steel. *Materials* **2021**, *14*, 2530. [CrossRef]
121. Yamada, Y.; Newman, J.C. Crack-Closure Behavior of 2324-T39 Aluminum Alloy near-Threshold Conditions for High Load Ratio and Constant Kmax Tests. *Int. J. Fatigue* **2009**, *31*, 1780–1787. [CrossRef]
122. Casperson, M.C.; Carroll, J.D.; Lambros, J.; Sehitoglu, H.; Dodds, R.H. Investigation of Thermal Effects on Fatigue Crack Closure Using Multiscale Digital Image Correlation Experiments. *Int. J. Fatigue* **2014**, *61*, 10–20. [CrossRef]
123. Pippan, R.; Hohenwarter, A. Fatigue Crack Closure: A Review of the Physical Phenomena. *Fatigue Fract. Eng. Mater. Struct.* **2017**, *40*, 471–495. [CrossRef]
124. Culliton, D.; Betts, A.J.; Kennedy, D. Impact of Intermetallic Precipitates on the Tribological and/or Corrosion Performance of Cast Aluminium Alloys: A Short Review. *Int. J. Cast Met. Res.* **2013**, *26*, 65–71. [CrossRef]
125. Koleva, D.A.; Hu, J.; Stroeven, P. Microstructure Alterations Underlying Electrochemical Process of Chloride-Induced Corrosion. In *Brittle Matrix Composites 8*; Brandt, A.M., Li, V.C., Marshall, I.H., Eds.; Woodhead Publishing: Sawston, UK, 2006; pp. 571–580.
126. Shrivastava, V.; Dubey, S.; Gupta, G.K.; Singh, I.B. Influence of Alpha Nanoalumina Reinforcement Content on the Microstructure, Mechanical and Corrosion Properties of Al6061-Al_2O_3 Composite. *J. Mater. Eng. Perform.* **2017**, *26*, 4424–4433. [CrossRef]
127. Kciuk, M.; Kurc, A.; Szewczenko, J. Structure and Corrosion Resistance of Aluminium AlMg2.5; AlMg5Mn and AlZn5Mg1 Alloys. *J. Achiev. Mater. Manuf. Eng.* **2010**, *41*, 74–81.
128. Zhu, Y.; Sun, K.; Frankel, G.S. Intermetallic Phases in Aluminum Alloys and Their Roles in Localized Corrosion. *J. Electrochem. Soc.* **2018**, *165*, C807–C820. [CrossRef]
129. Zarif, M.; Spacil, I.; Pabel, T.; Schumacher, P.; Li, J. Effect of Ca and P on the Size and Morphology of Eutectic Mg_2Si in High-Purity Al-Mg-Si Alloys. *Metals* **2023**, *13*, 784. [CrossRef]
130. Blau, P.J. Elevated-Temperature Tribology of Metallic Materials. *Tribol. Int.* **2010**, *43*, 1203–1208. [CrossRef]
131. Gain, A.K.; Zhang, L. Temperature and Humidity Effects on Microstructure and Mechanical Properties of an Environmentally Friendly Sn–Ag–Cu Material. *J. Mater. Sci.* **2019**, *54*, 12863–12874. [CrossRef]
132. Arrabal, R.; Pardo, A.; Merino, M.C.; Merino, S.; Mohedano, M.; Casajús, P. Corrosion Behaviour of Mg/Al Alloys in High Humidity Atmospheres. *Mater. Corros. Und. Korros.* **2011**, *62*, 326–334. [CrossRef]
133. Shyam, A.; Lara-Curzio, E. A Model for the Formation of Fatigue Striations and Its Relationship with Small Fatigue Crack Growth in an Aluminum Alloy. *Int. J. Fatigue* **2010**, *32*, 1843–1852. [CrossRef]
134. Laird, C. The Influence of Metallurgical Structure on the Mechanisms of Fatigue Crack Propagation. *ASTM STP 415 Am. Soc. Test. Mats* **1967**, *131*, 1–10.
135. Data, F.M. Linear Elastic Fracture Mechanics (LEFM): Part Two. Available online: https://www.totalmateria.com/page.aspx?ID=CheckArticle&site=kts&LN=IT&NM=299 (accessed on 3 September 2023).
136. Liu, Y.; Pan, Q.; Liu, B.; Yu, Q.; Li, G.; Pan, D. Effect of Aging Treatments on Fatigue Properties of 6005A Aluminum Alloy Containing Sc. *Int. J. Fatigue* **2022**, *163*, 107103. [CrossRef]
137. Zhu, M.; Yang, S.; Bai, Y.; Fan, C. Microstructure and Fatigue Damage Mechanism of 6082-T6 Aluminium Alloy Welded Joint. *Mater. Res. Express* **2021**, *8*, 056505. [CrossRef]
138. Anis, S.F.; Koyama, M.; Hamada, S.; Noguchi, H. Mode I Fatigue Crack Growth Induced by Strain-Aging in Precipitation-Hardened Aluminum Alloys. *Theor. Appl. Fract. Mech.* **2019**, *104*, 102340. [CrossRef]
139. Ma, L.; Liu, C.; Ma, M.; Wang, Z.; Wu, D.; Liu, L.; Song, M. Fatigue Fracture Analysis on 2524 Aluminum Alloy with the Influence of Creep-Aging Forming Processes. *Materials* **2022**, *15*, 3244. [CrossRef]
140. Doré, M.J.; Maddox, S.J. Accelerated Fatigue Crack Growth in 6082 T651 Aluminium Alloy Subjected to Periodic Underloads. *Procedia Eng.* **2013**, *66*, 313–322. [CrossRef]
141. Sharma, V.M.J.; Sree Kumar, K.; Nageswara Rao, B.; Pathak, S.D. Effect of Aging Treatments on the Fatigue Crack Growth and Threshold Behavior of AA 2219 Aluminium Alloy. *Trans. Indian Inst. Met.* **2010**, *63*, 535–540. [CrossRef]
142. Chapetti, M.D.; Gubeljak, N.; Kozak, D. Intrinsic Fatigue Limit and the Minimum Fatigue Crack Growth Threshold. *Materials* **2023**, *16*, 5874. [CrossRef]

143. Mathieu, F.; Hild, F.; Roux, S. Identification of a Crack Propagation Law by Digital Image Correlation. *Int. J. Fatigue* **2012**, *36*, 146–154. [CrossRef]
144. Kallien, Z.; Knothe-Horstmann, C.; Klusemann, B. Fatigue Crack Propagation in AA5083 Structures Additively Manufactured via Multi-Layer Friction Surfacing. *Addit. Manuf. Lett.* **2023**, *6*, 100154. [CrossRef]
145. Belan, J.; Kuchariková, L.; Tillová, E.; Závodská, D.; Chalupová, M. Effect of Fatigue Loading Mode on 718 Alloy Fatigue Properties. *Period. Polytech. Transp. Eng.* **2019**, *47*, 335–341. [CrossRef]
146. Ma, J.; Li, D.; Du, S.; Han, Z.; Luo, P.; Zhao, J. Comparison of Subcritical Crack Growth and Dynamic Fracture Propagation in Rocks under Double-Torsion Tests. *Int. J. Rock Mech. Min. Sci.* **2023**, *170*, 105481. [CrossRef]
147. Hana, N.; Umeda, M.; Akiyoshi, M.; Mitamura, K.; Amaya, K. Crack Identification by Digital Image Correlation Method Using Crack Shape as Prior Information. *ASME J. Press. Vessel Technol.* **2023**, *145*, 041601. [CrossRef]
148. Ermakova, A.; Ganguly, S.; Razavi, N.; Berto, F.; Mehmanparast, A. Experimental Investigation of the Fatigue Crack Growth Behavior in Wire Arc Additively Manufactured ER100S-1 Steel Specimens. *Fatigue Fract. Eng. Mater. Struct.* **2022**, *45*, 371–385. [CrossRef]
149. Milella, P.P. *Fatigue and Corrosion in Metals*; Springer Science & Business Media: Berlin/Heidelberg, Germany, 2012.
150. Arzaghi, E.; Abbassi, R.; Garaniya, V.; Binns, J.; Chin, C.; Khakzad, N.; Reniers, G. Developing a Dynamic Model for Pitting and Corrosion-Fatigue Damage of Subsea Pipelines. *Ocean Eng.* **2018**, *150*, 391–396. [CrossRef]
151. Guo, Y.; Shao, Y.; Gao, X.; Li, T.; Zhong, Y.; Luo, X. Corrosion Fatigue Crack Growth of Serviced API 5L X56 Submarine Pipeline. *Ocean Eng.* **2022**, *256*, 111502. [CrossRef]
152. Gangloff, R.P. Environmental Cracking Corrosion Fatigue. In *ASTM International: West Conshohocken*; ASTM International: West Conshohocken, PA, USA, 2005; pp. 302–321.
153. Wang, Q.Y.; Pidaparti, R.M.; Palakal, M.J. Comparative Study of Corrosion-Fatigue in Aircraft Materials. *AIAA J.* **2001**, *39*, 325–330. [CrossRef]
154. Luer, K.; DuPont, J.; Marder, A.; Skelonis, C. Corrosion Fatigue of Alloy 625 Weld Claddings in Combustion Environments. *Mater. High Temp.* **2001**, *18*, 11–19. [CrossRef]
155. Muñoz, A.F.; Buenhombre, J.L.M.; García-Diez, A.I.; Fabal, C.C.; Díaz, J.J.G. Fatigue Study of the Pre-Corroded 6082-T6 Aluminum Alloy in Saline Atmosphere. *Metals* **2020**, *10*, 1260. [CrossRef]
156. Correia, J.A.F.O.; De Jesus, A.M.P.; Alves, A.S.F.; Lesiuk, G.; Tavares, P.J.S.; Moreira, P.M.G.P. Fatigue Crack Growth Behaviour of the 6082-T6 Aluminium Using CT Specimens with Distinct Notches. *Procedia Struct. Integr.* **2016**, *2*, 3272–3279. [CrossRef]
157. Alireza Behvar, M.H. A Critical Review on Very High Cycle Corrosion Fatigue: Mechanisms, Methods, Materials, and Models. *J. Space Saf. Eng.* **2023**, *10*, 281–323. [CrossRef]
158. Saintier, N.; El May, M.; Odemer, G.; Hénaff, G.; Bosch, C.; Feaugas, X.; Couvant, T. Corrosion and Hydrogen Fatigue at Different Scales. In *Mechanics—Microstructure—Corrosion Coupling*; Christine Blanc, I.A., Ed.; Elsevier: Amsterdam, The Netherlands, 2019; pp. 385–411.
159. Mudali, U.K. Materials for Hostile Corrosive Environments. In *Materials Under Extreme Conditions*; Tyagi, A.K., Banerjee, S., Eds.; Elsevier: Amsterdam, The Netherlands, 2017; pp. 91–128.
160. Gong, K.; Wu, M.; Liu, X.; Liu, G. Nucleation and Propagation of Stress Corrosion Cracks: Modeling by Cellular Automata and Finite Element Analysis. *Mater. Today Commun.* **2022**, *33*, 104886. [CrossRef]
161. Chen, W.; Lu, W.; Gou, G.; Dian, L.; Zhu, Z.; Jin, J. The Effect of Fatigue Damage on the Corrosion Fatigue Crack Growth Mechanism in A7N01P-T4 Aluminum Alloy. *Metals* **2023**, *13*, 104. [CrossRef]
162. Ge, F.; Fan, L.; Liang, J.; Pang, K.; Li, H.; Wang, X.; Cui, Z. Corrosion Evolution of High-Strength Aluminum Alloys in the Simulated Service Environment of Amphibious Aircraft in the Presence of Chloride and Bisulfite. *Acta Metall. Sin. (Engl. Lett.)* **2021**, *34*, 1679–1694. [CrossRef]
163. Mahmood, S.; Gallagher, C.; Engelberg, D.L. Atmospheric Corrosion of Aluminum Alloy 6063 Beneath Ferric Chloride Corrosion Product Droplets. *Corrosion* **2020**, *76*, 985–994. [CrossRef]
164. Kumar, P.; Verma, B.B. Propagation of Corrosion Induced Fatigue Crack in Aluminum Alloy. *AIMS Mater. Sci.* **2022**, *9*, 512–521. [CrossRef]
165. Chen, H.; Chow, C.L.; Lau, D. Deterioration Mechanisms and Advanced Inspection Technologies of Aluminum Windows. *Materials* **2022**, *15*, 354. [CrossRef]
166. Shan, J.; Hou, D.; Zhang, J.; Xin, X.; Cao, G.; Genzhe, H. Effects of the Extrusion Ratio on the Intergranular Corrosion Behaviour of 6082 Aluminium Alloy. *Mater. Corros. Und. Korros.* **2018**, *69*, 365–375. [CrossRef]
167. Nagumo, M. *Fundamentals of Hydrogen Embrittlement*, 1st ed.; Springer: Singapore, 2018.
168. Sergeev, N.N.; Sergeev, A.N.; Kutepov, S.N.; Kolmakov, A.G.; Gvozdev, A.E. Mechanism of the Hydrogen Cracking of Metals and Alloys, Part II (Review). *Inorg. Mater. Appl. Res.* **2019**, *10*, 32–41. [CrossRef]
169. Albrecht, J.; Bernstein, I.M.; Thompson, A.W. Evidence for Dislocation Transport of Hydrogen in Aluminum. *Met. Mater. Trans. A* **1982**, *13*, 811–820. [CrossRef]
170. Bochkaryova, A.V.; Li, Y.V.; Barannikova, S.A.; Zuev, L.B. The Effect of Hydrogen Embrittlement on the Mechanical Properties of Aluminum Alloy. *IOP Conf. Ser. Mater. Sci. Eng.* **2015**, *71*, 012057. [CrossRef]
171. Ambat, R.; Dwarakadasa, E.S. Effect of Hydrogen in Aluminium and Aluminium Alloys: A Review. *Bull. Mater. Sci.* **1990**, *19*, 103–114. [CrossRef]

172. Xie, D.; Li, S.; Li, M.; Wang, Z.; Gumbsch, P.; Sun, J.; Ma, E.; Li, J.; Shan, Z. Hydrogenated Vacancies Lock Dislocations in Aluminium. *Nat. Commun.* **2016**, *7*, 13341. [CrossRef] [PubMed]
173. Liang, Y.; Ahn, D.; Sofronis, P.; Dodds, R.; Bammann, D. Effect of Hydrogen Trapping on Void Growth and Coalescence in Metals and Alloys. *Mech. Mater.* **2008**, *40*, 115–132. [CrossRef]
174. Wang, Y.; Sharma, B.; Xu, Y.; Shimizu, K.; Fujihara, H.; Hirayama, K.; Takeuchi, A.; Uesugi, M.; Cheng, G.; Toda, H. Switching Nanoprecipitates to Resist Hydrogen Embrittlement in High-Strength Aluminum Alloys. *Nat. Commun.* **2022**, *13*, 6860. [CrossRef] [PubMed]
175. Nielsen, C.; Amirkhizi, A.V.; Nemat-Nasser, S. An Empirical Model for Estimating Fracture Toughness Using the DCDC Geometry. *Int. J. Fract.* **2014**, *188*, 113–118. [CrossRef]
176. Begum, Y.; Bharath, K.N.; Doddamani, S.; Rajesh, A.M.; Kaleemulla, K.M. Optimization of Process Parameters of Fracture Toughness Using Simulation Technique Considering Aluminum–Graphite Composites. *Trans. Indian Inst. Met.* **2020**, *73*, 3095–3103. [CrossRef]
177. Abdellah, M.Y.; Hassan, M.K.; Hashem, A.M. Finite Element Computational Approach of Fracture Toughness in Composite Compact-Tension Specimen. *Int. J. Mech. Mechatron. Eng.* **2012**, *12*, 57–61.
178. Achebo, J. Development of a Predictive Model for Determining Mechanical Properties of AA 6061 Using Regression Analysis. *Prod. Manuf. Res.* **2015**, *3*, 169–184. [CrossRef]
179. Gander, M.J.; Hennicker, J.; Masson, R. Modeling and Analysis of the Coupling in Discrete Fracture Matrix Models. *SIAM J. Numer. Anal.* **2021**, *59*, 195–218. [CrossRef]
180. Zhong, Y.; Shao, Y.; Gao, X.; Luo, X.; Zhu, H. Fatigue Crack Growth of EH36 Steel in Air and Corrosive Marine Environments. *J. Constr. Steel Res.* **2023**, *210*, 108104. [CrossRef]
181. Vishnuvardhan, S.; Saravanan, M.; Gandhi, P.; Raghava, G. Fatigue Crack Growth Studies on Power Plant Piping Materials under Corrosive Environment. *Procedia Struct. Integr.* **2019**, *14*, 482–490. [CrossRef]
182. Brückner, A.; Munz, D. The Effect of Curve Fitting on the Prediction of Failure Probabilities from the Scatter in Crack Geometry and Fracture Toughness. *Reliab. Eng.* **1983**, *5*, 139–156. [CrossRef]
183. Guddhur, H.; Naganna, C.; Doddamani, S. Taguchi's Method of Optimization of Fracture Toughness Parameters of Al-SiCp Composite Using Compact Tension Specimens. *Int. J. Optim. Control. Theor. Appl.* **2021**, *11*, 152–157. [CrossRef]
184. Younis, H.B.; Kamal, K.; Sheikh, M.F.; Hamza, A. Prediction of Fatigue Crack Growth Rate in Aircraft Aluminum Alloys Using Optimized Neural Networks. *Theor. Appl. Fract. Mech.* **2022**, *117*, 103196. [CrossRef]
185. Kovalov, D.; Fekete, B.; Engelhardt, G.R.; Macdonald, D.D. Prediction of Corrosion Fatigue Crack Growth Rate in Alloys. Part II: Effect of Electrochemical Potential, NaCl Concentration, and Temperature on Crack Propagation in AA2024-T351. *Corros. Sci.* **2019**, *152*, 130–139. [CrossRef]
186. Zhang, T.; He, Y.; Li, C.; Zhang, T.; Zhang, S. Effect of Alternate Corrosion and Fatigue on Fatigue Crack Growth Characterization of 2024-T4 Aluminum Alloy. *Math. Probl. Eng.* **2020**, *2020*, 7314241. [CrossRef]
187. Bergner, F.; Zouhar, G. New Approach to the Correlation between the Coefficient and the Exponent in the Power Law Equation of Fatigue Crack Growth. *Int. J. Fatigue* **2000**, *22*, 229–239. [CrossRef]
188. Alqahtani, I.; Starr, A.; Khan, M. Investigation of the Combined Influence of Temperature and Humidity on Fatigue Crack Growth Rate in Al6082 Al-Loy in a Coastal Environment. *Materials* **2023**, *16*, 6833. [CrossRef]
189. He, F.; Khan, M.; Aldosari, S. Interdependencies between Dynamic Response and Crack Growth in a 3D-Printed Acrylonitrile Butadiene Styrene (ABS) Cantilever Beam under Thermo-Mechanical Loads. *Polymers* **2022**, *14*, 982. [CrossRef] [PubMed]
190. Yang, Y.; Chen, H.; Feng, W.; Xu, S.; Li, Y.; Zhang, R. Fatigue Life Analysis for 6061-T6 Aluminum Alloy Based on Surface Roughness. *PLoS ONE* **2021**, *16*, e0252772. [CrossRef] [PubMed]
191. Zai, B.A.; Khan, M.; Khan, S.Z.; Asif, M.; Khan, K.A.; Saquib, A.N.; Mansoor, A.; Shahzad, M.; Mujtaba, A. Prediction of Crack Depth and Fatigue Life of an Acrylonitrile Butadiene Styrene Cantilever Beam Using Dynamic Response. *J. Test. Eval.* **2020**, *48*, 1520–1536. [CrossRef]
192. Guo, M.; Li, D.; Rao, S.X.; Guo, B.L. Effect of Environmental Factors on the Corrosion of 2024T3 Aluminium Alloy. In Proceedings of the 9th International Conference on Aluminium Alloys, Brisbane, Australia, 2–5 August 2004; Nie, J.F., Morton, A.J., Muddle, B.C., Eds.; Institute of Materials Engineering Australasia Ltd.: Melbourne, Australia, 2004; pp. 433–438.
193. Ichitani, K. Effect of Experimental Humidity on Fatigue Fracture of 6XXX-Series Aluminum Alloys. *Mater. Sci. Eng.* **2012**, 15–21.
194. Bayoumi, M. Fatigue Behaviour of a Commercial Aluminium Ahoy in Sea Water at Different Temperatures. *Eng. Fract. Mech.* **1993**, *45*, 297–307. [CrossRef]
195. Khan, M.A.; Khan, S.Z.; Sohail, W.; Khan, H.S.M.; Nisar, S. Mechanical Fatigue in Aluminum at Elevated Temperature and Remaining Life Prediction Based on Natural Frequency Evolution. *Fatigue Fract. Eng. Mater. Struct.* **2015**, *38*, 897–903. [CrossRef]
196. Omar, I.K.M.; Starr, A. Comparative Analysis of Machine Learning Models for Predicting Crack Propagation under Coupled Load and Temperature. *Appl. Sci.* **2023**, *13*, 7212. [CrossRef]
197. He, F.; Khan, M. Effects of Printing Parameters on the Fatigue Behaviour of 3d-Printed Abs under Dynamic Thermo-Mechanical Loads. *Polymers* **2021**, *13*, 2362. [CrossRef]
198. Kamei, K.; Khan, M.A. Current Challenges in Modelling Vibrational Fatigue and Fracture of Structures: A Review. *J. Braz. Soc. Mech. Sci. Eng.* **2020**, *43*, 77. [CrossRef]

199. Agala, A.; Khan, M.; Starr, A. Degradation Mechanisms Associated with Metal Pipes and the Effective Impact of LDMs and LLMs in Water Transport and Distribution. *Proc. Inst. Mech. Eng. Part C J. Mech. Eng. Sci.* **2023**, *237*, 1855–1876. [CrossRef]
200. Khan, S.Z.; Khan, T.M.; Joya, Y.F.; Khan, M.A.; Ahmed, S.; Shah, A. Assessment of Material Properties of AISI 316L Stainless Steel Using Non-Destructive Testing. *Nondestruct. Test. Eval.* **2016**, *31*, 360–370. [CrossRef]
201. Gairola, S.; Rengaswamy, J.; Verma, R. A Study on XFEM Simulation of Tensile, Fracture Toughness, and Fatigue Crack Growth Behavior of Al 2024 Alloy through Fatigue Crack Growth Rate Models Using Genetic Algorithm. *Fatigue Fract. Eng. Mater. Struct.* **2023**, *46*, 2121–2138. [CrossRef]
202. Hareesha, G.; Chikkanna, N.; Doddamani, S. Finite Element Simulation of Fracture Toughness of Al6061 Reinforced with Silicon Carbide. *IOP Conf. Ser. Mater. Sci. Eng.* **2021**, *1065*, 012036. [CrossRef]
203. Saleemsab Doddamani, M.K. Indentation Fracture Toughness of Alumnum6061-Graphite Composites. *Int. J. Fract. Damage Mech.* **2016**, *1*, 40–46.
204. Zhou, Z.; Bhamare, S.; Qian, D. Ductile Fracture in Thin Sheet Metals: A FEM Study of the Sandia Fracture Challenge Problem Based on the Gurson–Tvergaard–Needleman Fracture Model. *Int. J. Fract.* **2014**, *186*, 185–200. [CrossRef]
205. Chen, Z.; Dai, Y.; Liu, Y. Life Prediction of Corrosion-Fatigue Based on a New Crack Growth Rate Model with Damage and the Extended Finite Element Method. *Eng. Fract. Mech.* **2023**, *289*, 109445. [CrossRef]
206. Huang, Y.; Ye, X.; Hu, B.; Chen, L. Equivalent Crack Size Model for Pre-Corrosion Fatigue Life Prediction of Aluminum Alloy 7075-T6. *Int. J. Fatigue* **2016**, *88*, 217–226. [CrossRef]
207. Hu, P.; Meng, Q.; Hu, W.; Shen, F.; Zhan, Z.; Sun, L. A Continuum Damage Mechanics Approach Coupled with an Improved Pit Evolution Model for the Corrosion Fatigue of Aluminum Alloy. *Corros. Sci.* **2016**, *113*, 78–90. [CrossRef]
208. Safyari, M.; Khossossi, N.; Meisel, T.; Dey, P.; Prohaska, T.; Moshtaghi, M. New Insights into Hydrogen Trapping and Embrittlement in High Strength Aluminum Alloys. *Corros. Sci.* **2023**, *223*, 111453. [CrossRef]
209. Gholami, M.D.; Hashemi, R.; Sedighi, M. The Effect of Temperature on the Mechanical Properties and Forming Limit Diagram of Aluminum Strips Fabricated by Accumulative Roll Bonding Process. *J. Mater. Res. Technol.* **2020**, *9*, 1831–1846. [CrossRef]
210. Mouritz, A.P. 19—Fracture Toughness Properties of Aerospace Materials. In *Introduction to Aerospace Materials*; Mouritz, A.P., Ed.; Woodhead Publishing: Sawston, UK, 2012; pp. 454–468. ISBN 9781855739468.

metals

MDPI

Review

The Corrosion Behavior in Different Environments of Austenitic Stainless Steels Subjected to Thermochemical Surface Treatments at Low Temperatures: An Overview

Francesca Borgioli

Department of Industrial Engineering, University of Florence, via S. Marta 3, 50139 Florence, Italy; francesca.borgioli@unifi.it; Tel.: +39-055-275-8734

Abstract: Low-temperature thermochemical treatments are particularly suitable for use in the surface hardening of austenitic stainless steels without impairing their corrosion resistance. In fact, when using treatment media rich in nitrogen and/or carbon at relatively low temperatures (<450 °C for nitriding, <550 °C for carburizing), it is possible to inhibit the formation of chromium compounds and obtain modified surface layers that consist mainly of a supersaturated solid solution, known as expanded austenite or S-phase. It has been observed that this hard phase allows the enhancement of corrosion resistance in chloride-ion-containing solutions, while the results were contradictory for chloride-free acidic solutions. This overview aims to discuss the corrosion behavior of low-temperature-treated austenitic stainless steels, taking into account the different microstructures and phase compositions of the modified layers, as well as the different test environments and conditions. In particular, the corrosion behavior in both chloride-ion-containing solutions and chloride-free solutions (sulfuric acid, sulfate and borate solutions) is discussed. The analysis of the international literature presents evidence that the microstructure and phase composition of the modified layers have key roles in corrosion resistance, especially in sulfuric acid solutions.

Keywords: stainless steels; low-temperature treatments; nitriding; carburizing; expanded austenite; S-phase; corrosion resistance

Citation: Borgioli, F. The Corrosion Behavior in Different Environments of Austenitic Stainless Steels Subjected to Thermochemical Surface Treatments at Low Temperatures: An Overview. *Metals* **2023**, *13*, 776. https://doi.org/10.3390/met13040776

Academic Editors: Nong Gao and Frank Czerwinski

Received: 28 February 2023
Revised: 30 March 2023
Accepted: 13 April 2023
Published: 15 April 2023

1. Introduction

Stainless steels owe their name to their capability to remain without apparent stain, i.e., without rust traces, in many environments due to their excellent resistance to general corrosion. The key element allowing to improve the corrosion resistance of these iron (Fe)-based alloys is chromium (Cr). Since the 19th century, it has been noted that Fe-Cr alloys have higher corrosion resistance in various environments in comparison with carbon steels, and that this resistance increases as the Cr content grows. However, the discovery of the first alloys, which nowadays are called stainless steels, can be dated to the period of 1905–1912 [1]. In stainless steels, Cr, added in a content of at least 10.5 wt.%, promotes the formation of a passive surface film. This is compact, adherent and self-healing if broken in the presence of oxygen, and it protects the substrate from corrosive environments [2,3]. Even if Cr oxide plays a fundamental role in the passive film, the structure and composition of the film are fairly complex and depend on the alloy elements of the stainless steel [4,5] and the conditions in which the film forms (for example, in air, in passivating solutions, with electrochemical treatments) [5–7]. Additionally, they may change when the steel is put in contact with the corrosive environment [8,9]. Therefore, the conditions over which passivity is maintained depend on both the stainless steel itself (composition, surface finishing) and the environment [3]. The passive behavior allows stainless steels to have very low rates of uniform corrosion in many environments [3,10]. The protective effect of the passive film may cease when the environment is too aggressive to allow for the maintenance of passivity. This fact occurs when the environment is too reducing, meaning that the formation of passive species is inhibited (dissolution of the film in the active state), or too oxidizing, with the result that the oxidized species become not stable (transpassive dissolution of the film) [3,10]. As a consequence, general (uniform) corrosion may be observed when the alloys are put in contact with aqueous solutions of acids as hydrochloric acid (HCl) or sulfuric acid (H_2SO_4). Furthermore, the passive film may be locally broken and not able to reform, as occurs when halide ions are present or the sensitization of the steel occurs, and therefore localized corrosion phenomena are produced as pitting, crevices and intergranular corrosion [3,4,10]. Pitting and crevice corrosion can be particularly harmful since, even if the weight loss caused is usually negligible, corrosion may penetrate deeply into a steel component, prejudicing its structural integrity. Chloride-ions (Cl^-) are the most common causes of pitting attack due to their role in weakening the stability of passive films and decreasing their protective effect [3,10–12]. The combination of a Cl^- environment and tensile stress may also cause stress corrosion cracking phenomena [3,10]. Intergranular corrosion usually occurs when Cr compounds, as Cr carbides, precipitate at grain boundaries as a consequence of a thermal treatment, an event which causes so-called sensitization. The regions adjacent to the precipitate are depleted in Cr up to values lower than the passivity limit, and they become susceptible to corrosion phenomena at grain boundaries in aggressive environments [3,10].

Besides Cr, which stabilizes the body-centered cubic (b.c.c.) structure of α-Fe, or ferrite, other alloy elements may be added to steel. These include molybdenum (Mo), silicon (Si), niobium (Nb) and titanium (Ti), which are all ferrite-stabilizers, as well as nickel (Ni), manganese (Mn), carbon (C) and nitrogen (N), that stabilize the face-centered cubic (f.c.c.) structure of γ-Fe, austenite [2]. As a consequence, ferrite, austenite and martensite lattices and their related microstructures can be obtained, and the stainless steels are usually classified on the basis of these microstructures: ferritic, austenitic, martensitic, duplex (ferritic-austenitic) and precipitation hardening [2]. As recalled previously, alloy elements influence not only the mechanical properties and the workability of the different grades, but also the chemical composition, stability and thickness of the passive film, and hence the corrosion resistance [4].

Stainless steels are employed in many fields, from mild environments, such as for architectural applications, furniture and kitchenware, to the more severe ones, as components for chemical and power engineering industries [2,3,10]. Among the different steel grades, austenitic stainless steels have high corrosion resistance. This has resulted in them becoming the material of choice for many applications as components obtained by both conventional manufacturing techniques and additive manufacturing processes [10,13]. The use of these alloys can be extended further by improving their low surface hardness, poor tribological properties or their resistance to localized corrosion. Among the different surface engineering strategies, in the last years there has been an increasing interest in the so-called low-temperature thermochemical treatments, that have been the subjects of intensive research. Nitriding, carburizing or nitrocarburizing processes, which are usually applied to low-alloy steels or tool steels in order to obtain a N and/or C surface alloying by means of diffusion, cannot be transferred directly to stainless steels. In fact, when the same temperatures employed for the nitriding (500–550 °C) [14] or carburizing (850–980 °C) [14] of low-alloy steels and tool steels are used for stainless steels, Cr forms hard compounds (nitrides, carbides); therefore, the Cr-depleted matrix cannot maintain a uniform and protective passive film, and corrosion can easily occur [4,15]. However, by using temperatures so low that interstitial atoms (N, C) can easily diffuse while significant Cr (substitutional) diffusion cannot occur, the formation of Cr compounds is inhibited. In these so-called para-equilibrium conditions [16], N and C atoms are retained in solid solution in austenite lattice above the solubility limit and thus a supersaturated solid solution, known as "expanded austenite", forms.

The formation of modified surface layers consisting mainly of expanded austenite is known to increase surface hardness, improve tribological properties and enhance fatigue resistance. Regarding corrosion behavior, it has been reported that the modified layers allow to maintain or even increase the corrosion resistance of austenitic stainless steels in Cl^--containing solutions. On the contrary, the studies on the corrosion resistance of low-temperature-treated austenitic stainless steels in Cl-free aqueous environments, such as sulfuric acid solutions, have given contradictory results. Low-temperature treatments have not only been performed on massive specimens, but also on coatings obtained with various techniques [17–19]. Recently, they have also been performed on powders that potentially are usable for thermal spraying and additive manufacturing [20,21]. From the first studies of the 1980s up to today, more than 1000 papers have been published on low-temperature treatments of stainless steels, with most of them concerning the treatment of austenitic stainless steels. Likewise, extended reviews [22–26] and briefs overviews have been produced on general [27–29] or on specific topics [30–32] regarded this steel grade, which has also been discussed in reviews regarding all the different stainless steel grades [33–35]. Although the effects of low-temperature treatments on the corrosion behavior of austenitic stainless steels are of paramount importance for the usability of these alloys and are the subjects of many studies, this topic has usually only been reported as a part of general reviews [22,24–27,33,36], being discussed without an in-depth analysis of the various factors which contribute to causing the success or the failure of the modified surface layers in a potential corrosive environment.

The aim of the present review is to report on and discuss the major studies on the characteristics of low-temperature-treated austenitic stainless steels and their corrosion behavior in different environments, in order to highlight the conditions which allow to maintain or even improve the corrosion resistance of the parent alloys. The review is organized into two main sections. The first section is devoted to discussing the solid solution of interstitial atoms in Fe and the formation of expanded austenite, as well as the effects of N and C on the passive film and corrosion resistance. The second section considers the characteristics of the modified surface layers which are formed in austenitic stainless steels and summarizes the corrosion behavior studies carried out in aqueous environments with and without Cl^- using different test conditions and methodologies, including electrochemical impedance spectroscopy (EIS) analysis, potentiodynamic and cyclic potentiodynamic polarization tests, galvanostatic tests, and immersion tests.

2. N and C Alloying of Fe and Its Effects on Corrosion Behavior

2.1. Interstitial Solid Solutions in Fe and Formation of Expanded Austenite

N and C atoms in Fe-based solid solutions are very effective strengtheners. In f.c.c. austenite interstitial solute atoms are usually accommodated in the larger octahedral sites (Figure 1a). According to Fe-N and Fe-C phase diagrams, the maximum solubility of N in f.c.c. γ-Fe is 10.3 at.% at 650 °C [37], while the maximum solubility of C is 9.23 at.% at 1147 °C [38].

Figure 1. Sketch of the f.c.c. lattice with the octahedral interstitial sites (red spheres) (**a**) and the tetrahedral interstitial sites (black spheres) (**b**).

The atomic radius of N is smaller than that of C, but the effective atom size of N in solid solutions is larger due to the influence on the density of electronic states at the Fermi level, which results in the f.c.c. lattice having larger distortion [39]. Moreover, in Fe-N solid solutions, a tendency of short-range ordering of N atoms is observed; conversely, in Fe-C solid solutions there is a tendency towards the clustering of C atoms [39]. In stainless steels, the precipitation of Cr carbides, which causes the so-called sensitization, is fairly easy and can be considered to be a consequence of the clustering of C atoms and the C-assisted clustering of Cr atoms. Instead, N atoms tend to hinder the clustering of Cr atoms and precipitation of Cr nitrides and shift the sensitization curve in the time-temperature-transformation diagram to the right of the time scale. As a consequence, in an austenitic stainless steel such as AISI 316, the C content is maintained at a level lower than 0.015 at.% to avoid carbide precipitation [40], while the equilibrium solubility of N in austenite is larger, being less than 0.65 at.% [22].

When low-temperature thermochemical treatments are carried out, the (interstitial) diffusion of N and C is several orders of magnitude higher than that of substitutional atoms (Cr, Mn, Mo, Ni) that remain fairly "immobile" in the lattice [41]. Thus, the formation of nitrides or carbides tends to be inhibited, even if it should be favored from a thermodynamic point of view, and the interstitial atoms are retained in solid solution with a content that can be up to several hundred times higher than the level of equilibrium solubility, producing the so-called "colossal" supersaturation [40,42]. As a consequence, a significant lattice expansion of the original Fe-based lattice is observed, and this is particularly high when N is solubilized due to the fact that up to about 0.61 occupancy of the octahedral interstitial sites (equivalent to the number of interstitial atoms per metal atom in a f.c.c. lattice) may be obtained [23]. As the N content increases, it does not become clear whether the N atoms are hosted only in the octahedral sites (Figure 1a), so that there is a tendency to a more ordered structure as the NaCl-type structure of γ'''-FeN, or if the tetrahedral sites are also occupied (Figure 1b), so that the ZnS type of γ''-FeN tends to be produced [33]. In austenitic stainless steels, a maximum N content of ~38 at.% [42] is observed in the so-called N-rich expanded austenite, while the maximum C content in C-rich expanded austenite is ~19 at.% [43].

2.2. Effects of N and C on the Characteristics of the Passive Film

As mentioned before, the passive film has a key role in allowing the very good corrosion resistance of stainless steels in many environments owing to its capability of slowing corrosion (dissolution) reactions by many orders of magnitude. However, once formed, the passive film cannot be considered as a rigid layer, but rather a dynamic system capable of adjusting its composition and thickness as a consequence of the change in environmental conditions [44].

The passive film is enriched in Cr(III) oxide [8,45–47], but the amount depends on the passivation conditions. As recently summarized by Maurice and Marcus [47], the dissolution rate of Cr(III) oxide is smaller than that of Fe(II)/Fe(III) oxides in acid aqueous environments, meaning that a strong Cr enrichment occurs in the passive range. In the transpassive range, Cr dissolves as Cr(VI), decreasing the Cr content. In alkaline solutions, Fe(II)/Fe(III) oxides are less soluble, and Cr enrichment in the passive film decreases. The thickness of the passive film is typically about 1–3 nm [8,47].

In situ and ex situ analysis of the passive film points out that it has usually a complex structure. The outer region, at the film–electrolyte interface, is rich in Fe oxides and hydroxides, and the inner region, in contact with the metallic substrate, is rich in Cr oxide [45,46,48] (Figure 2a).

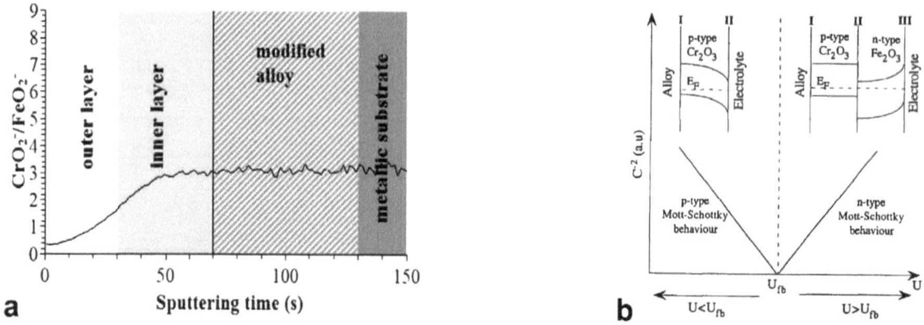

Figure 2. (**a**) CrO_2^- to FeO_2^- ratio in an air grown passive film (For further details, see Ref. [45]) (Reprinted from [45]. Reproduced under the terms of the CC-BY license. Copyright © 2019 Wang, Paschalidou, Seyeux, Zanna, Maurice and Marcus. Published by Frontiers Media S.A.). (**b**) Mott–Schottky plot and schematic representation of the electronic structure model of passive film (For further details, see Ref. [46]) (Reprinted with permission from Ref. [46]. Copyright 1995 Elsevier).

Surface alloying with N and/or C does not significantly change the structure of the passive film. However, these interstitial atoms remain "entrapped" in the film and contribute to the mitigation of corrosion phenomena. When N-rich expanded austenite (treated steel: 1Cr18Ni9Ti) is put in contact with a 3 wt.% NaCl solution, the passive film has an outer region rich in Fe hydroxide/oxide, which also has minor N content, and an inner Cr-rich hydroxide/oxide region in which N significantly increases, as can be detected by Auger electron spectroscopy (AES) and X-ray photoelectron spectroscopy (XPS) [49]. A similar structure, with an outer part enriched in Fe and the inner part enriched in Cr together with N, is observed when the passive layer formed in a borate buffer solution [50,51]. XPS analysis shows the presence of Cr-N and Fe-N bonds within the passive layer [49,50], and it is hypothesized that they are related to an $(Fe,Cr)_3(N,O)_4$ spinel present in the inner part of the passive layer [51]. High-resolution transmission electron microscopy (TEM) studies of the passive film, which forms on N-rich expanded austenite in a borate buffer solution, show that the film evolves when in contact with the solution. In particular, there is a change in the composition of the passive layer, and an increase in Cr and N content in the inner part of the film is observed after 3 h [51].

The analysis of passive films forming on C-rich expanded austenite by means of grazing incidence XPS shows that the film maintains a similar structure, with an inner layer rich in Cr-oxides and an outer layer in which comparable amounts of Fe and Cr are present [52]. Regarding C, a very high C content is detected at the surface, which is probably due to carbon soot, carbon contamination and carbon presence in the oxide film. At about 0.5 nm depth, C concentration markedly decreases after which it increases in the inner part of the passive film.

The passive films of stainless steels have semiconducting properties as their oxide layers are nonstoichiometric. By using Mott–Schottky analysis, it is observed that the outer Fe-rich oxide region has an *n*-type semiconductivity, while the inner Cr-rich oxide region has a *p*-type semiconductivity [46]. The *n*-type semiconductivity, caused by the presence of oxides such as Fe_2O_3, α, β and γ FeO(OH), MoO_3, is developed either by transport through interstitial diffusion or by anion diffusion inward toward the metal. Conversely, the *p*-type semiconductive behavior, caused by the presence of oxides such as Cr_2O_3, $FeCr_2O_4$, MoO_2, occurs owing to a deficiency in metal ions or excess of cation vacancies [53] (Figure 2b).

The N alloying influences this semiconductive behavior, and a decrease in donor and acceptor density is observed in AISI 316LN steel as the N content increases [53]. When a larger N content is present, such as for N-rich expanded austenite, a further significant reduction in the donor and acceptor density, up to about two orders of magnitude, is registered and a decrease in flat band potential, at which the transition from a *p*-type into an *n*-type semiconductive behavior occurs, is reported [50,54]. A similar effect is observed for C-rich expanded austenite, even if it is smaller [55,56]. This fact is particularly important, especially when taking into account that the dominant donor species for the outer *n*-type oxides is oxygen vacancy [55,57]. The reduction in the charge carrier density in the passive film decreases the conductivity of the film itself, and thus a better stability may be hypothesized [55,58].

2.3. *Effect of N and C on Corrosion Behavior*

2.3.1. Nitrogen

N is added as an alloy element in austenitic stainless steels [3,59]. It has been proposed that a N-alloyed austenitic stainless steel be considered "high N" when N content is higher than 0.4 wt.% [59,60]. In austenitic stainless steels, N is used as a small addition in traditional steels, as occurs in AISI 316LN (~0.1–0.16 wt.%), and in larger amounts in high-N austenitic stainless steels and Ni-free austenitic stainless steels (0.45–1.1 wt.%), in which it is the substitute of Ni together with Mn [60,61].

Besides its ability to improve tensile properties [59,60,62], N also has a remarkable effect on corrosion behavior [4,59,60,62]. A beneficial effect is recognized when localized corrosion phenomena occur [4,59,60]. N is able to promote passivity, widen the passive range, and enhance the resistance to stress corrosion cracking and to intergranular corrosion [4,60,63–65]. The protection mechanism is still not completely clear, and different theories have been proposed. The primary suggestions are the following.

1. When interstitial N atoms are released during the early stages of the corrosion process, they may react with H^+ and form ammonium (NH_4^+) ions, following the subsequent reaction [65]:

$$N + 4H^+ + 3e^- \rightarrow NH_4^+ \quad (1)$$

   The formation of ammonium ions consumes protons, and then it locally increases the pH value in incipient pits or crevices and promotes repassivation [63,65–69]. At fairly high potential, nitrates (NO_3^-) are reported to form [63,69] (Figure 3), and is hypothesized that they are adsorbed at the surface of the passive film and may act as inhibitors [70]. Moreover, it is suggested that ammonium ions may form nitrates [66,69], that they may react with free chlorine in chlorinated water and form species which are less effective oxidants [71], or they may produce a passive ammonium sulfate layer in sulfate solutions [72].
2. Negatively charged N ($N^{\delta-}$) may accumulate at the passive layer and may have a repulsive action towards Cl^-, promoting a faster repassivation [68,73]. It is also suggested that $N^{\delta-}$ may change into ammonium ions [73].
3. During passivation, an enrichment of N occurs at the passive film/substrate interface, preventing anion attack [53,67]. An incorporation of N and ammonium ions in the passive film is also observed [53].
4. First-principles calculations suggest that interstitial N causes a decrease in the electronic density of states (DOS) of Fe at and near the Fermi level, and thus that it has a stabilization effect on the electronic structure and tends to suppress electrochemical reactivity [69].

When N-rich expanded austenite forms, the local alkalization of the solution according to (1) is the main mechanism invoked to justify the increase in localized corrosion resistance [22,74,75]. The formation of NH_3 and NH_4^+ species is observed on the outermost surface of the passive layer formed on N-rich expanded austenite subjected to anodic polarization in a 3 wt.% NaCl solution [49]. Using XPS analysis, Lei and Zhu [49] detect N, which was bonded with Cr and Fe, in the passive film and hypothesize that, during the early stages of anodic polarization, N is released onto the passivating surface while stable oxides form, and thus it is able to form NH_3. It is supposed that, as N is released, two-step N hydrolysis to NH_4^+ may occur. In the first step, N hydrolyzes to NH_3 with a very fast reaction rate, which is affected by the concentration of released N and not by the pH value of the solution, and in the second step NH_3 further hydrolyzes to NH_4^+ with a slow reaction rate, one which becomes faster as the pH of the solution lowers [76].

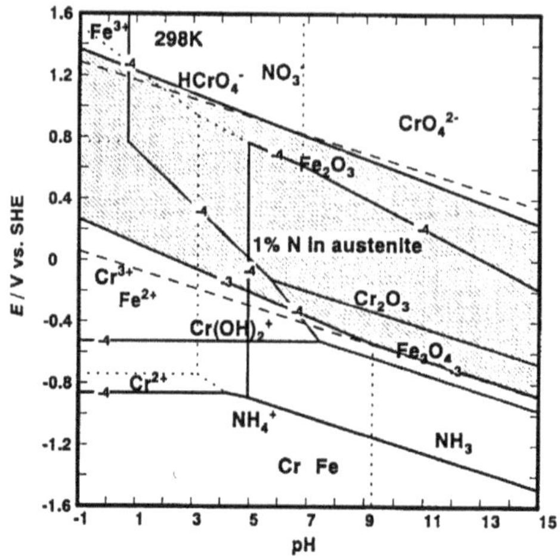

Figure 3. E–pH diagram for interstitial N, Fe and Cr. (For further details, see Ref. [63]) (Reprinted with permission from Ref. [63]. Copyright 2002 Elsevier).

Flis-Kabulska et al. [75] observe a local increase in pH due to N using in situ measurements for plasma-nitrided AISI 316L tested in deaerated solutions of 0.1 M Na_2SO_4 + 0.4 M NaCl acidified down to pH 2.4. The authors register an increase in pH at the surface of the nitrided samples and ascribe this alkalization to the formation of ammonium ions. It is hypothesized that the increase in corrosion resistance is related to a more efficient repassivation of incipient pits, which is occurring due in turn to the formation of oxides, promoted by the local alkalization, and to the formation of inhibiting NO_2^- anions, rather than due to a slow pit nucleation process.

It is also hypothesized that the strong bond between Cr and N may play a role in this process. In fact, as recalled before, in the passive film Cr and N enrichment is observed in the inner region, and thus the strong Cr-N interaction has the potential to restrict Cr migration [51].

A further effect may be due to the change in the semiconducting properties of the passive film formed on N-rich expanded austenite. The lower donor density in the outer *n*-type region of the passive layer, consisting mainly of Fe oxide/hydroxide, may decrease the pitting nucleation ability, hindering the Cl^- adsorption [54], and it allows to obtain more protective film [55,57].

Taking into account all the experimental results, it may be hypothesized that both the local alkalization of the solution and the changes in the characteristics of the passive film have roles in mitigating the corrosion phenomena.

When general corrosion is taken into account, both positive [59,77,78] and negative [60,79] effects of N alloying are reported.

2.3.2. Carbon

C improves the mechanical properties of stainless steels. However, from a corrosion resistance point of view, its use is considered detrimental [3,4]. Low C content is used in order to stabilize the passive layer and reduce the risk of sensitization [3,4].

The formation of C-rich expanded austenite highlights that C is able to improve resistance to localized corrosion, in particular increasing the corrosion and pitting potential. Different mechanisms are hypothesized as explaining this beneficial effect. A first hypothesis by Martin et al. [80] suppose that C atoms present in the passive film act as mobile interstitials and enhance/maintain the oxygen vacancy motion. Heuer et al. [52] suggest that the improvement in corrosion resistance is related to the lower thickness of the passive film at a given potential, thus reducing the detrimental effect of the thickness undulation of the oxide film, which can cause dielectric breakdown and pitting nucleation. A recent hypothesis suggests that the carburized layer decreases the mobility of charge carriers in the passive film, since a decrease of the oxygen vacancy concentration occurs as the C content is higher, which is due to the strong Cr-C bonds forming at the metal-passive film interface [81]. This hypothesis is in accordance with the lower donor density observable in carburized specimens, which in turn is related to a decrease in the density of oxygen vacancy in the passive film and thus to a more stable and protective film [55,56]. Li et al. [82] suggest that C may weaken the passive film, but that, owing to the formation of localized covalent bonds with the metal atoms, it tends to immobilize metal atoms and thus metal dissolution becomes more difficult.

3. Austenitic Stainless Steels

The formation of an expanded phase was firstly recognized in austenitic stainless steels, as outlined by Borgioli [24] for low-temperature nitriding and by Christiansen and Somers [29] for carburizing treatments. Today, the most widely used names for the expanded phase formed in this stainless steel type are "expanded austenite", "S-phase" (with reference to the shift in its diffraction peaks in comparison with those of the untreated austenite), γ_N for the N-rich phase and γ_C for the C-rich phase (for all the used names, see [24] and the reference therein). The main characteristics of N- and C-rich expanded austenite, formed by means of nitriding or carburizing, are reported in Table 1.

Table 1. Main characteristics of the modified surface layers containing expanded austenite, obtained by nitriding or carburizing austenitic stainless steels.

Properties	Nitriding	Carburizing
Formation temperature (without formation of Cr compounds) (°C)	300–450	300–550
Max. interstitial content in expanded austenite (at.%)	38	19
Max. surface hardness	1450	1000
Hardness profile	abrupt change	gradual change

3.1. Microstructure and Characteristics of the Modified Surface Layers

The "colossal" supersaturation of interstitial atoms in the austenite lattice has an initial important effect in that it produces localized plastic deformations, which are observable starting from the surface. Slip bands are present inside the grains, crossing them in different directions depending on crystallographic orientations. Subsequently, a swelling and a rotation of the grains occur, causing the presence of reliefs at grain boundaries and the leaning forward of some grain boundaries onto adjacent grains [83–87] (Figure 4a). This peculiar surface morphology was observed in samples treated with both plasma- [83–86,88] and gas- [89,90] based processes, and on nitrided [83–86,88,89], carburized [91–93] and nitrocarburized [94] samples. As a consequence, even when starting from polished untreated samples, a surface roughness increase is obtained after the treatment [85].

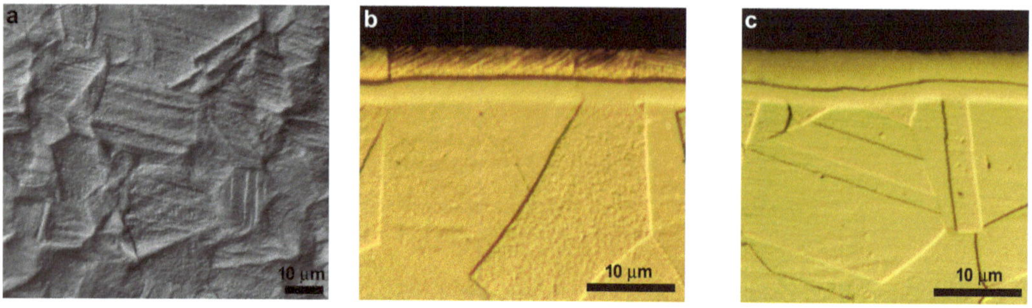

Figure 4. Surface morphology (**a**) and cross-sectional micrographs of a nitrided AISI 202 (**b**) and of nitrided AISI 316L (**c**) austenitic stainless steels (etchant: acetic glyceregia). (Nitriding conditions: 380 °C, 340 Pa, 5 h) (For further experimental details, see Ref. [95]).

Further interesting features can be observed at the cross sections and tapered sections of the treated samples. The high corrosion resistance of the modified surface layer consisting of expanded austenite can also be inferred by the fact that some of the chemical etchants usually employed for delineating the microstructure of stainless steels, such as Marble's reagent are not able to etch onto it [96,97]. The presence of a continuous and fairly homogeneous but almost featureless layer, separated from the matrix by a strong etched line, suggested that the "new" phase could be a nitride [24]. However, when using etchants such as glyceregia, additional features can be observed. When nitriding [95,98–103] and nitrocarburizing [28,94,104] are performed, the modified layer consists of a two-layer microstructure; conversely, in carburized samples, only one layer is present [28,55,105–107]. It was hypothesized that the strong etched lines, which separated the two modified layers and the substrate, were due to the sudden change in N concentration and as a response to chemical etching, and that they were not thermodynamic or crystallographic interfaces [102]. Grain boundaries were observable in the modified layer, and they were the continuation of those found in the austenite matrix [83]. In the two-layer microstructure, thin lines were observed in the outer layer, and it was suggested that they were slip lines occurring as the consequence of local plastic deformations induced by the solubilization of a large amount of N [108]. The number and the extension in the substrate of these slip lines depend on the nitriding conditions and steel composition [95,99,108]. As an example, in Figure 4b, the cross-sectional microstructure of the low-Ni CrMn AISI 202 steel, nitrided at 380 °C for 3 h, is depicted; as a comparison, the microstructure of the CrNi-grade AISI 316L, nitrided in the same conditions, is shown in Figure 4c.

The main constituent of the modified layer is expanded austenite, which is usually regarded as a supersaturated solid solution of interstitial atoms in the f.c.c. austenite lattice. The X-ray diffraction patterns of this phase have a peculiar feature since the peaks are shifted towards lower angles in comparison with those of austenite, with an apparently anisotropic expansion [24,28,33,84,109] (Figure 5). The nature of this expanded phase is still an open question. Many studies carried out with different analytical techniques suggested that a preferential bonding of N with Cr atoms occurred [110–112] and that Cr-N short-range order (SRO) regions were present together with regions having a γ'-Fe_4N-like long-range order (LRO) region, indicated as γ_N' [113–116]. Further details on the structure of the expanded austenite are reported in [22,24,33].

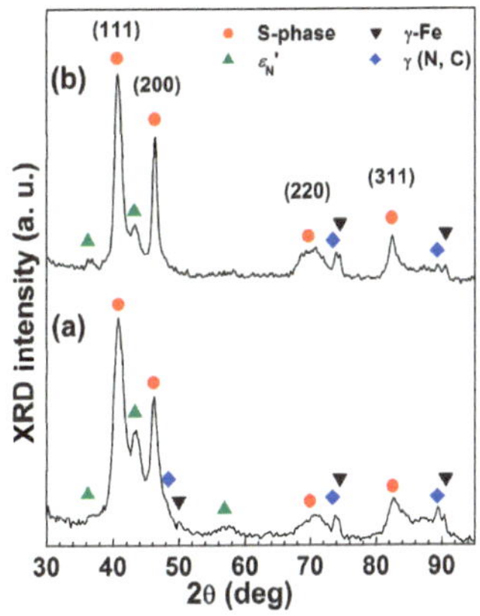

Figure 5. X-ray diffraction patterns of nitrided AISI 202 (**a**) and AISI 316L (**b**) austenitic stainless steels (Nitriding conditions: 380 °C, 340 Pa, 5 h) (For further experimental details, see Ref. [95]).

Together with expanded austenite, another N-rich phase is detected in the outer modified layer of nitrided samples, the so-called N-induced h.c.p. martensite, ε_N', which is a solid solution of N in the h.c.p. martensite, ε' [117]. The formation of this phase was hypothesized to be related to the local plastic deformations. Tao et al. [118] suggested that ε_N' formed from expanded austenite through a martensitic shear transformation which occurred above a critical N concentration. On the contrary, Tong et al. [119] ascribed these h.c.p. regions to "clustered" stacking faults instead of plates of N-rich h.c.p. ε' martensite. The formation of ε_N' depends on both solubilized N content and steel composition, and is promoted in steels with a low stacking fault energy, as well as in those with a high Mn content [88,95,99]. As an example, in Figure 5 the X-ray diffraction patterns of high-Mn low-Ni AISI 202 and low-Mn CrNi AISI 316L austenitic stainless steels, which were nitrided at 380 °C for 3 h, are depicted. The ε_N' phase can also be the precursor to h.c.p. ε-$Fe_{2\text{-}3}N$ nitride, which may have formed from it as a result of the ordered arrangement of N atoms and the distortion of the lattice [120].

The formation of ε_N' and the γ_N' LRO phases, which can be regarded as the precursors of ε and γ' Fe-nitrides, is not peculiar to nitrided samples, as these phases were also detected on nitrocarburized specimens [121].

The presence of an inner modified layer is assessed well when the nitrocarburizing treatment is performed, and it is ascribed to the formation of a C-rich expanded austenite [28,94,104,122]. When nitrided samples were analyzed, an inner layer was not detected by all the authors. Borgioli et al. [108] observed an inner modified layer on AISI 202 samples which had been glow-discharge-nitrided at 380 °C. By performing X-ray diffraction analysis at different incident angles, the authors assessed that this layer could be related to the presence of a f.c.c. phase as it had a lattice parameter slightly larger than that of austenite. The formation of this phase, which can be regarded as a solid solution of interstitial atoms (N, C) in austenite, γ(N, C), was observed both in steels having both a fairly high C content, such as 202 and Ni-free steel, and a low C content, such as AISI 316L [95,108,123–125], so that the hypothesis that this is only related to a local increase in C atoms [126], pushed ahead by diffusing N atoms, is not completely convincing. Williamson et al. [98] suggested that it may be due to the high residual stress, induced between the expanded austenite layer and the substrate, while Christiansen et al. [102] hypothesized that it may be related to the N concentration profile. Recently, in 316Ti steel subjected to N low-energy ion implantation in order to study the initial phase of nitriding, Manova and Mändl [127] observed the formation of a low-expansion f.c.c. phase, that subsequently transformed into expanded austenite when the N content increased.

Nitrides are usually observable in the outer modified layer as strong etched zones; these are present especially along the slip lines and grain boundaries. When CrN is able to form, the loss of solubilized Cr from the matrix has the potential to cause a transformation of f.c.c. austenite into b.c.c. α-Fe [128]. The maximum temperature at which Cr-compounds do not form is usually 450 °C for nitriding [27,60] and 550 °C for carburizing [27]. However, the temperature, at which nitride precipitation occurs, depends on steel composition, and tends to decrease as the Mn-content increases and Ni-content decreases [99,123,129,130]. Moreover, the nitride precipitation also depends on the treatment duration and N feeding [84,123,125,129,131]. Similarly, C solubilization and carbide formation also depend on alloy composition [93,107].

The concentration profile of interstitial atoms in the modified layers depends on the treatment used. In nitrided samples, N concentration has high values at the surface, a steep decrease to lower values, followed by a nearly constant plateau, and then a steep decrease to matrix values [22,24,28,102]. In carburized samples, C concentration decreases steeply near the surface and then more smoothly to the matrix values [22,24,28,55]. In nitrocarburized samples, the formation of an outer layer consisting of N-rich expanded austenite and an inner layer, in which C-rich expanded austenite is present, produces a smoother profile, as compared to those of nitrided steels [24,28,132]. The maximum N content solubilized in expanded austenite is 38 at.% [42], while the maximum C content is 19 at.% [43].

The huge interstitial atom solubilization causes an increase in surface hardness. A maximum surface hardness of ~1450 $kg_f\ mm^{-2}$ is registered for nitrided steels [131], and a maximum surface hardness of ~1000 $kg_f\ mm^{-2}$ is observed for carburized ones [133]. The microhardness profiles resemble those of interstitial atom concentration. For the nitrided samples, the microhardness profiles usually show an abrupt decrease from the high hardness values near the surface to the matrix values. Conversely, for the carburized specimens, a more gradual decrease in hardness values is registered [24,104]. Nitrocarburizing allows to obtain high microhardness values at the surface together with a smoother hardness profile [24,104].

Further details on the characteristics of the modified layers are summarized in [22,24,26,33].

3.2. Corrosion Behavior in Presence of Cl^-

The presence of Cl^- ions in an aqueous environment may be harmful for austenitic stainless steels. since these ions are able to weaken the protective effect of the passive film and cause localized corrosion phenomena, such as pitting and crevice, and promote stress corrosion cracking when a tensile stress is applied. General corrosion may occur in HCl solutions. The corrosion resistance characteristics of the modified surface layers, obtained on austenitic stainless steels by means of nitriding or carburizing, are summarized in Table 2.

Table 2. Corrosion resistance in Cl^--containing solutions of the modified surface layers obtained on austenitic stainless steels by means of nitriding or carburizing.

Solution	Modified Layer Type	Nitriding	Carburizing
NaCl	- expanded austenite only - with martensite - with nitrides (low amount)	very good very good–good acceptable	good - -
Body fluid-like	- expanded austenite only - with martensite	good good	good -
NaCl + Na_2SO_4	- with martensite/nitrides [1]	good	-
$FeCl_3$	- expanded austenite	good	good
HCl	- expanded austenite	good	good

[1] Hypothesized on the basis of nitriding conditions and X-ray diffraction pattern.

3.2.1. Electrochemical Impedance Spectroscopy (EIS) Analysis

EIS analysis is a powerful tool for investigating the general corrosion resistance of modified surface layers. For stainless steels, the Nyquist plot is characterized by the presence of a single semicircle [134,135] or two connected semicircles [135–138], which are characteristic of one time constant or two time constants, depending on the test conditions. By modeling the experimental data with an electrical equivalent circuit (EEC), it is possible to obtain a physical picture of the corrosion phenomena. The semicircle trend can be modeled using a Randles circuit, i.e., a solution resistance element in series with the parallel combination of a resistance element, R, and capacitance, C. These are identified with the charge transfer resistance and double-layer capacitance [139]. When two connected semicircles are observed, the data can be modeled with two parallel combinations of R and C, either hierarchically connected [78,135,140–142] or in series [143]. When a model with two time constants is required, one constant is usually connected to the charging/discharging processes that take place at the interface between the electrode and the electrolyte, whereas the other is connected to the charge transfer and mass transport processes that occur in the oxide phase. However, their attribution to the low or high time constant is not univocal [78,135,138,141]. The sum of the resistance elements (charge transfer resistance and oxide-related resistance) is usually indicated as polarization resistance and it can be considered a measure of the surface total resistance to general corrosion [138].

When low-temperature nitriding produced modified surface layers consisting essentially of expanded austenite, for tests in NaCl solutions, in the Nyquist plot the semicircle of nitrided samples was usually wider than that observed for untreated alloys, suggesting a better corrosion resistance [54,95,124,144]. When the experimental data were depicted in a Bode plot, higher impedance values were reported and, in the analysis of a phase angle plot, the presence of a second inflection at a lower frequency was registered, meaning that two time constants could be hypothesized [54,95,123,124,145]. As examples, the Nyquist and Bode plots of untreated and nitrided AISI 202 samples are shown in Figure 6.

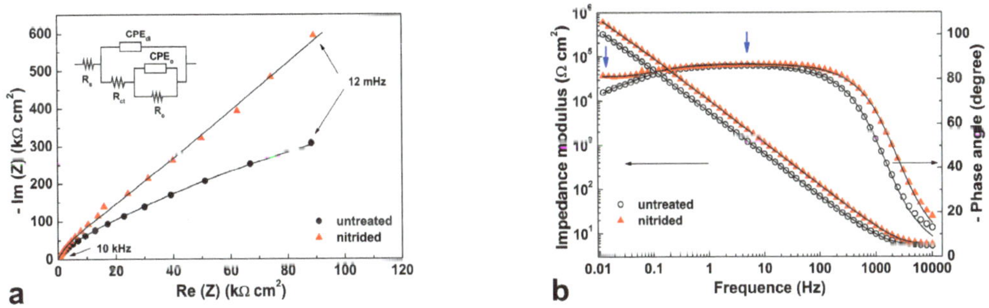

Figure 6. Nyquist plots (**a**) and Bode plots (**b**) of untreated and nitrided (380 °C, 340 Pa, 3 h) AISI 202 samples, recorded at the respective open circuit potentials (symbols: experimental data; lines: modeled data using the equivalent electrical circuit drawn in (**a**)). For the circuit: R_s, electrolyte resistance; R_{ct}, charge transfer resistance, CPE_{dl}, double layer/space charge capacitance, R_o, resistance of the oxide layer, and CPE_o capacitance/mass transfer processes in the oxide layer) (solution: 5 wt.% NaCl, aerated). (The blue arrows in the phase angle plots indicate the two inflections of the plot.) (For further details, see Ref. [95]).

As was observed for the untreated alloys, the attribution of the different processes to the two time constant elements is not univocal. Borgioli et al. [95,123], who tested untreated and nitrided austenitic stainless steel samples in 5 wt.% NaCl solution, fitted the EIS data using two parallel R–C elements, hierarchically connected, and they related the high-frequency (HF) time constant to the faster charging/discharging processes occurring at the electrode/electrolyte interface, while the low-frequency (LF) time constant was linked to the slower processes taking place in the oxide phase. Pure capacitance elements were substituted with constant phase elements (CPE). It was observed that, for the HF time constant, the CPE exponent, n, was usually very close to unity [95,145], so that it was hypothesized that the behavior of the double layer deviated from that of a pure capacitance structure ($n = 1$) due to surface heterogeneities at the atomic (surface disorder as dislocations and steps, chemical inhomogeneities) and micrometric (roughness, polycrystalline structure) scales and due to adsorption phenomena [136,141]. For the LF time constant, a tendency to a significant decrease in CPE exponent was observed in comparison with that of untreated steels [95]. It was usually suggested that this fact may be due to surface heterogeneities [136,141], diffusion phenomena in the oxide film [136], and variation in oxide composition, each of which caused a distribution of resistivity and dielectric constant [146]. A reduction in the capacitive contribution of the oxide film was observed in the nitrided samples [95], as was a marked increase in the resistance of the oxide film, which enhanced corrosion resistance [95,145]. A similar interpretation of the two time constants was given by Zhu et al. [54], who modeled the EIS spectra using a solution resistance element in series with two parallel combinations of R and CPE and ascribed the HF time constant to the charge transfer process at the expanded austenite layer/electrolyte interface and the LF time constant to the occurrence of the diffusion process through the passive film. It has to be noted that for the nitrided steel, with this attribution to the two time constants, the resistance ascribed to the oxide film was higher than that related to the charge transfer resistance at the electrode/electrolyte interface [54,95,123]. However, the overall polarization resistance increased for the nitrided samples, in comparison with the untreated steels [54,95,123].

The formation of low amounts of N-rich h.c.p. martensite, ε_N', together with expanded austenite does not significantly affect the corrosion behavior of nitrided samples. Studies carried out on different austenitic stainless steels (AISI 316L, AISI 202 and Ni-free P558), nitrided at 360 and 380 °C, showed that similar trends of impedance data were observed when samples were tested in 5 wt.% NaCl aerated solution [95]. However, when the ε_N' amount was fairly high and the h.c.p.-based ε-nitride could form, as for P558 Ni-free austenitic stainless steel samples nitrided at 380 °C, three inflections were observed in the phase angle plots, suggesting the presence of three constant phase elements [95,123]. Models with three hierarchically connected time constants were used for inhomogeneous surfaces [147] or porous oxide films [148]. As a result, it was hypothesized that the (1) element was related with a heterogeneous and/or porous oxide layer, that the element (2) with the processes taking place at the electrode/electrolyte interface, and that the (3) element with the processes occurring in the inner compact oxide layer. Nevertheless, a fairly high polarization resistance was observed for these samples.

Luiz et al. [149] studied the change in EIS spectra over a timeframe ranging from 1 h to 30 days for a nitrided UNS S31254 super-austenitic stainless steel, which the authors tested in a 3.5 wt.% NaCl aerated solution. In the Nyquist plot, the semicircles became wider as the test duration reached 7 days. Then, after 15 and 30 days, a slight decrease was observed. The authors used an EEC with two parallel R–C elements hierarchically connected, but related the HF time constant to the passive film formation, the LF time constant to the charge and the mass transfer phenomena, occurring through the passive film, to the double layer. With this model, the higher resistance values were related to the charge transfer resistance, while the passive film resistance had lower values. The polarization resistance was at a maximum for 7 days; then, it slightly decreased, and it was comparable to that of the untreated steel. The estimated passive film capacitance was slightly lower than that of the untreated alloy and did not change significantly, suggesting the formation of a highly stable film which did not undergo relevant dissolution processes.

When significant amounts of nitrides formed, narrower semicircles in Nyquist plots are observed, meaning that a decrease in corrosion resistance may be supposed [123].

Nitrocarburized samples have a behavior similar to that observed for nitrided ones, with wider semicircles in the Nyquist plots and higher impedance values in the Bode plots when mainly expanded austenite is able to form [94].

A higher corrosion resistance is also hypothesized for C-rich expanded austenite, since wider semicircles in the Nyquist plots [56,81] and higher impedance values in the Bode plots [55,56] are observed.

3.2.2. Open Circuit Potential Measurements

Even if there was usually a delay before electrochemical tests were performed to allowing the sample to reach an equilibrium condition in the solution, data regarding open circuit potential (OCP) measurements, which allow researchers to evaluate the behavior in free corroding conditions, were seldom reported.

The OCP of the untreated alloys usually shows sharp decreases and increases due to metastable pit nucleation and repassivation phenomena [30,150]. As the delay duration increases, different trends are observable, showing either increases [131] or decreases [95] in OCP values, as depicted in Figure 7. The increasing trend is typical of passive materials, which may be related to the passivation effect in the solution on freshly polished samples [151]. However, for fairly long delay durations, a decrease is also observed, which may be ascribed to a degradation of the air-formed passive layer due to the absorption of Cl^- [8,130,152,153]. A subsequent potential increase such as that observed by Borgioli et al. [95] may be ascribed to repassivation phenomena.

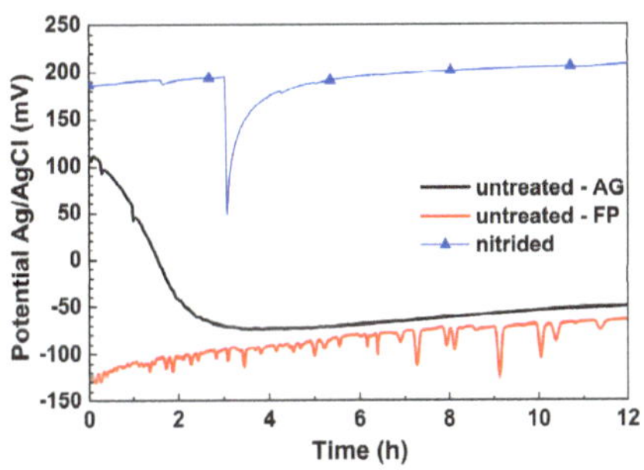

Figure 7. Open circuit potential vs. time for AISI 316L samples: untreated with air-grown passive film (AG) and freshly polished (FP); nitrided at 380 °C for 3 h (air-grown passive film) (for further details on the treatment, see Ref. [95]) (solution: 5 wt.% NaCl, aerated).

When expanded austenite is able to form, an increasing trend of OCP values is registered in NaCl aerated solutions, and the potential values are higher than those of the untreated alloys [95,131,150] (Figure 7). Fluctuations are also observed due to metastable pit formation and repassivation. However, the pit nucleation events are fewer than those of the untreated alloy [131,150], and the pit repassivation kinetic tends to be slower than that of the alloy [95,131,150]. The formation of a considerable amount of nitrides causes a decrease in OCP values, probably due to the occurrence of many corrosion events on the surface [150].

The study of the OCP values up to 30 days for the nitrided UNS S31254 superaustenitic stainless steel tested in a 3.5 wt.% NaCl solution showed that, after a fast increase during the first day, OCP tended to increase more slowly and had a maximum value at 15 days of immersion in the probe solution and then experienced a slight decrease [149]. It was hypothesized that a slow dissolution process of the passive film took place for immersion periods longer than 15 days.

The experimental data present in the literature suggest that equilibrium conditions may be attained by the treated alloys with fairly long durations. This fact should be taken into account when the delay time is chosen before electrochemical tests.

3.2.3. Localized Corrosion Phenomena in NaCl Solutions

NaCl solutions are the most commonly used test solutions to assess the localized corrosion behavior of alloys, and electrochemical techniques are usually employed, such as potentiodynamic polarization, cyclic potentiodynamic polarization, potentiostatic and galvanostatic measurements, together with crevice tests. It has to be taken into account that the different experimental results presented in the literature depend on both the test conditions and the characteristics of the modified layers, and so a direct comparison cannot be straightforwardly performed.

When low-temperature nitriding treatments are carried out and modified surface layers formed, consisting of N-rich expanded austenite, the polarization curves possess the typical features showed by N-containing austenitic stainless steels, as depicted in Figure 8a. When compared to the untreated alloy, the nitrided samples usually have higher corrosion potential, lower anodic current density in the passive branch and higher pitting potential [22,30,54,74,95,129,149,154,155]. As a consequence, after the tests it was found that the surface of the nitrided samples showed fewer pits than the untreated steel, or seemed fairly untouched, with micrometric or submicrometric pits that could be observed only using microscopic techniques [95,108,150] (Figure 8b,c). A slight change in color was sometimes observed, which could be ascribed to the transpassive dissolution of the oxide film, due to the high potential values reached in the test [95]. It has to be recalled that the localized corrosion phenomena produced by potentiodynamic polarization tests may cause both pits and crevices, which may occur in the area shielded by the gasket.

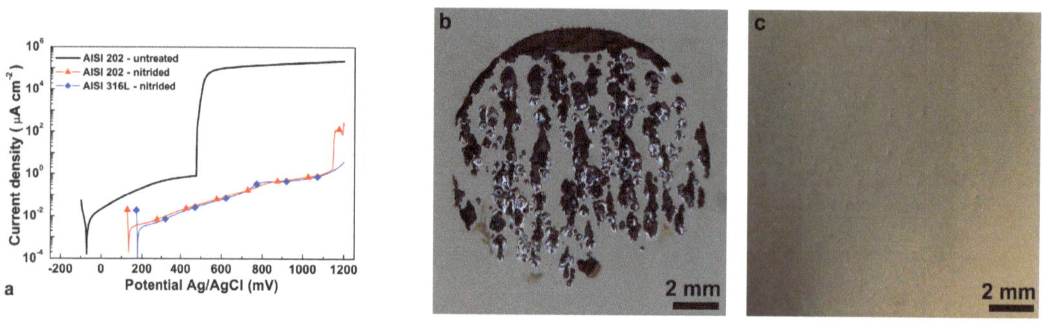

Figure 8. Corrosion behavior of AISI 202 (untreated and nitrided) and AISI 316L (nitrided) (nitriding conditions: 380 °C, 340 Pa, 3 h): polarization curve (**a**), details of the surface morphology after corrosion test of untreated (**b**) and nitrided (**c**) AISI 202 samples (solution: 5 wt.% NaCl, aerated). (For further experimental details, see Ref. [95]).

The protection effect of expanded austenite and its ability in hindering and delaying the occurrence of corrosion phenomena tend to increase as the N content in this phase is higher, the precipitation of significant number of nitrides is avoided, and the modified surface layer is thicker. Thus, treatment temperature, duration and N feeding have to be chosen in order to produce a N-rich, nitride-free and thick modified layers [83,84,95,123,124,129,131,150,156]. It has to be noted that the localized plastic deformations, which cause the increase in surface roughness and the formation of N-induced h.c.p. martensite, ε_N', do not significantly impair the protectiveness of the modified layer. The presence of deformation bands in N-alloyed austenitic stainless steel, present due to cold working conditions, was reported to increase the number of sites for pitting nucleation and reduce corrosion resistance [157]. On the contrary, even in nitrided austenitic stainless steels showing a large amount of localized plastic deformations in the modified surface layers, such as AISI 202, an improvement in corrosion resistance was registered due to expanded austenite [99,108,129]. A similar observation was made by Lei et al. [158], who obtained comparable (high) corrosion resistance in a 1 wt.% NaCl solution for modified layers of nitrided 1Cr18Ni9Ti steel which consisted of expanded austenite with and without ε_N'. Similarly, Tao et al. [130] observed that the formation of N-induced expanded martensite did not have a detrimental influence on the corrosion resistance of the nitrided (400 °C, 20 h) high-Mn Staballoy® AG 17 steel tested in 3.5 wt.% NaCl solution.

A peculiar behavior was registered by Borgioli et al. [95,123] on nitrided Ni-free austenitic stainless steels, either as the treatment temperature increased (380 °C) or at lower temperatures (360 °C) for longer durations (5 h). With these treatment conditions, the modified surface layers had a high density of slip lines, even if no detectable amount of nitride precipitates was observed. These samples had a higher corrosion potential when compared to that of the untreated alloy, as well as lower anodic current density values in the passive branch up to approximately +630 mV (Ag/AgCl). For higher potential values, an increase in anodic current density was registered, up to a maximum at about +960 mV (Ag/AgCl), and then a decrease occurred down to values of a second passive branch, suggesting the occurrence of oxidation phenomena. It was hypothesized that oxidation phenomena might be promoted by the heterogeneous structure formed by expanded austenite with a fairly large amount of h.c.p. phase (ε_N' and/or submicrometric nitride precipitates), which might hinder repassivation. However, a very small anodic current density peak was observed at about the same potential values for other nitrided austenitic stainless steel types, such as AISI 316L [95,129], AISI 304L [99] and AISI 202 [99,129]. On the basis of the Pourbaix diagrams of the main alloy elements [63,159] and taking into account an acidic solution, as the one expected for pitting or crevice phenomena, it was hypothesized that this anodic current peak was related to the oxidation of Cr(III), mainly in form of Cr_2O_3, into Cr(VI).

The effect of inclusions, such as MnS, on the corrosion resistance of the modified surface layers, consisting of N-rich expanded austenite, was recently studied [160]. The inclusions appeared to be pitting initiation sites, but they did not cause pitting or crevice in the expanded austenite. In fact, localized corrosion occurred in the untreated substrate, beneath the modified layer, where the electrolyte was able to penetrate due to the partial dissolution of the inclusions. Thus, as observed before, the protective effect of the expanded austenite was higher when thicker modified layers formed.

The effect of alloy elements on the corrosion behavior of nitrided austenitic stainless steels was also studied [95,99,129,161]. The comparison of the potentiodynamic curves of AISI 316L, AISI 202 and Ni-free P558 steel samples, nitrided at 360 and 380 °C and tested in a 5 wt.% NaCl aerated solution, suggested that the corrosion potential and anodic current density values were influenced mainly by the solubilized N, and that the pitting potential depended on both the N content and alloy elements [95]. In particular, Mo-containing alloys had an improved corrosion resistance after nitriding due to the synergy of Mo and N in hindering corrosion phenomena [95,99,161].

The influence of surface finishing on the corrosion behavior was also assessed [124]. It is well known that a smoother surface has a higher corrosion resistance since the possible sites for pits are more open, preventing the accumulation of Cl^- and H^+, so that metastable pits are less capable of propagating and a higher potential is required to cause a stable pitting [162]. However, surface roughness comparable to that obtained with a 2D finishing (Ra~0.2 μm) maintained a very good corrosion resistance when the samples were subjected to low temperature nitriding [124]. However, it has to be recalled that localized plastic deformations occur when expanded austenite forms, and thus even a very smooth surface of an untreated steel increases its roughness after nitriding, but that the rougher surface does not impair the improvement in corrosion resistance due to N solubilization [95,124].

Nitriding treatment is also able to improve corrosion resistance in combination with a surface mechanical attrition treatment (SMAT). Li and Wang [163] carried out a duplex treatment consisting of a first SMAT step, which produced a nanostructured layer, and then a nitriding treatment (400 °C, 6 h). The duplex treated samples showed an improved corrosion resistance in comparison with the samples subjected to SMAT only.

The ability of the expanded austenite modified surface layers to repassivate is usually investigated using the cyclic potentiodynamic method. Different results were obtained with this technique. The repassivating potential was higher than that of the untreated steel [149], or it was lower [158,164], suggesting that repassivation did not occur, or was not measured [74,158], due to the transpassive dissolution of the passive film. These results may be ascribed to the different extent of the damage to the modified layer caused by this technique, which may hinder repassivation. The galvanostatic technique may be another useful technique for assessing repassivation, or protection, potential [165]. Galvanostatic tests highlight that, after the occurrence of localized corrosion phenomena, potential values below which corrosion did not occur were higher than those of corrosion potential, so that it was hypothesized that repassivation could occur when the depth of pits or crevices was not too large [95].

By using cyclic polarization tests, Zhu and Lei [74] obtained potential–pH diagrams for AISI 1Cr18Ni9Ti steel, untreated and nitrided at 380 °C for 4 h (N concentration in expanded austenite: 32 at.%), and tested in 3 wt.% NaCl solutions with pH values ranging from 0.4 to 13 (Figure 9). The formation of expanded austenite extended the immunity and perfect passivity zones and reduced the imperfect passivity and pitting regions in comparison with the untreated alloy. For pH 0.4–3, an active–passive behavior was observed for both untreated and nitrided specimens, with a change from general corrosion to imperfect passivity–perfect passivity–imperfect passivity–pitting corrosion, even if the pitting potential values of nitrided steel were higher than those of the untreated alloy. For pH 4–13, the untreated steel had the typical passive–transpassive transition due to pitting phenomena, while for nitrided samples the perfect passivity region was widened and pitting was not observed. Above pH 11, both sample types were resistant to pitting corrosion. On the basis of these results, it was hypothesized that the neutralizing effect of NH_4^+ ions in acidic pits was significantly only above a certain range of pH, and, in particular, that there was a significant improvement in pitting corrosion resistance in 3 wt.% NaCl solutions at pH values ranging from 4 to 11.

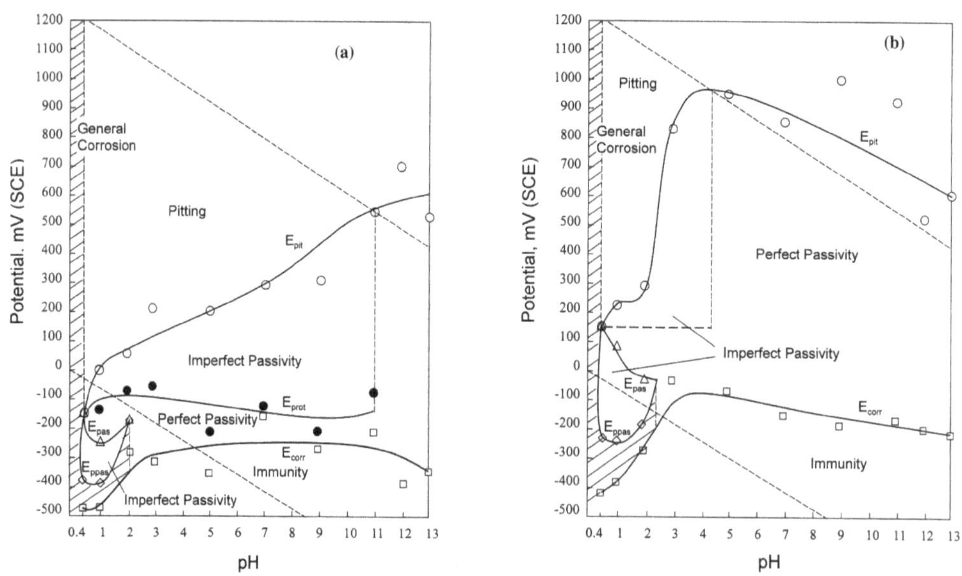

Figure 9. Potential–pH diagrams in 3 wt.% NaCl solutions for 1Cr18Ni9Ti: (**a**) untreated steel; (**b**) N-rich expanded austenite layer (For further details, see Ref. [74]). (Reprinted with permission from Ref. [74]. Copyright 2000 Elsevier).

The crevice tests highlight that low-temperature-nitrided austenitic stainless steels also have an excellent resistance to this type of localized corrosion. AISI 316L samples, plasma-nitrided at 430 and 400 °C for 5 h, did not have any observable crevice corrosion event after 60 days when tested in a 10 wt.% NaCl solution at 55 °C; meanwhile, for the untreated steel, the first corrosion attack occurred after only 3 days [150] (Figure 10). When the samples were nitrided at 500 °C, and a large amount of Cr nitride was able to form, the resistance to crevice corrosion markedly decreased.

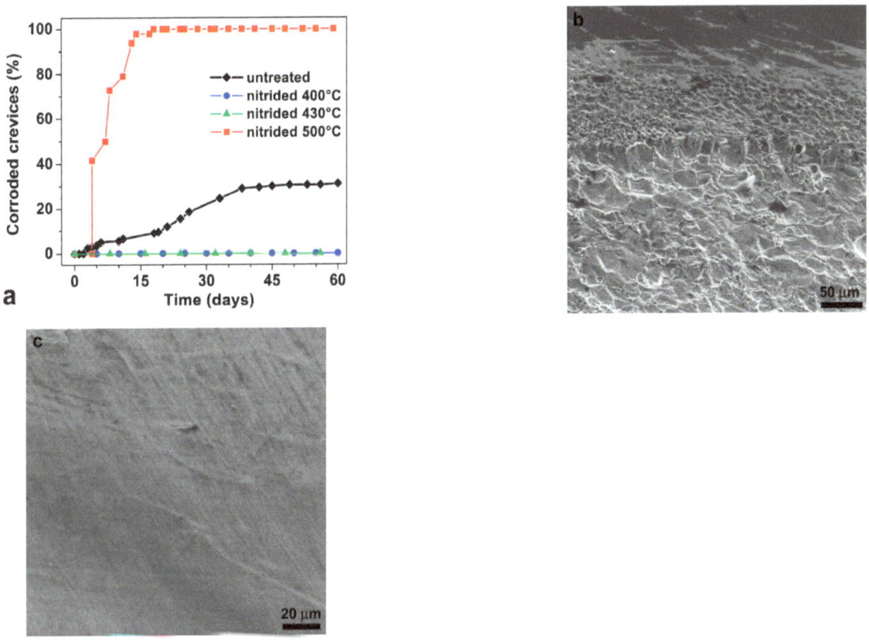

Figure 10. (**a**) Percentages of crevices, in which corrosion process are active, vs. time for AISI 316L samples untreated and nitrided as indicated (nitriding time: 5 h). Surfaces of AISI 316L samples untreated (**b**) and nitrided at 400 °C (**c**) after 60 days (solution: 10 wt.% NaCl, aerated, at 55 °C). (For further details, see Ref. [150]).

According to Olzon-Dionysio et al. [155], the improvement in corrosion resistance of nitrided samples might be due not only to N-rich expanded austenite, but also to the presence of ε and γ' nitrides, which formed up to a depth of 0.1 µm and were detected by means of Mössbauer spectroscopy analysis. However, it is not clear whether these phases were really nitrides or they were the h.c.p. N-rich martensite, ε_N', and the LRO γ_N' phase, which were expected to form in the outermost layer where the N content was higher.

The formation of a small number of nitrides, together with expanded austenite, do not hinder the modified surface layers to exert their protective effect. Even if lower corrosion potential and slightly higher anodic current density values in the passive branch were observed, the pitting potential was usually higher and the damage on the surface was smaller than that of the untreated steel [84,96,99,108,124,150]. This significant improvement was not observed when crevice tests were performed [150]. When a large amount of Cr-rich nitrides forms, a marked decrease in corrosion resistance is registered [83,89,108,123,125,150].

For the nitrided AISI 316L steel, an increase in the critical pitting temperature from 40 to 65 °C was reported [166].

Nitrocarburizing is effective in improving corrosion resistance, increasing the corrosion potential and pitting potential values and decreasing the anodic current density in the passive branch [94,104,122,167,168]. A slower pit formation was observed in the modified surface layer obtained on active-screen nitrocarburized AISI 316L, in comparison with that of both untreated and nitrided specimens, suggesting that C, in combination with N, acted in inhibiting the corrosion process [168]. The formation of N-rich Fe-based phases, ε_N' and the LRO γ_N', tended to cause an increase in the anodic current density in the passive branch, even if the passive branch was wider than that of the untreated steel [121].

Low-temperature carburizing causes an increase of corrosion resistance in NaCl solutions [25,55,56,106]. A significant increase in pitting potential, in comparison with that of the untreated alloy, is observed. A very good ability to repassivate is also registered when using cyclic potentiodynamic tests [25,55].

Sun [55] studied the corrosion behavior of carburized AISI 316L in a 0.5 M NaCl solution and pointed out that the high potential value (+1200 mV(SCE)) at which anodic current density increased was related to crevice corrosion phenomena, and not to stable pit formation. It has to be noted that this very good result was obtained by removing the first microns of the modified layer, in which a carbon soot and carbides might form. Moreover, by progressively removing the modified layer and repeating the cyclic potentiodynamic tests, similar polarization curves were registered both in the forward scan and in the reverse scan. In particular, it was observed that there existed a critical C concentration of ~0.25 wt.% (~1.1 at.%), above which C-rich expanded austenite had an excellent corrosion resistance without a significant dependence on the C content (Figure 11a). The very good corrosion resistance of the carburized layer was also assessed by performing potentiodynamic tests on the cross section (Figure 11b). The growth of pits near the modified layer was hindered; meanwhile, in the untreated substrate, pitting formed freely. Using potentiostatic tests, it was observed that metastable pit formation tended to be inhibited on carburized specimens. Very good corrosion resistance of the carburized steel was observed also in a 0.5 M NaCl + 0.5 M HCl solution.

For the carburized AISI 316, an increase in the critical pitting temperature was reported from 16.9 to 79.1 °C [25].

When treatment conditions allow the formation of carbides at the surface, a decrease in corrosion resistance is observed, with a smaller passive branch having higher anodic current density values [56].

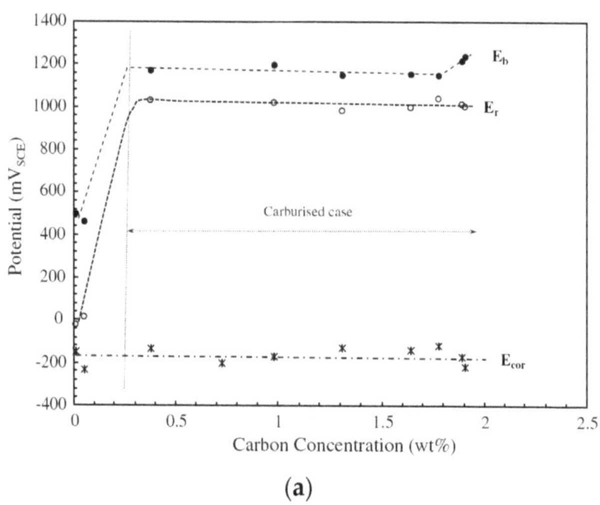

(**a**)

Figure 11. *Cont.*

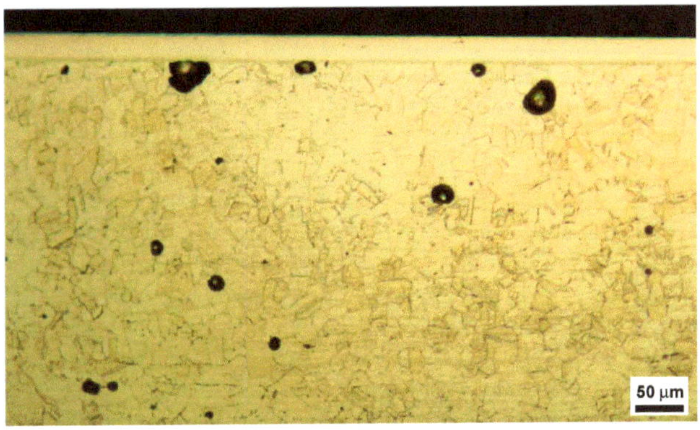

(**b**)

Figure 11. (**a**) Variation in corrosion potential (E_{cor}), breakdown potential (E_b) and repassivation potential (E_r) with C concentration for a carburized AISI 316L stainless steel. (**b**) Particular of the surface after potentiodynamic test on the cross section (chemical etching was performed after the test for delineating the microstructure) (solution: 0.5 M NaCl, aerated). (For further details, see Ref. [55]). (Reprinted with permission from Ref. [55]. Copyright 2010 Elsevier).

3.2.4. Localized Corrosion Phenomena in Mixed Saline Solutions

Electrochemical tests are also performed on low-temperature-treated austenitic stainless steels using solutions which contain NaCl together with other salts, such as those used for the simulation of body fluids.

An increase in corrosion resistance is observed for nitrided stainless steels tested in phosphate-buffered saline (PBS) solution [169–172]. The nitrided samples usually have higher corrosion potential and lower anodic current density values than the untreated ones [169,170,172]. Regarding the potential value at which the passive–transpassive transition occurs, higher values are not always registered for nitrided samples. For example, Martinesi et al. [170] observed comparable potential values for the passive–transpassive transition for AISI 316L samples, untreated or nitrided at 400 °C, 5 h, and tested in PBS, while similar samples registered a marked difference in pitting potential when tested in 5 wt.% NaCl [129] (Figure 12). It was hypothesized that the phosphates present in the PBS solution were adsorbed to the stainless steel surface and exerted their anodic inhibitor effect, thus delaying pit initiation [173].

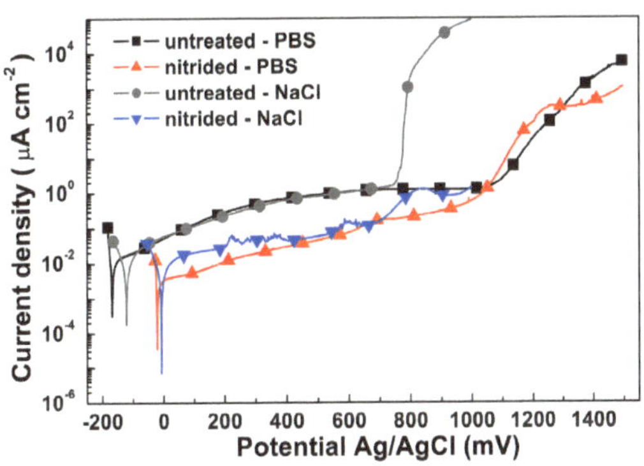

Figure 12. Potentiodynamic polarization curves of AISI 316L untreated and nitrided tested in PBS and 5 wt.% NaCl solutions (aerated). (For further details, see Ref. [170] (PBS) and Ref. [129] (NaCl solution)).

Lei and Zhu [174] tested untreated and nitrided (380 °C, 4 h) AISI 316L and 1Cr18Ni9Ti steels in Ringer's solution at pH values of 3.5–7.2 at 37 °C. When the solution was between pH 7.2 and 5.5, the nitrided samples did not suffered pitting corrosion due to the transpassive dissolution of the passive films on the surface. Conversely, for a pH of 3.5, a transition from passivation to pitting corrosion was observed, even if the pitting potential values of the nitrided steels were higher than those of the untreated alloys. The increase in corrosion resistance of plasma nitrided AISI 316L (350 and 400 °C, 2 h) was also observed when the samples were tested by means of potentiodynamic tests and EIS analysis in a simulated body fluid (SBF) solution [175].

The formation of a low amount of Fe-based nitrides did not impair the corrosion resistance of nitrided AISI 316L in Ringer's solution and a more extended passive branch was observed with lower anodic current density values in comparison with the untreated steel [96].

For the food and medical industries bacterial colonization is a major concern, together with corrosion resistance, and so antibacterial textures are employed. Dashtbozorg et al. [176] studied the effects of ultrashort pulsed laser texturing on the corrosion resistance of N-rich expanded austenite in Ringer's solution. The authors reported that the modified layers maintained a high corrosion resistance when femtoseconds laser pulses were employed. Conversely, when using laser pulses within the nanosecond regime the thermal decomposition of expanded austenite occurred and a worsening of corrosion resistance was observed.

When tests were carried out in a 0.4 M NaCl + 0.1 M Na_2SO_4 solution at pH 3, low-temperature-nitrided AISI 304L (425 °C, 30 h) had a higher resistance to pitting corrosion than the untreated alloy. N accelerated anodic oxidation in the initial stages of polarization, and promoted the subsequent passivation, mainly due to the precipitation of corrosion products [177]. It has to be pointed out that the used nitriding conditions might promote the formation of ε_N' martensite and nitrides precipitates, which might be responsible for the observed accelerated anodic oxidation in the initial stages.

An improvement in corrosion resistance is registered in Ringer's solution for nitrocarburized samples, whereby the treatment conditions inhibited the formation of precipitates [178], and also for carburized samples [178,179].

3.2.5. Localized Corrosion Phenomena in $FeCl_3$ Solutions

$FeCl_3$ solutions are often used for evaluating localized corrosion phenomena (see, for example, ASTM G48 standard—"Test Methods for Pitting and Crevice Corrosion Resistance of Stainless Steels and Related Alloys by Use of Ferric Chloride Solution").

Bottoli et al. [90] tested gas-nitrided AISI 316 (430 °C, 20 h) in a 3 wt.% $FeCl_3$ solution at pH 1 applying Teflon shaped washers in order to evaluate the resistance to crevice corrosion. After 168 h immersion duration (1 week) the weight loss of this sample type was significantly lower than that of the untreated alloy, confirming the very good resistance of low-temperature-nitrided austenitic stainless steels to crevice corrosion.

Crevice corrosion resistance on carburized 316LVM samples was assessed using an acidified ferric chloride solution [179]. A significant improvement in crevice corrosion resistance, after a 72 h test, was observed, in particular on active screen treated samples, with a reduction in weight loss of 98% in comparison with the untreated steel. Similar tests were performed on AISI 304L and Ni-free austenitic stainless steels, which were subjected to a carburizing treatment by means of Kolsterizing process [106]. An improvement was registered in crevice corrosion resistance.

3.2.6. Corrosion Phenomena in HCl Solutions

The good corrosion resistance of low-temperature-treated austenitic stainless steels in HCl solutions can also be inferred by the fairly unetched appearance of the modified surface layers in metallographic analysis since HCl is present in many chemical etchants usually used for stainless steels, such as glyceregia and Marble's reagent.

Li and Bell [180] tested the corrosion resistance of AISI 316, which was treated by active screen plasma nitriding at 420 °C for 20 h, in a 10 vol.% HCl water solution for a duration up to 120 h. The weight loss was comparable to that of the untreated steel, but the SEM analysis showed that the dominant corrosion mechanism changed from localized pitting corrosion for the untreated steel to a more general corrosion for the nitrided samples, even if dissolution preferably occurred along slip lines and grain boundaries. A marked increase in the weight loss was observed when a high content of nitrides was able to form, i.e., when the samples were treated at 500 °C. A good corrosion resistance is reported also for carburized stainless steels. Sun [181] tested a carburized cast AISI 316 steel (480 °C, 20 h) in a 2 vol.% HCl solution, and registered a significantly reduction in weight loss up to 6 days when compared to that of the untreated alloy. Potentiodynamic polarization tests showed a higher corrosion resistance of carburized AISI 316L in a 2.5 vol.% HCl solution [182].

3.3. *Corrosion Behavior in Presence of Cl-Free Solutions*

Austenitic stainless steels have a good corrosion resistance in many environments which do not contain Cl^-. These alloys have a good resistance in very diluted or highly concentrated sulfuric acid (H_2SO_4) solutions, while corrosion occurs more or less easily at intermediate concentrations depending on steel composition. Borate solutions are not usually employed in industrial applications; however, due to their benign character, a stable passive layer can form, and therefore they are commonly used to study the characteristics of this film. Regarding these environments, the corrosion resistance characteristics of the modified surface layers obtained on austenitic stainless steels by means of nitriding or carburizing are summarized in Table 3.

Table 3. Corrosion resistance in Cl-free solutions of the modified surface layers obtained on austenitic stainless steels by means of nitriding or carburizing.

Solution	Modified Layer Type	Nitriding	Carburizing
H_2SO_4	- expanded austenite only	good	good
	- with martensite	poor	-
	- with nitrides (low amount)	poor	-
Na_2SO_4 with/without H_2SO_4	- expanded austenite only	good [1]	
	- with martensite/nitrides	poor [1]/good [2]	-
Borate	- expanded austenite	good	-

[1] pH 3; [2] pH 6, 9

3.3.1. Corrosion Behavior in H_2SO_4 and Sulfate Solutions

The study of the corrosion behavior of low-temperature-nitrided austenitic stainless steels in H_2SO_4 solutions is particularly interesting since these surface modified alloys have been proposed as prospective materials for bipolar plates in proton exchange membrane fuel cells (PEMFC) [183–185]. Lin et al. [183] observed typical active–passive behavior with a slight increase in corrosion potential and decrease in corrosion current density for AISI 316 subjected to active screen plasma nitriding (370, 410 and 450 °C, 7 h), when tested in a 0.05 M H_2SO_4 solution, while the anodic current density values in the passive branch tended to be higher than those of the untreated steel. Moreover, higher corrosion resistance was observed in active screen-nitrided (450 °C) specimens, in comparison with the DC plasma-nitrided samples. Tests carried out on plasma-nitrided AISI 316L (370 °C, 2 h) in a H_2SO_4 solution containing 2 ppm F^- (pH = 1–5), used to simulate a PEMFC environment, showed that this sample type had a nobler corrosion potential and lower anodic current density in the passive branch than those of the untreated steel [184]. For pH 1 the corrosion current density of nitrided samples was slightly lower and for pH 2–5 it was comparable to that of the untreated alloy. Further studies on the corrosion behavior of nitrided AISI 316L (450 °C, durations from 3 to 12 h) in a PEFMC environment (0.5 M H_2SO_4 + 2 ppm

HF at 70 °C) showed that the corrosion resistance depended on the phases formed in the modified layers [185]. Both EIS analysis and potentiodynamic polarization tests registered an improvement in the corrosion resistance in comparison with that of the untreated steel. However, as treatment duration was longer and small amounts of nitrides were able to form together with expanded austenite, the corrosion resistance tended to decrease.

The corrosion resistance in H_2SO_4 appears to be especially sensitive to the phases present in the modified layers, and, in particular, to the formation of heterogeneous microstructures. Lei and Zhu [156] studied the corrosion behavior in a 0.05 M H_2SO_4 solution of the expanded austenite with two different N concentrations at the surface, 26 and 32 at.% (steel: 1Cr18Ni9Ti). The potentiodynamic polarization curves of the nitrided samples were comparable to those of the untreated alloy, even if with a slightly higher corrosion resistance, suggesting that the resistance to general corrosion in this environment was equivalent to that of the untreated steel (Figure 13). However, when Lei et al. [158] tested 1Cr18Ni9Ti stainless steel with modified layers consisting of N-rich expanded austenite and expanded austenite + ε_N' in a 0.5 M H_2SO_4 solution, they observed that a single expanded austenite layer had a general corrosion resistance comparable to that of the untreated steel, while the presence of a heterogeneous structure (expanded austenite + N-induced hc.p. martensite) caused an increase in passive current values of 1-2 order of magnitude, which differed from what was observed in a NaCl solution. The precipitation of nitrides decreased the corrosion resistance further on. A similar result was found by Spies et al. [186], who observed that the corrosion resistance in 0.05 M H_2SO_4 depended on nitriding parameters and it was improved only up to a treatment performed at 420 °C for 36 h.

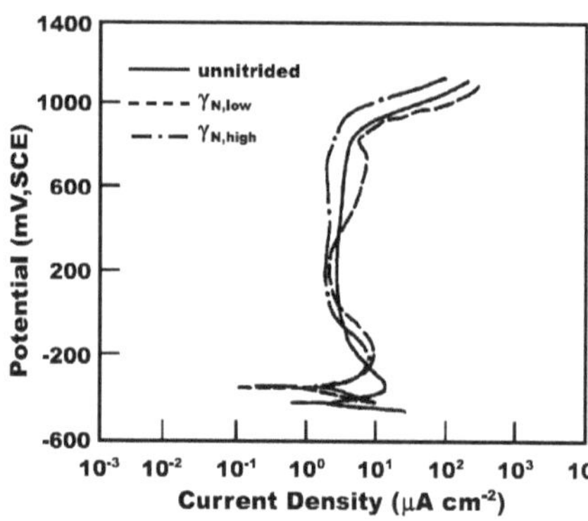

Figure 13. Potentiodynamic polarization curves of 1Cr18Ni9Ti austenitic stainless steel untreated and nitrided with two different N content in expanded austenite ($\gamma_{N,low}$: 26 at.%, $\gamma_{N,high}$: 32 at.%) (solution: 0.05 M H_2SO_4). (For further details, see Ref. [156]). (Reprinted with permission from Ref. [156]. Copyright 2005 Elsevier).

Flis and Gajek [187] tested nitrided AISI 304L steel in a 0.1 M Na_2SO_4 solution acidified with H_2SO_4 to pH 3 by using potentiodynamic method and evaluated the impedance by repeating EIS analysis up to 240 min. All the tests were performed after holding the samples at −1.4 V (MSE) for 2 min, so it has to be hypothesized that at least a part of the passive film formed in the air was destroyed. Using these initial conditions nitrided samples had a higher anodic reactivity than the untreated steel, with a decrease in corrosion resistance. Regarding EIS analysis, using an EEC analogous to that used for electrolyte–polymer–metal system, the authors suggested that the porosity of the film on the nitrided steel was higher than that of the untreated steel. On the basis of steel type, treatment conditions and reported X-ray diffraction patterns, it has to be supposed that N-induced h.c.p. martensite, ε_N', was able to form together with expanded austenite, meaning that the corrosion behavior might be influenced by both a heterogeneous surface layer and a thin passive film, which was not able to properly grow in the probe solution. Further studies showed that when nitrided

AISI 304L (425 °C, 30 h) was put in contact with 0.1 M Na_2SO_4 solution at pH 6 or 9, significantly lower anodic current density values were registered, especially if compared to those of the untreated steel, and a higher corrosion resistance was observed [177].

The influence of N content on the corrosion resistance in a 0.1 M Na_2SO_4 solution acidified to pH 3 was studied by Kuczynska-Wydorska and Flis [188] on nitrided AISI 304L and AISI 316L (425 °C, 30 h). The outer layers, having a higher N concentration (~7–14 wt.%, i.e., ~22–35.7 at.%) and presumably also containing the N-induced h.c.p. martensite and nitrides precipitates due to the nitriding conditions, showed a lower corrosion resistance. Meanwhile, for deeper regions that had a N content less than 7 wt.% (~22 at.%) and presumably consisted of expanded austenite only, the anodic current density values were slightly higher or even lower in the passive and transpassive states in comparison with those of the untreated alloys. An initial acceleration of anodic oxidation was suggested for these samples.

Regarding carburizing, Sun [189] evaluated the corrosion behavior of C-rich expanded austenite formed on AISI 316L steel in a 1 M H_2SO_4 solution. Besides a thin outermost layer, which was due to carbon soot and redeposition of sputtered materials and had a poor electrochemical behavior, the C-rich expanded austenite of the inner part of the modified layer showed an improved corrosion resistance. While the untreated alloy had an active–passive behavior with a clear anodic peak, in the polarization curves of the carburized specimen, related to different depths and thus different C concentrations, very small anodic peaks were observed; moreover, the corrosion potential tended to increase and the corrosion current density tended to decrease as the C content rose. General passivity and transpassivity were not influenced by C concentration and were comparable to those of the untreated alloy. However, in the early stage of potentiostatic polarization experiments a faster film growth kinetic was observed for the carburized samples, and thus it was supposed than there was an enhancement of the initial corrosion resistance in the test solution for this sample type. Carburized AISI 316 showed also a very good corrosion resistance in boiling 16 vol.% H_2SO_4 solution, with a significant reduction in weight loss [190].

The comparison of the corrosion behavior of nitrided (415 °C, 15 h) and carburized (470 °C, 15 h) AISI 316L samples in a 0.5 M H_2SO_4 solution highlighted that the nitrided specimens were more prone to corrosion phenomena, which was probably due to the precipitation of N-induced h.c.p. martensite and/or nitrides, while the carburized samples had a higher corrosion resistance in comparison with that of the untreated steel [191].

3.3.2. Corrosion Behavior in Borate Solutions

A study of the corrosion behavior in a borate buffer solution with pH 8.4 for N-rich expanded austenite, formed on AISI 304L steel, also showed that the modified surface layers maintained good corrosion resistance with these environmental conditions [50]. EIS analysis evidenced higher impedance values and a higher polarization resistance for the nitrided samples than those of the untreated steel. Additionally, in the polarization curves, the nitrided steel had a slightly higher corrosion potential and significantly lower anodic current density values in the passive branch if compared to the untreated alloy. Additional studies, performed in a borate buffer solution with pH increasing from 6.5 to 8.9 [76] and at pH 7.7 [51], also showed that in these conditions N-rich expanded austenite had a stronger tendency to passivation, with an increase in corrosion potential and decrease in anodic current density if compared to the untreated alloy.

4. Conclusions

Low-temperature thermochemical treatments are becoming the treatments of choice for enhancing surface hardness, wear and fatigue resistance without impairing the corrosion resistance of austenitic stainless steels. The main characteristics of the modified surface layers obtained on austenitic stainless steels by means of nitriding or carburizing are summarized in Table 4.

Table 4. Main characteristics of the modified surface layers obtained on austenitic stainless steels by means of nitriding or carburizing.

Properties		Nitriding	Carburizing
Formation temperature (without formation of Cr compounds) (°C)		300–450	300–550
Max. interstitial content in expanded austenite (at.%)		38	19
Max. surface hardness (kg_f mm^{-2})		1450	1000
Localized corrosion resistance in Cl^--containing solutions	- expanded austenite only - with martensite - with nitrides (low amount)	very good very good-good acceptable	good - -
Corrosion resistance in H_2SO_4 solutions	- expanded austenite only - with martensite/nitrides	good poor	good -
Corrosion resistance in borate buffer solutions	- expanded austenite	good	-

On the basis of the studies presented in the international literature, the following main conclusions can be drawn.

- The protection effect of expanded austenite and its ability in hindering and delaying the occurrence of corrosion phenomena tend to increase as the interstitial content in this phase is high, the formation of a significant amount of precipitates is avoided and the modified surface layer is thicker. Thus, treatment temperature, duration and N and/or C feeding have to be chosen as functions of the steel composition in order to obtain an interstitial-rich, precipitates-free and thick modified layer.
- Regarding Cl^--containing solutions, both N- and C-rich expanded austenite have improved resistance to localized corrosion (pitting, crevice), delaying or even inhibiting the occurrence of localized corrosion phenomena. The formation of N-induced h.c.p. martensite and/or small amounts of Fe-based nitride precipitates does not impair this resistance. Nitrided samples tested in a 3 wt.% NaCl solution showed an enhanced corrosion resistance in the pH range from 0.4 to 11, with a significant improvement in the pH range 4–11. Carburized samples also showed a good corrosion resistance in NaCl solution at a low pH.
- For nitrided samples, the improvement in corrosion resistance is usually ascribed to the local alkalization of the incipient pits due to the formation of ammonium ions, while for carburized samples the mechanism is still unclear. Taking into account that, in the passive layer, the concentration of both N and C tends to increase in the inner part, where Cr content is usually higher, it may be hypothesized that the preferential Cr-N and Cr-C bonds tend to hinder Cr solubilization and allow to have a more stable and protective film. The change in the semiconductive properties of the film, with a decrease in donor and acceptor density and in the flat band potential value, may also contribute to increasing the protective effect of the passive film.
- In the presence of aqueous environments containing H_2SO_4, the corrosion resistance is comparable or even higher than that of the untreated steel when the modified surface layers consist of expanded austenite only. The corrosion resistance is worse when N-induced h.c.p. martensite and/or Fe-based nitride precipitates are able to form in nitrided samples, or when carbides form in carburized samples. Thus, treatment conditions should be chosen as functions of the steel which are to be treated in order to avoid the formation of a heterogeneous layer at the surface. It has to be pointed out that it is not easy to avoid the formation of h.c.p. martensite in nitrided samples since the solubilization of a high amount of N in expanded austenite induces local plastic deformations, which may be accompanied by the formation of extended stacking faults, i.e., h.c.p. martensite-like zones. On the other hand, modified surface layers, consisting of carbide-free C-rich expanded austenite, can be obtained more easily.

Thus, for applications in H_2SO_4-containing environments, the possible choices seem to be either carburized austenitic stainless steels or nitrided austenitic stainless steels with lower N content than that able to induce the formation of h.c.p. martensite.

- In borate solutions, the good corrosion resistance of austenitic stainless steels is improved further on when N-rich expanded austenite forms.

Thus, the formation of modified surface layers consisting of expanded austenite allows to further extend the applications of austenitic stainless steels. However, a careful choice of the treatment parameters should be made, these being functions of alloy composition, in order to obtain the best performances possible in the service environment.

Funding: This research received no external funding.

Data Availability Statement: Data sharing is not applicable.

Conflicts of Interest: The author declares no conflict of interest.

References

1. Cobb, H.M. *The History of Stainless Steel*; ASM International: Materials Park, OH, USA, 2010; ISBN 978-1-61503-011-8.
2. Washko, S.D.; Aggen, G. Wrought Stainless Steels. In *ASM Handbook Vol. 1*; ASM International: Materials Park, OH, USA, 1997; pp. 841–907.
3. Grubb, J.F.; DeBold, T.; Fritz, J.D. Corrosion of Wrought Stainless Steels. In *ASM Handbook Vol. 13B*; Cramer, S.D., Covino, B.S., Jr., Eds.; ASM International: Materials Park, OH, USA, 2005; pp. 54–77. [CrossRef]
4. Sun, J.; Tang, H.; Wang, C.; Han, Z.; Li, S. Effects of Alloying Elements and Microstructure on Stainless Steel Corrosion: A Review. *Steel Res. Int.* **2022**, *93*, 2100450. [CrossRef]
5. Asami, K.; Hashimoto, K. Importance of initial surface film in the degradation of stainless steels by atmospheric exposure. *Corros. Sci.* **2003**, *45*, 2263–2283. [CrossRef]
6. Habibzadeh, S.; Li, L.; Shum-Tim, D.; Davis, E.C.; Omanovic, S. Electrochemical polishing as a 316L stainless steel surface treatment method: Towards the improvement of biocompatibility. *Corros. Sci.* **2014**, *87*, 89–100. [CrossRef]
7. Łyczkowska-Widłak, E.; Lochyński, P.; Nawrat, G. Electrochemical Polishing of Austenitic Stainless Steels. *Materials* **2020**, *13*, 2557. [CrossRef] [PubMed]
8. Olsson, C.O.A.; Landolt, D. Passive films on stainless steels-Chemistry, structure and growth. *Electrochim. Acta* **2003**, *48*, 1093–1104. [CrossRef]
9. Jiang, R.; Wang, Y.; Wen, X.; Chen, C.; Zhao, J. Effect of time on the characteristics of passive film formed on stainless steel. *Appl. Surf. Sci.* **2017**, *412*, 214–222. [CrossRef]
10. McGuire, M.F. *Stainless Steels for Design Engineers*; ASM International: Materials Park, OH, USA, 2008; ISBN 978-0-87170-717-8.
11. Saadi, S.A.; Yi, Y.; Cho, P.; Jang, C.; Beeley, P. Passivity breakdown of 316L stainless steel during potentiodynamic polarization in NaCl solution. *Corros. Sci.* **2016**, *111*, 720–727. [CrossRef]
12. Wang, Z.; Seyeux, A.; Zanna, S.; Maurice, V.; Marcus, P. Chloride-induced alterations of the passive film on 316L stainless steel and blocking effect of pre-passivation. *Electrochim. Acta* **2020**, *329*, 135159. [CrossRef]
13. Astafurov, S.; Astafurova, E. Phase Composition of Austenitic Stainless Steels in Additive Manufacturing: A Review. *Metals* **2021**, *11*, 1052. [CrossRef]
14. Schneider, M.J.; Chatterjee, M.S. Introduction to Surface Hardening of Steels. In *ASM Handbook Vol. 4A*; Dosset, J.L., Totten, G.E., Eds.; ASM International: Materials Park, OH, USA, 2013; pp. 389–397. [CrossRef]
15. Davis, J.R. Surface Engineering of Stainless Steels. In *ASM Metal Handbook Vol. 5*; ASM International: Materials Park, OH, USA, 1994; pp. 741–761. [CrossRef]
16. Michal, G.M.; Gu, X.; Jennings, W.D.; Kahn, H.; Ernst, F.; Heuer, A.H. Paraequilibrium Carburization of Duplex and Ferritic Stainless Steels. *Metall. Mater. Trans. A* **2009**, *40*, 1781–1790. [CrossRef]
17. Adachi, S.; Ueda, N. Formation of Expanded Austenite on a Cold-Sprayed AISI 316L Coating by Low-Temperature Plasma Nitriding. *J. Therm. Spray Technol.* **2015**, *24*, 1399–1407. [CrossRef]
18. Adachi, S.; Yamaguchi, T.; Ueda, N. Formation and Properties of Nitrocarburizing S-Phase on AISI 316L Stainless Steel-Based WC Composite Layers by Low-Temperature Plasma Nitriding. *Metals* **2021**, *11*, 1538. [CrossRef]
19. Lindner, T.; Kutschmann, P.; Löbel, M.; Lampke, T. Hardening of HVOF-Sprayed Austenitic Stainless-Steel Coatings by Gas Nitriding. *Coatings* **2018**, *8*, 348. [CrossRef]
20. Lindner, T.; Löbel, M.; Lampke, T. Phase Stability and Microstructure Evolution of Solution-Hardened 316L Powder Feedstock for Thermal Spraying. *Metals* **2018**, *8*, 1063. [CrossRef]
21. Qadri, S.A.R.; Sasidhar, K.N.; Meka, S.R. High nitrogen alloying of AISI 316 L stainless steel powder by nitriding. *Powder Technol.* **2021**, *390*, 456–463. [CrossRef]
22. Dong, H. S-phase surface engineering of Fe-Cr, Co-Cr and Ni-Cr alloys. *Int. Mater. Rev.* **2010**, *55*, 65–98. [CrossRef]

23. Christiansen, T.L.; Somers, M.A.J. Low-temperature gaseous surface hardening of stainless steel: The current status. *Int. J. Mater. Res.* **2009**, *100*, 1361–1377. [CrossRef]
24. Borgioli, F. From Austenitic Stainless Steel to Expanded Austenite-S Phase: Formation, Characteristics and Properties of an Elusive Metastable Phase. *Metals* **2020**, *10*, 187. [CrossRef]
25. Collins, S.R.; Williams, P.C.; Marx, S.V.; Heuer, A.; Ernst, F.; Kahn, H. Low-Temperature Carburization of Austenitic Stainless Steels. In *ASM Handbook Vol. 4D*; Dosset, J., Totten, G.E., Eds.; ASM International: Materials Park, OH, USA, 2014; pp. 451–460.
26. Casteletti, L.C.; Neto, A.L.; Totten, G.E. Nitriding of Stainless Steels. *Metallogr. Microstruct. Anal.* **2014**, *3*, 477–508. [CrossRef]
27. Bell, T. Surface engineering of austenitic stainless steel. *Surf. Eng.* **2002**, *18*, 415–422. [CrossRef]
28. Czerwiec, T.; He, H.; Marcos, G.; Thiriet, T.; Weber, S.; Michel, H. Fundamental and Innovations in Plasma Assisted Diffusion of Nitrogen and Carbon in Austenitic Stainless Steels and Related Alloys. *Plasma Process. Polym.* **2009**, *6*, 401–409. [CrossRef]
29. Somers, M.A.J.; Christiansen, T.L. Low temperature surface hardening of stainless steel. In *Thermochemical Surface Engineering of Steels*; Mittemeijer, E.J., Somers, M.A.J., Eds.; Woodhead Publishing: Oxford, UK, 2015; pp. 557–579. ISBN 978-0-85709-592-3. [CrossRef]
30. Fossati, A.; Galvanetto, E.; Bacci, T.; Borgioli, F. Improvement of corrosion resistance of austenitic stainless steels by means of glow-discharge nitriding. *Corros. Rev.* **2011**, *29*, 209–221. [CrossRef]
31. Luo, Q.; Yang, S. From Micro to Nano Scales -Recent Progress in the Characterization of Nitrided Austenitic Stainless Steels. *Int. J. Nanomed. Nanosurgery* **2015**, *1*, 1–11. [CrossRef]
32. Somers, M.; Kücükyildiz, Ö.; Ormstrup, C.; Alimadadi, H.; Hattel, J.; Christiansen, T.; Winther, G. Residual Stress in Expanded Austenite on Stainless Steel; Origin, Measurement, and Prediction. *Mater. Perform. Charact.* **2018**, *7*, 693–716. [CrossRef]
33. Borgioli, F. The "Expanded" Phases in the Low-Temperature Treated Stainless Steels: A Review. *Metals* **2022**, *12*, 331. [CrossRef]
34. Cardoso, R.P.; Mafra, M.; Brunatto, S.F. Low-temperature Thermochemical Treatments of Stainless Steels–An Introduction. In *Plasma Science and Technology-Progress in Physical States and Chemical Reactions*; Mieso, T., Ed.; InTech: Rijeka, Croatia, 2016; pp. 107–130. ISBN 978-953-51-2280-7. [CrossRef]
35. Tschiptschin, A.P.; Pinedo, C.E. Surface Hardening of Stainless Steel. In *Stainless Steels*; Singh, A., Ed.; IntechOpen: Rijeka, Croatia, 2022; pp. 94–174. ISBN 978-1-80355-133-3. [CrossRef]
36. Spies, H.-J. Corrosion behaviour of nitrided, nitrocarburised and carburised steels. In *Thermochemical Surface Engineering of Steel*; Mittemeijer, E.J., Somers, M.A.J., Eds.; Woodhead Publishing: Oxford, UK, 2015; pp. 267–309. ISBN 978-0-85709-592-3. [CrossRef]
37. Wriedt, H.A.; Gokcen, N.A.; Nafziger, R.H. The Fe-N (Iron-Nitrogen) system. *Bull. Alloy Phase Diagr.* **1987**, *8*, 355–377. [CrossRef]
38. Okamoto, H. The C-Fe (carbon-iron) system. *J. Phase Equilibria* **1992**, *13*, 543–565. [CrossRef]
39. Gavriljuk, V.G. Carbon and nitrogen in iron-based austenite and martensite: An attempt at comparative analysis. *J. Phys. IV Fr.* **2003**, *112*, 51–59. [CrossRef]
40. Cao, Y.; Ernst, F.; Michal, G.M. Colossal carbon supersaturation in austenitic stainless steels carburized at low temperature. *Acta Mater.* **2003**, *51*, 4171–4181. [CrossRef]
41. Williamson, D.L.; Ozturk, O.; Wei, R.; Wilbur, P.J. Metastable phase formation and enhanced diffusion in f.c.c. alloys under high dose, high flux nitrogen implantation at high and low ion energies. *Surf. Coat. Technol.* **1994**, *65*, 15–23. [CrossRef]
42. Christiansen, T.; Somers, M.A.J. Controlled dissolution of colossal quantities of nitrogen in stainless steel. *Metall. Mater. Trans. A* **2006**, *37*, 675–682. [CrossRef]
43. Christiansen, T.L.; Ståhl, K.; Brink, B.K.; Somers, M.A.J. On the Carbon Solubility in Expanded Austenite and Formation of Hägg Carbide in AISI 316 Stainless Steel. *Steel Res. Int.* **2016**, *87*, 1395–1405. [CrossRef]
44. Schmuki, P. From Bacon to barriers: A review on the passivity of metals and alloys. *J. Solid State Electrochem.* **2002**, *6*, 145–164. [CrossRef]
45. Wang, Z.; Paschalidou, E.-M.; Seyeux, A.; Zanna, S.; Maurice, V.; Marcus, P. Mechanisms of Cr and Mo Enrichments in the Passive Oxide Film on 316L Austenitic Stainless Steel. *Front. Mater.* **2019**, *6*, 232. [CrossRef]
46. Hakiki, N.B.; Boudin, S.; Rondot, B.; Da Cunha Belo, M. The electronic structure of passive films formed on stainless steels. *Corros. Sci.* **1995**, *37*, 1809–1822. [CrossRef]
47. Maurice, V.; Marcus, P. Progress in corrosion science at atomic and nanometric scales. *Prog. Mater. Sci.* **2018**, *95*, 132–171. [CrossRef]
48. Maurice, V.; Marcus, P. Current developments of nanoscale insight into corrosion protection by passive oxide films. *Curr. Opin. Solid State Mater. Sci.* **2018**, *22*, 156–167. [CrossRef]
49. Lei, M.K.; Zhu, X.M. Role of Nitrogen in Pitting Corrosion Resistance of a High-Nitrogen Face-Centered-Cubic Phase Formed on Austenitic Stainless Steel. *J. Electrochem. Soc.* **2005**, *152*, B291–B295. [CrossRef]
50. Wang, K.S.; Tong, S.; Lei, M.K. Corrosion and Passivation of High Nitrogen Face-Centered-Cubic Phase Formed on AISI 304L Austenitic Stainless Steel in Borate Buffer Solution. *J. Electrochem. Soc.* **2015**, *162*, C601–C609. [CrossRef]
51. Tong, S.; Che, H.L.; Lei, M.K. High-resolution TEM characterization of epitaxial passivation for a high nitrogen face-centered-cubic phase formed on AISI 304L austenitic stainless steel in borate buffer solution. *Electrochim. Acta* **2021**, *393*, 139075. [CrossRef]
52. Heuer, A.H.; Kahn, H.; Ernst, F.; Michal, G.M.; Hovis, D.B.; Rayne, R.J.; Martin, F.J.; Natishan, P.M. Enhanced corrosion resistance of interstitially hardened stainless steel: Implications of a critical passive layer thickness for breakdown. *Acta Mater.* **2012**, *60*, 716–725. [CrossRef]

53. Ningshen, S.; Mudali, U.K.; Mittal, V.K.; Khatak, H.S. Semiconducting and passive film properties of nitrogen-containing type 316LN stainless steels. *Corros. Sci.* **2007**, *49*, 481–496. [CrossRef]
54. Zhu, X.M.; Guo, Y.; Xing, Z.Q.; Lei, M.K. Effect of Nitrogen on Semiconducting Properties of Passive Films of a High Nitrogen Face-Centered-Cubic Phase Formed on Austenitic Stainless Steel. *J. Electrochem. Soc.* **2012**, *159*, C319–C325. [CrossRef]
55. Sun, Y. Corrosion behaviour of low temperature plasma carburised 316L stainless steel in chloride containing solutions. *Corros. Sci.* **2010**, *52*, 2661–2670. [CrossRef]
56. Liu, H.Y.; Che, H.L.; Gao, J.Y.; Li, G.B.; Lei, M.K. Low-pressure hollow cathode plasma source carburizing of AISI 304L austenitic stainless steel at low temperature. *Surf. Coat. Technol.* **2022**, *442*, 128548. [CrossRef]
57. Paredes, E.C.; Bautista, A.; Alvarez, S.M.; Velasco, F. Influence of the forming process of corrugated stainless steels on their corrosion behaviour in simulated pore solutions. *Corros. Sci.* **2012**, *58*, 52–61. [CrossRef]
58. Cheng, Y.F.; Luo, J.L. Electronic structure and pitting susceptibility of passive film on carbon steel. *Electrochim. Acta* **1999**, *44*, 2947–2957. [CrossRef]
59. Hänninen, H.; Romu, J.; Ilola, R.; Tervo, J.; Laitinen, A. Effects of processing and manufacturing of high nitrogen-containing stainless steels on their mechanical, corrosion and wear properties. *J. Mater. Process. Technol.* **2001**, *117*, 424–430. [CrossRef]
60. Lo, K.H.; Shek, C.H.; Lai, J.K.L. Recent developments in stainless steels. *Mater. Sci. Eng. R Rep.* **2009**, *65*, 39–104. [CrossRef]
61. Sumita, M.; Hanawa, T.; Teoh, S.H. Development of nitrogen-containing nickel-free austenitic stainless steels for metallic biomaterials—Review. *Mater. Sci. Eng. C* **2004**, *24*, 753–760. [CrossRef]
62. Chen, S.; Wang, Q.; Yang, H.; Yang, K. High-Nitrogen Nickel-Free Stainless Steel: An Attractive Material with Potential for Biomedical Application. *Steel Res. Int.* **2022**, 2200355. [CrossRef]
63. Baba, H.; Kodama, T.; Katada, Y. Role of nitrogen on the corrosion behavior of austenitic stainless steels. *Corros. Sci.* **2002**, *44*, 2393–2407. [CrossRef]
64. Speidel, M.O. Corrosion Science of Stainless Steels. In *Stainless Steels '91. Proceedings of the International Conference on Stainless Steels, Chiba, Japan, 10–13 June 1991*; Iron and Steel Institute of Japan: Tokyo, Japan, 1991; Volume 1, pp. 25–35.
65. Jargelius-Pettersson, R.F.A. Electrochemical investigation of the influence of nitrogen alloying on pitting corrosion of austenitic stainless steels. *Corros. Sci.* **1999**, *41*, 1639–1664. [CrossRef]
66. Baba, H.; Katada, Y. Effect of nitrogen on crevice corrosion in austenitic stainless steel. *Corros. Sci.* **2006**, *48*, 2510–2524. [CrossRef]
67. Olefjord, I.; Wegrelius, L. The influence of nitrogen on the passivation of stainless steels. *Corros. Sci.* **1996**, *38*, 1203–1220. [CrossRef]
68. Vehovar, L.; Vehovar, A.; Metikoš-Huković, M.; Tandler, M. Investigations into the stress corrosion cracking of stainless steel alloyed with nitrogen. *Mater. Corros.* **2002**, *53*, 316–327. [CrossRef]
69. Kadowaki, M.; Saengdeejing, A.; Muto, I.; Chen, Y.; Frankel, G.S.; Doi, T.; Kawano, K.; Sugawara, Y.; Hara, N. Roles of Interstitial Nitrogen, Carbon, and Boron in Steel Corrosion: Generation of Oxyanions and Stabilization of Electronic Structure. *J. Electrochem. Soc.* **2020**, *167*, 81503. [CrossRef]
70. Baba, H.; Katada, Y. Effect of Nitrogen on Crevice Corrosion and Repassivation Behavior of Austenitic Stainless Steel. *Mater. Trans.* **2008**, *49*, 579–586. [CrossRef]
71. Ives, M.B.; Lu, Y.C.; Luo, J.L. Cathodic reactions involved in metallic corrosion in chlorinated saline environments. *Corros. Sci.* **1991**, *32*, 91–102. [CrossRef]
72. Clayton, C.R.; Rosenzweig, L.; Oversluizen, M.; Lu, Y.C. The influence of nitrogen on the passivity of 18-8 (0.24%N) stainless steels. In *Surfaces, Inhibition and Passivation*; McCafferty, E., Brodd, R.J., Eds.; The Electrochemical Society: Pennington, NJ, USA, 1986; pp. 323–339.
73. Bayoumi, F.M.; Ghanem, W.A. Effect of nitrogen on the corrosion behavior of austenitic stainless steel in chloride solutions. *Mater. Lett.* **2005**, *59*, 3311–3314. [CrossRef]
74. Zhu, X.M.; Lei, M.K. Pitting corrosion resistance of high nitrogen f.c.c. phase in plasma source ion nitrided austenitic stainless steel. *Surf. Coat. Technol.* **2000**, *131*, 400–403. [CrossRef]
75. Flis-Kabulska, I.; Sun, Y.; Flis, J. Monitoring the near-surface pH to probe the role of nitrogen in corrosion behaviour of low-temperature plasma nitrided 316L stainless steel. *Electrochim. Acta* **2013**, *104*, 208–215. [CrossRef]
76. Tong, S.; Che, H.L.; Wang, K.S.; Lei, M.K. Passivation kinetics of a high nitrogen face-centered-cubic phase formed on AISI 304L austenitic stainless steel in borate buffer solutions by photo- and electrochemical methods. *Electrochim. Acta* **2021**, *394*, 139110. [CrossRef]
77. Metikoš-Huković, M.; Babić, R.; Grubač, Z.; Petrović, Ž.; Lajçi, N. High corrosion resistance of austenitic stainless steel alloyed with nitrogen in an acid solution. *Corros. Sci.* **2011**, *53*, 2176–2183. [CrossRef]
78. Sun, S.; Wei, S.; Wang, G.; Jiang, Z.; Lian, J.; Ji, C. The Synthesis and Electrochemical Behavior of High-Nitrogen Nickel-Free Austenitic Stainless Steel. *J. Mater. Eng. Perform.* **2014**, *23*, 3957–3962. [CrossRef]
79. Wu, X.Q.; Xu, S.; Huang, J.B.; Han, E.H.; Ke, W.; Yang, K.; Jiang, Z.H. Uniform corrosion and intergranular corrosion behavior of nickel-free and manganese alloyed high nitrogen stainless steels. *Mater. Corros.* **2008**, *59*, 676–684. [CrossRef]
80. Martin, F.J.; Lemieux, E.J.; Newbauer, T.M.; Bayles, R.A.; Natishan, P.M.; Kahn, H.; Michal, G.M.; Ernst, F.; Heuer, A.H. Carburization-induced passivity of 316 L austenitic stainless steel. *Electrochem. Solid-State Lett.* **2007**, *10*, C76–C78. [CrossRef]
81. Niu, W.; Lillard, R.S.; Li, Z.; Ernst, F. Properties of the Passive Film Formed on Interstitially Hardened AISI 316L Stainless Steel. *Electrochim. Acta* **2015**, *176*, 410–419. [CrossRef]

82. Li, T.; Chien, S.-C.; Ren, Z.; Windl, W.; Ernst, F.; Frankel, G.S. Understanding the efficacy of concentrated interstitial carbon in enhancing the pitting corrosion resistance of stainless steel. *Acta Mater.* **2021**, *221*, 117433. [CrossRef]
83. Borgioli, F.; Fossati, A.; Galvanetto, E.; Bacci, T. Glow-discharge nitriding of AISI 316L austenitic stainless steel: Influence of treatment temperature. *Surf. Coat. Technol.* **2005**, *200*, 2474–2480. [CrossRef]
84. Borgioli, F.; Fossati, A.; Galvanetto, E.; Bacci, T.; Pradelli, G. Glow discharge nitriding of AISI 316L austenitic stainless steel: Influence of treatment pressure. *Surf. Coat. Technol.* **2006**, *200*, 5505–5513. [CrossRef]
85. Borgioli, F.; Galvanetto, E.; Bacci, T. Influence of surface morphology and roughness on water wetting properties of low temperature nitrided austenitic stainless steels. *Mater. Charact.* **2014**, *95*, 278–284. [CrossRef]
86. Stinville, J.C.; Cormier, J.; Templier, C.; Villechaise, P. Modeling of the lattice rotations induced by plasma nitriding of 316L polycrystalline stainless steel. *Acta Mater.* **2015**, *83*, 10–16. [CrossRef]
87. Czerwiec, T.; Tsareva, S.; Andrieux, A.; Bruyère, S.; Marcos, G. Effects of surface topography at different scales on the dispersion of the wetting data for sessile water droplets on nitrided austenitic stainless steels. *Surf. Coat. Technol.* **2022**, *441*, 128510. [CrossRef]
88. Tao, X.; Liu, X.; Matthews, A.; Leyland, A. The influence of stacking fault energy on plasticity mechanisms in triode-plasma nitrided austenitic stainless steels: Implications for the structure and stability of nitrogen-expanded austenite. *Acta Mater.* **2019**, *164*, 60–75. [CrossRef]
89. Baranowska, J.; Arnold, B. Corrosion resistance of nitrided layers on austenitic steel. *Surf. Coat. Technol.* **2006**, *200*, 6623–6628. [CrossRef]
90. Bottoli, F.; Jellesen, M.S.; Christiansen, T.L.; Winther, G.; Somers, M.A.J. High temperature solution-nitriding and low-temperature nitriding of AISI 316: Effect on pitting potential and crevice corrosion performance. *Appl. Surf. Sci.* **2018**, *431*, 24–31. [CrossRef]
91. Farrell, K.; Specht, E.D.; Pang, J.; Walker, L.R.; Rar, A.; Mayotte, J.R. Characterization of a carburized surface layer on an austenitic stainless steel. *J. Nucl. Mater.* **2005**, *343*, 123–133. [CrossRef]
92. Gallo, S.C.; Dong, H. EBSD and AFM observations of the microstructural changes induced by low temperature plasma carburising on AISI 316. *Appl. Surf. Sci.* **2011**, *258*, 608–613. [CrossRef]
93. Maistro, G.; Yao, Y.; Klement, U.; Nyborg, L.; Cao, Y. On surface carbides in low-temperature carburized austenitic stainless steels. *Mater. Charact.* **2020**, *167*, 110462. [CrossRef]
94. Cisquini, P.; Ramos, S.V.; Viana, P.R.P.; Lins, V.d.F.C.; Franco, A.R.; Vieira, E.A. Effect of the roughness produced by plasma nitrocarburizing on corrosion resistance of AISI 304 austenitic stainless steel. *J. Mater. Res. Technol.* **2019**, *8*, 1897–1906. [CrossRef]
95. Borgioli, F.; Galvanetto, E.; Bacci, T. Corrosion behaviour of low temperature nitrided nickel-free, AISI 200 and AISI 300 series austenitic stainless steels in NaCl solution. *Corros. Sci.* **2018**, *136*, 352–365. [CrossRef]
96. De Las Heras, E.; Ybarra, G.; Lamas, D.; Cabo, A.; Dalibon, E.L.; Brühl, S.P. Plasma nitriding of 316L stainless steel in two different N_2-H_2 atmospheres-Influence on microstructure and corrosion resistance. *Surf. Coat. Technol.* **2017**, *313*, 47–54. [CrossRef]
97. Li, G.Y.; Lei, M.K. Microstructure and Properties of Plasma Source Nitrided AISI 316 Austenitic Stainless Steel. *J. Mater. Eng. Perform.* **2017**, *26*, 418–423. [CrossRef]
98. Williamson, D.L.; Davis, J.A.; Wilbur, P.J. Effect of austenitic stainless steel composition on low-energy, high-flux, nitrogen ion beam processing. *Surf. Coat. Technol.* **1998**, *103*, 178–184. [CrossRef]
99. Borgioli, F.; Galvanetto, E.; Bacci, T. Low temperature nitriding of AISI 300 and 200 series austenitic stainless steels. *Vacuum* **2016**, *127*, 51–60. [CrossRef]
100. Czerwiec, T.; Renevier, N.; Michel, H. Low-temperature plasma-assisted nitriding. *Surf. Coat. Technol.* **2000**, *131*, 267–277. [CrossRef]
101. Czerwiec, T.; He, H.; Weber, S.; Dong, C.; Michel, H. On the occurrence of dual diffusion layers during plasma-assisted nitriding of austenitic stainless steel. *Surf. Coat. Technol.* **2006**, *200*, 5289–5295. [CrossRef]
102. Christiansen, T.; Dahl, K.V.; Somers, M.A.J. Nitrogen diffusion and nitrogen depth profiles in expanded austenite: Experimental assessment, numerical simulation and role of stress. *Mater. Sci. Technol.* **2008**, *24*, 159–167. [CrossRef]
103. Michler, T. Influence of plasma nitriding on hydrogen environment embrittlement of 1.4301 austenitic stainless steel. *Surf. Coat. Technol.* **2008**, *202*, 1688–1695. [CrossRef]
104. Sun, Y. Hybrid plasma surface alloying of austenitic stainless steels with nitrogen and carbon. *Mater. Sci. Eng. A* **2005**, *404*, 124–129. [CrossRef]
105. Christiansen, T.; Somers, M.A.J. Low temperature gaseous nitriding and carburising of stainless steel. *Surf. Eng.* **2005**, *21*, 445–455. [CrossRef]
106. Formosa, D.; Hunger, R.; Spiteri, A.; Dong, H.; Sinagra, E.; Buhagiar, J. Corrosion behaviour of carbon S-phase created on Ni-free biomedical stainless steel. *Surf. Coat. Technol.* **2012**, *206*, 3479–3487. [CrossRef]
107. Werner, K.V.; Che, H.L.; Lei, M.K.; Christiansen, T.L.; Somers, M.A.J. Low Temperature Carburizing of Stainless Steels and the Development of Carbon Expanded Austenite. *HTM J. Heat Treat. Mater.* **2022**, *77*, 3–15. [CrossRef]
108. Borgioli, F.; Fossati, A.; Matassini, G.; Galvanetto, E.; Bacci, T. Low temperature glow-discharge nitriding of a low nickel austenitic stainless steel. *Surf. Coat. Technol.* **2010**, *204*, 3410–3417. [CrossRef]
109. Fewell, M.P.; Priest, J.M. High-order diffractometry of expanded austenite using synchrotron radiation. *Surf. Coat. Technol.* **2008**, *202*, 1802–1815. [CrossRef]

110. Zhidkov, I.S.; Kukharenko, A.I.; Makarov, A.V.; Savrai, R.A.; Gavrilov, N.V.; Cholakh, S.O.; Kurmaev, E.Z. XPS characterization of surface layers of stainless steel nitrided in electron beam plasma at low temperature. *Surf. Coat. Technol.* **2020**, *386*, 125492. [CrossRef]
111. Martinavičius, A.; Abrasonis, G.; Scheinost, A.C.; Danoix, R.; Danoix, F.; Stinville, J.C.; Talut, G.; Templier, C.; Liedke, O.; Gemming, S.; et al. Nitrogen interstitial diffusion induced decomposition in AISI 304L austenitic stainless steel. *Acta Mater.* **2012**, *60*, 4065–4076. [CrossRef]
112. Oddershede, J.; Christiansen, T.L.; Ståhl, K.; Somers, M.A.J. Extended X-ray absorption fine structure investigation of nitrogen stabilized expanded austenite. *Scr. Mater.* **2010**, *62*, 290–293. [CrossRef]
113. Che, H.L.; Tong, S.; Wang, K.S.; Lei, M.K.; Somers, M.A.J. Co-existence of γ_N' phase and γ_N phase on nitrided austenitic Fe–Cr–Ni alloys- I. experiment. *Acta Mater.* **2019**, *177*, 35–45. [CrossRef]
114. Che, H.L.; Lei, M.K. Microstructure of perfect nitrogen-expanded austenite formed by unconstrained nitriding. *Scr. Mater.* **2021**, *194*, 113705. [CrossRef]
115. Che, H.L.; Lei, M.K.; Somers, M.A.J. A simple model for nitrogen-induced lattice expansion of γ_N' and γN phases in Fe–Cr–Ni alloys with different chromium contents. *Philos. Mag. Lett.* **2020**, *100*, 435–441. [CrossRef]
116. Che, H.L.; Christiansen, T.L.; Lei, M.K.; Somers, M.A.J. Co-existence of γ_N' phase and γ_N phase in nitrided austenitic Fe-Cr-Ni alloys-II: A pragmatic modeling approach. *Acta Mater.* **2022**, *235*, 118094. [CrossRef]
117. Lei, M.K. Phase transformations in plasma source ion nitrided austenitic stainless steel at low temperature. *J. Mater. Sci.* **1999**, *34*, 5975–5982. [CrossRef]
118. Tao, X.; Qi, J.; Rainforth, M.; Matthews, A.; Leyland, A. On the interstitial induced lattice inhomogeneities in nitrogen-expanded austenite. *Scr. Mater.* **2020**, *185*, 146–151. [CrossRef]
119. Tong, K.; Ye, F.; Che, H.; Lei, M.K.; Miao, S.; Zhang, C. High-density stacking faults in a supersaturated nitrided layer on austenitic stainless steel. *J. Appl. Crystallogr.* **2016**, *49*, 1967–1971. [CrossRef]
120. Lei, M.K.; Huang, Y.; Zhang, Z.L. In situ Transformation of Nitrogen-induced h.c.p. Martensite in Plasma Source Ion–nitrided Austenitic Stainless Steel. *J. Mater. Sci. Lett.* **1998**, *17*, 1165–1167. [CrossRef]
121. Sah, J.V.; Dwivedi, P.K.; Mukherjee, S.; Jhala, G.; Joseph, A. Influence of γ_N' and ε_N' phases on the properties of AISI 304L after low-temperature plasma nitrocarburizing. *J. Vac. Sci. Technol. A* **2023**, *41*, 33101. [CrossRef]
122. Sah, J.; Joseph, A.; Jhala, G.; Mukherjee, S. On the Effects of H2 and Ar on Dual Layer Formed by Plasma Nitrocarburizing on Austenitic Stainless Steels. *J. Mater. Eng. Perform.* **2022**, *31*, 2664–2677. [CrossRef]
123. Borgioli, F.; Galvanetto, E.; Bacci, T. Surface Modification of a Nickel-Free Austenitic Stainless Steel by Low-Temperature Nitriding. *Metals* **2021**, *11*, 1845. [CrossRef]
124. Borgioli, F.; Galvanetto, E.; Bacci, T. Effects of surface modification by means of low temperature plasma nitriding on wetting and corrosion behavior of austenitic stainless steel. *Coatings* **2020**, *10*, 98. [CrossRef]
125. Borgioli, F.; Galvanetto, E.; Bacci, T. Surface modification of austenitic stainless steel by means of low pressure glow-discharge treatments with nitrogen. *Coatings* **2019**, *9*, 604. [CrossRef]
126. Tsujikawa, M.; Yamauchi, N.; Ueda, N.; Sone, T.; Hirose, Y. Behavior of carbon in low temperature plasma nitriding layer of austenitic stainless steel. *Surf. Coat. Technol.* **2005**, *193*, 309–313. [CrossRef]
127. Manova, D.; Mändl, S. Initial phase formation during nitriding of austenitic stainless steel. *Surf. Coat. Technol.* **2023**, *456*, 129258. [CrossRef]
128. Collins, G.A.; Hutchings, R.; Short, K.T.; Tendys, J.; Li, X.; Samandi, M. Nitriding of austenitic stainless steel by plasma immersion ion implantation. *Surf. Coat. Technol.* **1995**, *74*, 417–424. [CrossRef]
129. Borgioli, F.; Fossati, A.; Raugei, L.; Galvanetto, E.; Bacci, T. Low temperature glow-discharge nitriding of stainless steels. In Proceedings of the 7th European Stainless Steel Conference: Science and Market, Como, Italy, 21–23 September 2011; Associazione Italiana di Metallurgia: Milan, Italy, 2011.
130. Tao, X.; Li, X.; Dong, H.; Matthews, A.; Leyland, A. Evaluation of the sliding wear and corrosion performance of triode-plasma nitrided Fe-17Cr-20Mn-0.5N high-manganese and Fe-19Cr-35Ni-1.2Si high-nickel austenitic stainless steels. *Surf. Coat. Technol.* **2021**, *409*, 126890. [CrossRef]
131. Fossati, A.; Borgioli, F.; Galvanetto, E.; Bacci, T. Glow-discharge nitriding of AISI 316L austenitic stainless steel: Influence of treatment time. *Surf. Coat. Technol.* **2006**, *200*, 3511–3517. [CrossRef]
132. Wu, D.; Ge, Y.; Kahn, H.; Ernst, F.; Heuer, A.H. Diffusion profiles after nitrocarburizing austenitic stainless steel. *Surf. Coat. Technol.* **2015**, *279*, 180–185. [CrossRef]
133. Sun, Y.; Li, X.; Bell, T. Low temperature plasma carburising of austenitic stainless steels for improved wear and corrosion resistance. *Surf. Eng.* **1999**, *15*, 49–54. [CrossRef]
134. Revilla, R.I.; Wouters, B.; Andreatta, F.; Lanzutti, A.; Fedrizzi, L.; De Graeve, I. EIS comparative study and critical Equivalent Electrical Circuit (EEC) analysis of the native oxide layer of additive manufactured and wrought 316L stainless steel. *Corros. Sci.* **2020**, *167*, 108480. [CrossRef]
135. Marcelin, S.; Pébère, N.; Régnier, S. Electrochemical characterisation of a martensitic stainless steel in a neutral chloride solution. *Electrochim. Acta* **2013**, *87*, 32–40. [CrossRef]
136. Omanovic, S.; Roscoe, S.G. Electrochemical Studies of the Adsorption Behavior of Bovine Serum Albumin on Stainless Steel. *Langmuir* **1999**, *15*, 8315–8321. [CrossRef]

137. Perumal, G.; Grewal, H.S.; Arora, H.S. Enhanced durability, bio-activity and corrosion resistance of stainless steel through severe surface deformation. *Colloids Surf. B Biointerfaces* **2020**, *194*, 111197. [CrossRef] [PubMed]
138. Bou-Saleh, Z.; Shahryari, A.; Omanovic, S. Enhancement of corrosion resistance of a biomedical grade 316LVM stainless steel by potentiodynamic cyclic polarization. *Thin Solid Film.* **2007**, *515*, 4727–4737. [CrossRef]
139. Harrington, D.A.; van den Driessche, P. Mechanism and equivalent circuits in electrochemical impedance spectroscopy. *Electrochim. Acta* **2011**, *56*, 8005–8013. [CrossRef]
140. Tandon, V.; Patil, A.P. On the Influence of Cold Working and Electrochemical Nitridation on the Corrosion Behaviour of 316L Austenitic Stainless Steel in Acidic Environment. *Surf. Eng. Appl. Electrochem.* **2020**, *56*, 63–70. [CrossRef]
141. Sheng, X.; Ting, Y.P.; Pehkonen, S.O. The influence of sulphate-reducing bacteria biofilm on the corrosion of stainless steel AISI 316. *Corros. Sci.* **2007**, *49*, 2159–2176. [CrossRef]
142. Wang, L.; Tian, H.; Gao, H.; Xie, F.; Zhao, K.; Cui, Z. Electrochemical and XPS analytical investigation of the accelerative effect of bicarbonate/carbonate ions on AISI 304 in alkaline environment. *Appl. Surf. Sci.* **2019**, *492*, 792–807. [CrossRef]
143. Shahryari, A.; Omanovic, S.; Szpunar, J.A. Enhancement of biocompatibility of 316LVM stainless steel by cyclic potentiodynamic passivation. *J. Biomed. Mater. Res. Part A* **2009**, *89A*, 1049–1062. [CrossRef]
144. Wang, X.; Liu, Z.; Chen, Y.; Sun, J.; He, Q.; Liu, Q.; Liu, G.; Xie, K. Abrasive resistance and corrosion properties of AISI 316 sieve via low-temperature gaseous nitriding. *Surf. Coat. Technol.* **2019**, *361*, 349–356. [CrossRef]
145. Abreu, C.M.; Cristóbal, M.J.; Merino, P.; Nóvoa, X.R.; Pena, G.; Pérez, M.C. Electrochemical behaviour of an AISI 304L stainless steel implanted with nitrogen. *Electrochim. Acta* **2008**, *53*, 6000–6007. [CrossRef]
146. Hirschorn, B.; Orazem, M.E.; Tribollet, B.; Vivier, V.; Frateur, I.; Musiani, M. Constant-Phase-Element Behavior Caused by Resistivity Distributions in Films. *J. Electrochem. Soc.* **2010**, *157*, C458–C463. [CrossRef]
147. Jüttner, K. Electrochemical impedance spectroscopy (EIS) of corrosion processes on inhomogeneous surfaces. *Electrochim. Acta* **1990**, *35*, 1501–1508. [CrossRef]
148. Bai, H.; Wang, F. Protective Properties of High Temperature Oxide Films on Ni-based Superalloys in 3.5% NaCl Solution. *J. Mater. Sci. Technol.* **2007**, *23*, 541–546.
149. Luiz, L.A.; Kurelo, B.C.E.S.; Souza, G.B.; Andrade, J.; Marino, C.E.B. Effect of nitrogen plasma immersion ion implantation on the corrosion protection mechanisms of different stainless steels. *Mater. Today Commun.* **2021**, *28*, 102655. [CrossRef]
150. Fossati, A.; Borgioli, F.; Galvanetto, E.; Bacci, T. Corrosion resistance properties of glow-discharge nitrided AISI 316L austenitic stainless steel in NaCl solutions. *Corros. Sci.* **2006**, *48*, 1513–1527. [CrossRef]
151. Örnek, C.; Leygraf, C.; Pan, J. Passive film characterisation of duplex stainless steel using scanning Kelvin probe force microscopy in combination with electrochemical measurements. *NPJ Mater. Degrad.* **2019**, *3*, 8. [CrossRef]
152. Rybalka, K.V.; Shaldaev, V.S.; Beketaeva, L.A.; Malofeeva, A.N.; Davydov, A.D. Development of pitting corrosion of stainless steel 403 in sodium chloride solutions. *Russ. J. Electrochem.* **2010**, *46*, 196–204. [CrossRef]
153. Jin, S.; Atrens, A. Passive films on stainless steels in aqueous media. *Appl. Phys. A Solids Surf.* **1990**, *50*, 287–300. [CrossRef]
154. Schibicheski Kurelo, B.C.E.; de Souza, G.B.; Serbena, F.C.; Lepienski, C.M.; Chuproski, R.F.; Borges, P.C. Improved saline corrosion and hydrogen embrittlement resistances of superaustenitic stainless steel by PIII nitriding. *J. Mater. Res. Technol.* **2022**, *18*, 1717–1731. [CrossRef]
155. Olzon-Dionysio, M.; Olzon-Dionysio, D.; Campos, M.; Shigeyosi, W.T.; de Souza, S.D.; de Souza, S. Corrosion resistance of AISI 316L plasma nitrided at different temperatures and times. *Hyperfine Interact.* **2019**, *240*, 26. [CrossRef]
156. Lei, M.K.; Zhu, X.M. Plasma-based low-energy ion implantation of austenitic stainless steel for improvement in wear and corrosion resistance. *Surf. Coat. Technol.* **2005**, *193*, 22–28. [CrossRef]
157. Kamachi Mudali, U.; Shankar, P.; Ningshen, S.; Dayal, R.K.; Khatak, H.S.; Raj, B. On the pitting corrosion resistance of nitrogen alloyed cold worked austenitic stainless steels. *Corros. Sci.* **2002**, *44*, 2183–2198. [CrossRef]
158. Lei, M.K.; Zhang, Z.L.; Zhu, X.M. Effects of nitrogen-induced hcp martensite formation on corrosion resistance of plasma source ion nitrided austenitic stainless steel. *J. Mater. Sci. Lett.* **1999**, *18*, 1537–1538. [CrossRef]
159. Chao, K.L.; Liao, H.Y.; Shyue, J.J.; Lian, S.S. Corrosion behavior of high nitrogen nickel-free Fe-16Cr-Mn-Mo-N stainless steels. *Metall. Mater. Trans. B Process Metall. Mater. Process. Sci.* **2014**, *45*, 381–391. [CrossRef]
160. Escalada, L.; Dalibon, E.L.; Brühl, S.P.; Manova, D.; Mändl, S.; Simison, S. Influence of Inclusions in the Corrosion Behavior of Plasma Nitrided Stainless Steel. *Adv. Eng. Mater.* **2022**, 2201112. [CrossRef]
161. Saravanan, P.; Raja, V.S.; Mukherjee, S. Effect of alloyed molybdenum on corrosion behavior of plasma immersion nitrogen ion implanted austenitic stainless steel. *Corros. Sci.* **2013**, *74*, 106–115. [CrossRef]
162. Burstein, G.T.; Vines, S.P. Repetitive Nucleation of Corrosion Pits on Stainless Steel and the Effects of Surface Roughness. *J. Electrochem. Soc.* **2001**, *148*, 504–516. [CrossRef]
163. Li, N.; Wang, N. The effect of duplex Surface mechanical attrition and nitriding treatment on corrosion resistance of stainless steel 316L. *Sci. Rep.* **2018**, *8*, 8454. [CrossRef]

164. Picard, S.; Memet, J.B.; Sabot, R.; Grosseau-Poussard, J.L.; Rivière, J.P.; Meilland, R. Corrosion behaviour, microhardness and surface characterisation of low energy, high current ion implanted austenitic stainless steel. *Mater. Sci. Eng. A* **2001**, *303*, 163–172. [CrossRef]
165. Frangini, S.; De Cristofaro, N. Analysis of the galvanostatic polarization method for determining reliable pitting potentials on stainless steels in crevice-free conditions. *Corros. Sci.* **2003**, *45*, 2769–2786. [CrossRef]
166. Spies, H.-J.; Eckstein, C.; Biermann, H.; Franke, A. Corrosion behaviour of stainless steels after low temperature thermochemical treatment. *Materwiss. Werksttech.* **2010**, *41*, 133–141. [CrossRef]
167. Sun, Y. Enhancement in corrosion resistance of austenitic stainless steels by surface alloying with nitrogen and carbon. *Mater. Lett.* **2005**, *59*, 3410–3413. [CrossRef]
168. Borowski, T. Enhancing the Corrosion Resistance of Austenitic Steel Using Active Screen Plasma Nitriding and Nitrocarburising. *Materials* **2021**, *14*, 3320. [CrossRef]
169. Martinesi, M.; Bruni, S.; Stio, M.; Treves, C.; Bacci, T.; Borgioli, F. Biocompatibility evaluation of surface-treated AISI 316L austenitic stainless steel in human cell cultures. *J. Biomed. Mater. Res. Part A* **2007**, *80*, 131–145. [CrossRef]
170. Martinesi, M.; Stio, M.; Treves, C.; Borgioli, F. Biocompatibility studies of low temperature nitrided and collagen-I coated AISI 316L austenitic stainless steel. *J. Mater. Sci. Mater. Med.* **2013**, *24*, 1501–1513. [CrossRef] [PubMed]
171. Stio, M.; Martinesi, M.; Treves, C.; Borgioli, F. In vitro response of human peripheral blood mononuclear cells to AISI 316L austenitic stainless steel subjected to nitriding and collagen coating treatments. *J. Mater. Sci. Mater. Med.* **2015**, *26*, 100. [CrossRef] [PubMed]
172. Stio, M.; Martinesi, M.; Treves, C.; Borgioli, F. Cultures and co-cultures of human blood mononuclear cells and endothelial cells for the biocompatibility assessment of surface modified AISI 316L austenitic stainless steel. *Mater. Sci. Eng. C* **2016**, *69*, 1081–1091. [CrossRef]
173. Valero Vidal, C.; Igual Muñoz, A. Electrochemical characterisation of biomedical alloys for surgical implants in simulated body fluids. *Corros. Sci.* **2008**, *50*, 1954–1961. [CrossRef]
174. Lei, M.K.; Zhu, X.M. In vitro corrosion resistance of plasma source ion nitrided austenitic stainless steels. *Biomaterials* **2001**, *22*, 641–647. [CrossRef]
175. Yazıcı, M.; Çomaklı, O.; Yetim, T.; Yetim, A.F.; Çelik, A. The effect of plasma nitriding temperature on the electrochemical and semiconducting properties of thin passive films formed on 316L stainless steel implant material in SBF solution. *Surf. Coat. Technol.* **2015**, *261*, 181–188. [CrossRef]
176. Dashtbozorg, B.; Li, X.; Romano, J.; Garcia–Giron, A.; Sammons, R.L.; Dimov, S.; Dong, H. A study on the effect of ultrashort pulsed laser texturing on the microstructure and properties of metastable S phase layer formed on AISI 316L surfaces. *Appl. Surf. Sci.* **2020**, *511*, 145557. [CrossRef]
177. Flis, J.; Kuczynska, M. Effect of Low-Temperature Plasma Nitriding on Corrosion of 304L Stainless Steel in Sulfate and Chloride Solutions. *J. Electrochem. Soc.* **2004**, *151*, B573–B580. [CrossRef]
178. Buhagiar, J.; Dong, H. Corrosion properties of S-phase layers formed on medical grade austenitic stainless steel. *J. Mater. Sci. Mater. Med.* **2012**, *23*, 271–281. [CrossRef] [PubMed]
179. Buhagiar, J.; Spiteri, A.; Sacco, M.; Sinagra, E.; Dong, H. Augmentation of crevice corrosion resistance of medical grade 316LVM stainless steel by plasma carburising. *Corros. Sci.* **2012**, *59*, 169–178. [CrossRef]
180. Li, C.X.; Bell, T. Corrosion properties of active screen plasma nitrided 316 austenitic stainless steel. *Corros. Sci.* **2004**, *46*, 1527–1547. [CrossRef]
181. Sun, Y. Response of cast austenitic stainless steel to low temperature plasma carburizing. *Mater. Des.* **2009**, *30*, 1377–1380. [CrossRef]
182. Ceschini, L.; Chiavari, C.; Lanzoni, E.; Martini, C. Low-temperature carburised AISI 316L austenitic stainless steel: Wear and corrosion behaviour. *Mater. Des.* **2012**, *38*, 154–160. [CrossRef]
183. Lin, K.; Li, X.; Sun, Y.; Luo, X.; Dong, H. Active screen plasma nitriding of 316 stainless steel for the application of bipolar plates in proton exchange membrane fuel cells. *Int. J. Hydrog. Energy* **2014**, *39*, 21470–21479. [CrossRef]
184. Tian, R.; Sun, J.; Wang, L. Plasma-nitrided austenitic stainless steel 316L as bipolar plate for PEMFC. *Int. J. Hydrog. Energy* **2006**, *31*, 1874–1878. [CrossRef]
185. Tandon, V.; Patil, A.P.; Rathod, R.C. Influence of Time on Low Temperature Salt Bath Nitriding and its Corrosion Behavior of 316L ASS in PEMFC Environment. *Prot. Met. Phys. Chem. Surf.* **2020**, *56*, 772–779. [CrossRef]
186. Spies, H.-J.; Eckstein, C.; Zimdars, H. Structure and corrosion behaviour of stainless steels after plasma and gas nitriding. *Surf. Eng.* **2002**, *18*, 459–460. [CrossRef]
187. Flis, J.; Gajek, A. Impedance parameters of nitrided 304L stainless steel in an acidified sulphate solution. *J. Electroanal. Chem.* **2001**, *515*, 82–90. [CrossRef]
188. Kuczynska-Wydorska, M.; Flis, J. Corrosion and passivation of low-temperature nitrided AISI 304L and 316L stainless steels in acidified sodium sulphate solution. *Corros. Sci.* **2008**, *50*, 523–533. [CrossRef]
189. Sun, Y. Depth-profiling electrochemical measurements of low temperature plasma carburised 316L stainless steel in 1M H_2SO_4 solution. *Surf. Coat. Technol.* **2010**, *204*, 2789–2796. [CrossRef]

190. Corujeira Gallo, S.; Dong, H. Corrosion behaviour of direct current and active screen plasma carburised AISI 316 stainless steel in boiling sulphuric acid solutions. *Corros. Eng. Sci. Technol.* **2011**, *46*, 8–16. [CrossRef]
191. Sun, Y.; Bailey, R. Comparison of Wear Performance of Low Temperature Nitrided and Carburized 316L Stainless Steel under Dry Sliding and Corrosive-Wear Conditions. *J. Mater. Eng. Perform.* **2023**, *32*, 1238–1247. [CrossRef]

metals

MDPI

Article

Effects of Solid-Solution Carbon and Eutectic Carbides in AISI 316L Steel-Based Tungsten Carbide Composites on Plasma Carburizing and Nitriding

Shinichiro Adachi *, Takuto Yamaguchi, Keigo Tanaka, Takashi Nishimura and Nobuhiro Ueda

Osaka Research Institute of Industrial Science and Technology, 2-7-1 Ayumino, Izumi 594-1157, Osaka, Japan
* Correspondence: shinadachi@orist.jp; Tel.: +81-725-51-2648; Fax: +81-725-51-2749

Abstract: AISI 316L stainless-steel-based tungsten carbide composite layers fabricated via laser metal deposition are used for additive manufacturing. Heat treatment practices such as low-temperature plasma carburizing and nitriding improve the hardness and corrosion resistance of austenitic stainless steels via the formation of expanded austenite, known as the S phase. In the present study, practices to enhance the hardness and corrosion resistances of the stainless-steel parts in the composite layers have been investigated, including single plasma carburizing for 4 h and continuous plasma nitriding for 3.5 h following carburizing for 0.5 h at 400 and 450 °C. The as-deposited composite layers contain solid-solution carbon and eutectic carbides owing to the thermal decomposition of tungsten carbide during the laser metal deposition. The eutectic carbides inhibit carbon diffusion, whereas the original solid-solution carbon contributes to the formation of the S phase, resulting in a thick S phase layer. Both the single carburizing and continuous processes are effective in improving the Vickers surface hardness and corrosion resistance of the composite layers despite containing the solid-solution carbon and eutectic carbides.

Keywords: plasma carburizing; plasma nitriding; laser metal deposition; stainless steel; tungsten carbide; hardness; corrosion resistance

Citation: Adachi, S.; Yamaguchi, T.; Tanaka, K.; Nishimura, T.; Ueda, N. Effects of Solid-Solution Carbon and Eutectic Carbides in AISI 316L Steel-Based Tungsten Carbide Composites on Plasma Carburizing and Nitriding. *Metals* **2023**, *13*, 1350. https://doi.org/10.3390/met13081350

Academic Editor: Catalin Constantinescu

Received: 21 June 2023
Revised: 20 July 2023
Accepted: 25 July 2023
Published: 27 July 2023

1. Introduction

Additive manufacturing using molten metal and laser-beam heating is widely used for the fabrication of three-dimensional objects, particularly parts with complex shapes or highly functional materials that are difficult to machine. Recently, additive manufacturing technology has been rapidly developed and is now applied in the aerospace industry and medical field, as well as to fabricate general prototypes and machinery parts. Additive manufacturing includes methods such as directed energy deposition, whereby a molten metal powder or wire is deposited on a substrate to form an object via repeated layering, and powder bed fusion, wherein a laser beam is repeatedly irradiated onto a bed of powder to form an object [1–6].

Steel materials, particularly austenitic stainless steels such as AISI 316L, which are readily available and exhibit excellent corrosion resistance, are actively used for additive manufacturing. The metallurgical and mechanical properties of additively manufactured AISI 316L stainless steels have been extensively studied [7–13]. Additive manufacturing via laser metal deposition has been used to improve the wear resistance of stainless-steel materials via compositing with tungsten carbide particles [14–18]. Tungsten carbide can improve the hardness of stainless-steel materials via dispersion. Furthermore, eutectic carbides are formed by the reaction of pyrolyzed tungsten carbide particles with the metal elements in stainless steel; these eutectic carbides enhance the hardness of stainless steel. However, unlike hard materials, such as ceramics and cermet materials, the stainless-steel parts in tungsten carbide composite layers cannot withstand severe friction and wear environments.

Low-temperature nitriding of austenitic stainless steels at temperatures below 450 °C supersaturates the nitrogen atoms in the austenite crystal lattice, thereby forming an expanded austenite phase, namely the S phase, which improves their hardness and wear resistance. The S phase is a solid solution of nitrogen (not a nitride compound), which does not affect the passive-film-forming ability of chromium in stainless steel. Thus, high corrosion resistance is maintained, and pitting corrosion in chlorinated environments is inhibited [19–28]. In a previous study, AISI 316L stainless-steel-based tungsten carbide composite layers fabricated via laser metal deposition were subjected to low-temperature plasma nitriding, and the formation of the S phase was confirmed at temperatures of 400–450 °C and a processing time of 4 h [29]. The hardness of the stainless-steel parts in the composite layers increased from 500–700 HV to 1200–1400 HV. The as-deposited composite layers exhibited poor corrosion resistance in NaCl because the solid-solution chromium formed eutectic carbides and weakened the passive film. However, the plasma nitriding treatment significantly increased the pitting resistance equivalent number (PREN) with the dissolution of nitrogen, and consequently, the corrosion resistance of the composite layers increased. In addition, the metallurgical structure of the composite layers was significantly different from that of general stainless-steel materials. Solid-solution carbon was produced by the thermal decomposition of the tungsten carbide particles during the laser metal deposition. The composite layers, when subjected to low-temperature plasma nitriding, formed a dual S phase layer, i.e., a supersaturated nitrogen S phase (nitriding S phase) on the outside and a supersaturated carbon S phase (carburizing S phase) inside. These results suggest that the original solid-solution carbon played a major role in the formation of the S phase.

Low-temperature carburizing of stainless-steel materials using carburizing gases, such as methane, propane, and butane, is a common method for producing a carburizing S phase with supersaturated carbon. In this method, carbon diffuses from the surface to form an expanded austenitic phase, thus improving wear resistance similar to the nitriding S phase [30–37]. Furthermore, compared to only carburizing or nitriding, a continuous or simultaneous treatment that combines carburizing and nitriding can increase the thickness, hardness, and wear resistance of the S phase [38–45]. Therefore, low-temperature carburizing and combined treatments for AISI 316L stainless-steel-based tungsten carbide composite layers can be used to improve wear and corrosion resistance.

A few papers have reported low-temperature nitriding and carburizing for austenitic stainless-steel fabrication via directed energy deposition and powder bed fusion [22,24,32,46,47]. However, to the best of our knowledge, none of the studies published to date investigate tungsten carbide composites except the previous study [29]. The effect of original solid-solution carbon present in stainless-steel plates on low-temperature nitriding has been well-researched in previous studies, establishing the formation of thin carburizing S phase layers at the interface between the nitriding layer and substrate [21,26,48,49]. In contrast, the effect on low-temperature carburizing has not been investigated yet because the amount of original solid-solution carbon in stainless-steel materials is negligible compared with that of carbon diffused via carburizing. However, AISI 316L stainless-steel-based tungsten carbide composite layers contain large amounts of solid-solution carbon and eutectic carbides; therefore, their effect on the S phase cannot be ignored, unlike that in ordinary stainless-steel plates. Moreover, the original solid-solution carbon in the composite layers can influence the formation and properties of the carburizing S phase more than those of the nitriding S phase because the diffusing element is carbon.

In this study, AISI 316L stainless-steel-based tungsten carbide composite layers were fabricated via laser metal deposition and followed by treatment with two methods, a single plasma carburizing process, and a continuous plasma nitriding process after carburizing at temperatures below 450 °C. The microstructure, X-ray diffraction (XRD) patterns, and depth profiles of carbon and nitrogen were examined to reveal the formation mechanism of the S phase and evaluate the influence of the original solid-solution carbon and eutectic carbides. In addition, the S phase properties were evaluated via Vickers hardness and anodic polarization measurements. The effects of the tungsten carbide composition and plasma process conditions were investigated to develop the optimal treatment method for the composite layers.

2. Experimental Section

Composite layers of AISI 316L stainless-steel-based tungsten carbide with a thickness of approximately 2 mm were fabricated using a laser metal deposition machine equipped with a continuous-wave diode laser (wavelength, 940 nm; LDM-2000-60, Laserline, GmbH, Mulheim-Karlich, Germany) and attached coaxial powder feed head (COAX12, Fraunhofer, Munchen, Germany). The output power of the laser beam was 1.4 kW, and a single layer was deposited on the AISI 304 stainless-steel plates with a size of 65 mm × 65 mm × 10 mm and a composition of 0.04 wt.% C, 0.3 wt.% Si, 1.7 wt.% Mn, 0.04 wt.% P, 0.01 wt.% S, 8.5 wt.% Ni, 18.2 wt.% Cr and Fe (balance) by scanning the coaxial powder feed head at a transverse speed of 4 mm s^{-1} and length of 50 mm at 3 mm intervals. The AISI 316L stainless-steel powder with a particle size of −212/+63 μm and a composition of 0.03 wt.% C, 0.7 wt.% Si, 0.9 wt.% Mn, 0.03 wt.% P, 0.02 wt.% S, 12.4 wt.% Ni, 17.4 wt.% Cr, 2.1 wt.% Mo and Fe (balance) was mixed with 20 and 40 wt.% tungsten carbide powder with a particle size of −180/+53 μm and a carbon content of about 5 wt.%, which was delivered in the coaxial direction of the laser beam through the powder feed head with mass flow rates of 17–19 g min^{-1}; Ar was used as the carrier gas.

The low-temperature plasma treatments at 400 and 450 °C were performed using two processes: single carburizing for 4 h and continuous nitriding for 3.5 h after carburizing for 0.5 h, using a DC plasma ion treatment machine (FECH-1N, Fuji Electronics Industry, Osaka, Japan). Before the plasma treatments, the surfaces of the composite layers were polished to remove any oxide layer and flattened like a mirror because an oxide layer prevents the diffusion of carbon and nitrogen. The polished samples and a copper plate in a glass chamber were connected to the cathode and anode electrodes, respectively. CH_4:Ar:H_2 (5:50:45) and N_2:H_2 (80:20) with a flow rate of 1 L min^{-1} were used as the plasma gases in the chamber to maintain the pressure at 667 Pa for the carburizing and nitriding processes, respectively. The crystal structure was examined using XRD (SmartLab, Rigaku, Tokyo, Japan) via a conventional θ–2θ scan of Cu-Kα radiation at 40 kV and 150 mA with a scan rate of 10-degree min^{-1}. Compounds were identified from XRD peaks using the JCPDS cards. The cross-sectional microstructure was observed using an optical microscope (ECLIPSE MA100N; Nikon, Tokyo, Japan). The depth profiles of carbon and nitrogen were examined through glow-discharge optical emission spectroscopy (GDOES; GDA750 system Rigaku, Tokyo, Japan). The surface atomic composition was evaluated using X-ray photoelectron spectroscopy (XPS; PHI Quantera II instrument, Ulvac-PHI, Kanagawa, Japan). The XPS data was collected by a monochromatic Al Kα radiation with an X-ray beam diameter of 200 μm. Ar^+ was used for the sputtering with a rate of 2 nm min^{-1} for the oxide layer of SiO_2.

The surface hardness of the stainless-steel parts of the composite layers was measured using a Vickers tester (HM-220D, Mitutoyo, Tokyo, Japan) with a test load of 0.245 N. The Vickers tests were repeated at least eight times for each sample, and the maximum and minimum values were deleted. The hardness depth profiles of the cross-sections were measured using a nanoindentation tester (ENT-1100a, Elionix, Tokyo, Japan) with a test load of 1 mN and a loading rate of 1×10^{-4} N s^{-1}. The nanoindentation tests were repeated four times, and the data with an indentation depth of less than 50 nm were deleted.

The anodic polarization curves were obtained using a potentiostat (HSV-110, Hokuto Denko Corporation, Tokyo, Japan) with an Ag/AgCl reference electrode and platinum wire counter electrode. The samples were directly exposed to polarization tests, and the measurement voltage was applied between −0.6 V and 1 V with a scan rate of 1 mV s^{-1}—a 3.5 wt.% NaCl solution at a temperature of 30 °C was used for the measurements. Since the dissolved oxygen in the solution may affect the polarization curves, oxygen was removed by bubbling nitrogen gas for at least 0.5 h prior to the measurements. After the measurements, the surface morphologies were examined using a digital microscope (Dino-Lite, AnMo Electronics Corporation, Taipei, Taiwan).

3. Results and Discussion

3.1. Formation Mechanism of the S Phase during a Single Carburizing Process

Figure 1 shows the XRD patterns of the AISI 316L stainless-steel-based tungsten carbide composite layers. The as-deposited composite layers contained the austenite (γ) and a small amount of ferrite (α) phase in stainless steel with the eutectic carbides WC, W_2C, and M_6C. $M_{12}C$ was found only in the 40 wt.% tungsten carbide layer. After the single carburizing process had been conducted for 4 h, the diffraction peaks of the austenite phase shifted to lower angles. The etched cross-sectional microstructure in Figure 2 shows bright-contrast layers on the surface. These results indicate that the crystal lattice of the austenite phase was expanded by the supersaturated solid solution of carbon, that is, the S phase formed on the composite layers.

Figure 3 shows the S phase thickness obtained by measuring the bright-contrast layer thickness, as shown in Figure 2. The average S phase thicknesses of the 20 wt.% composite layers with single carburizing temperatures of 400 and 450 °C were 7.3 and 19.9 μm respectively, and those of the 40 wt.% composite layers were 7 and 14.5 μm respectively. Figure 4a,b show the depth profiles of carbon obtained by GDOES. As indicated by the arrows, the carbon content in the 20 wt.% composite layers increased at depths of 9 and 21 μm, respectively, at 400 and 450 °C, whereas the carbon content in the 40 wt.% composite layers increased at depths of 8.5 and 15.5 μm, respectively. These depths with increased carbon contents were marginally larger than the thickness of the S phase (difference of less than 2 μm) for both 20 and 40 wt.% composite layers. Considering that eutectic carbide existed in the as-deposited composite layers, high content of the original solid-solution carbon was expected before the carburizing process. Therefore, only a slight increase in the solid-solution carbon content due to the carburizing process can immediately lead to the formation of the S phase.

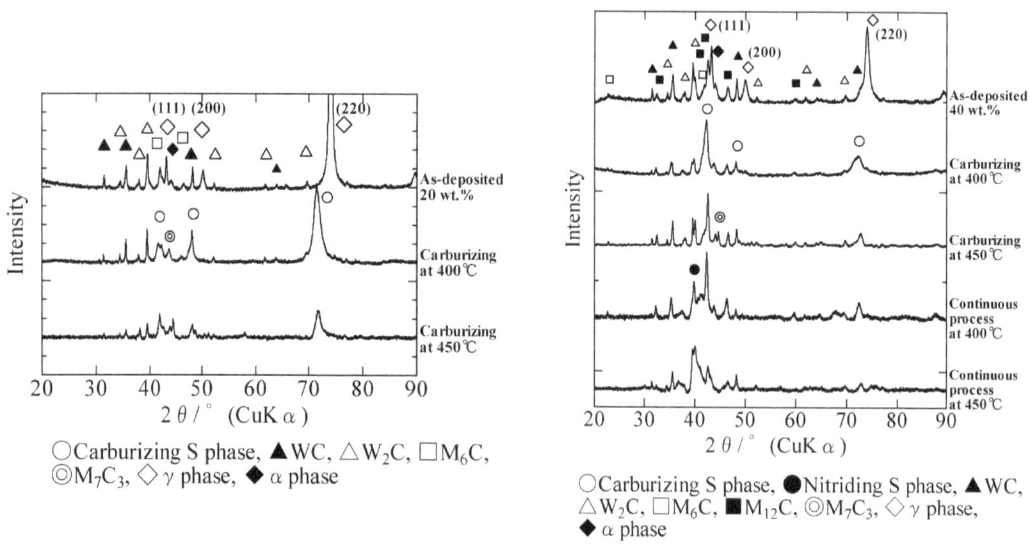

(a) (b)

Figure 1. XRD patterns of the AISI 316L steel-based composite layers at tungsten carbide composition of (**a**) 20 wt.% and (**b**) 40 wt.%.

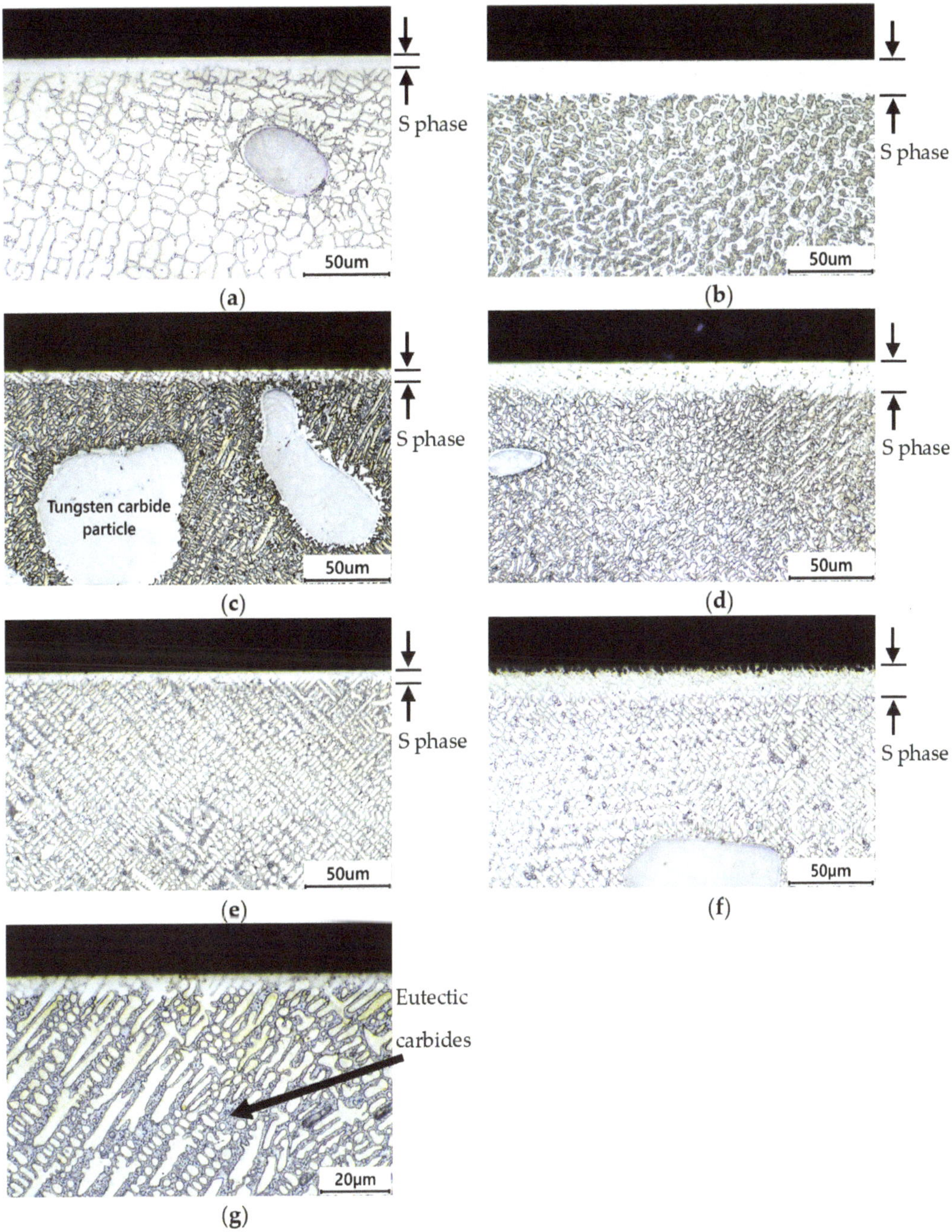

Figure 2. Cross-sectional micrographs of the AISI 316L steel-based composite layers obtained after the plasma processes at tungsten carbide compositions of (**a**) 20 wt.% carburized at 400 °C, (**b**) 20 wt.% carburized at 450 °C, (**c**) 40 wt.% carburized at 400 °C, (**d**) 40 wt.% carburized at 450 °C, (**e**) 40 wt.% continuously processed at 400 °C, (**f**) 40 wt.% continuously processed at 450 °C, and (**g**) enlarged view of 40 wt.% continuously processed at 400 °C.

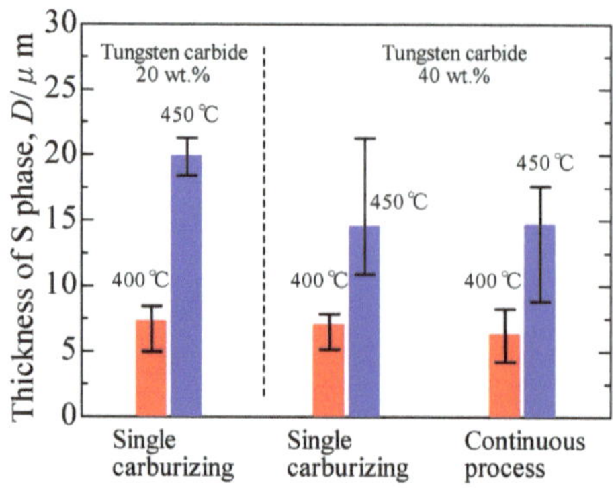

Figure 3. Thickness of the S phase on the AISI 316L steel-based composite layers at various tungsten carbide compositions after the plasma processes.

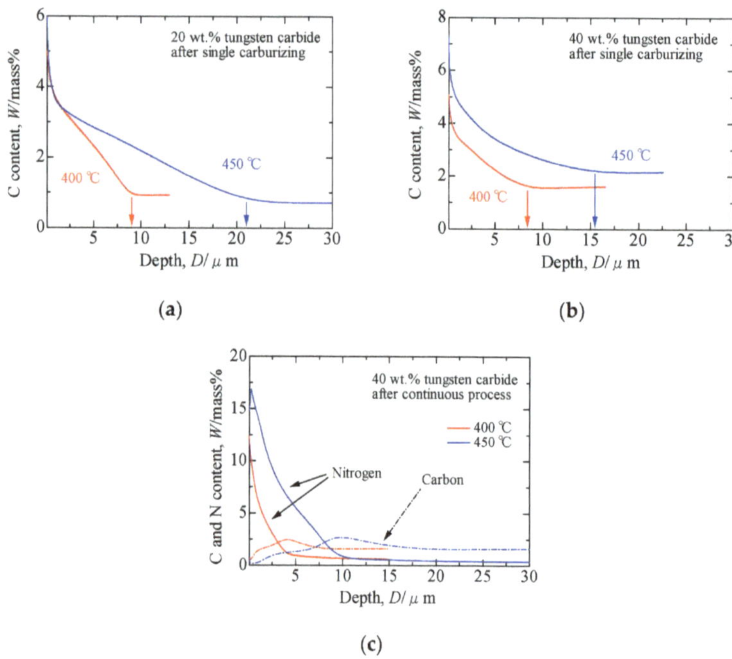

Figure 4. Carbon and nitrogen depth profiles of the AISI 316L steel-based composite layers after the plasma processes at tungsten carbide compositions of (**a**) 20 wt.% after carburizing, (**b**) 40 wt.% after carburizing, and (**c**) 40 wt.% after continuous process.

The depth profiles of carbon exhibit a smooth downward trend in general carburizing treatments [50]. This shape possibly originates from the dissolved chromium, which restricts carbon diffusion in stainless steel with binding carbon [51]. Figure 4a shows that a decrease in carbon content in the 20 wt.% composite layers follows a typical S phase profile. In contrast, in the 40 wt.% composite layers, the carbon content gradually decreases toward the interior, according to Fick's second law of diffusion (Figure 4b). This implies that the synthesis of eutectic carbides, such as M_6C and $M_{12}C$, in the as-deposited composite layers reduced the dissolved chromium content. Because the original solid-solution carbon is already bound to the dissolved chromium, its restriction on carbon diffusion weakens.

However, eutectic carbides can inhibit carbon diffusion. The XRD pattern in Figure 1 shows that the M_7C_3 carbide was formed by the single carburizing process. A prolonged low-temperature gas-phase carburizing of the AISI 316L stainless steel can cause the intragranular precipitation of M_7C_3 [52]. In this study, the diffused carbon was trapped by the eutectic carbides M_6C and $M_{12}C$, and the trapped carbon formed M_7C_3, with a high carbon equivalent, via the following reactions:

$$7M_{12}C + 29C \rightarrow 12M_7C_3, \quad (1)$$

$$7M_6C + 11C \rightarrow 6M_7C_3. \quad (2)$$

As shown in Figure 2g, the gray areas of the microstructure are eutectic carbides. Figure 2a–f reveals the presence of eutectic carbides as dendrite-like metallographic structures inside the composite layers. Evidently, the size of the equiaxed grains (equiaxed crystals) in the dendritic structures in the 40 wt.% composite layers is smaller than that of grains in the 20 wt.% composite layers. This result indicates that more densely distributed eutectic carbides were formed because of the higher tungsten carbide composition. The S phase in the 40 wt.% composite layer at 450 °C is thinner than that in the 20 wt.% layer (Figure 3). Further, carbon diffusion in the 40 wt.% composite layer at 450 °C is shallower than in the 20 wt.% composite layer (Figure 4). These results suggest that the eutectic carbides act as carbon diffusion barriers. In addition, no difference in the thickness of the S phase at 400 °C is observed. This suggests that the effect of the eutectic carbides could be negligible at 400 °C because the diffusion rate at 400 °C is slower than at 450 °C. However, the thickness of the S phase in the composite layers was larger than that on AISI 316L steel plate (12 μm) after 4 h plasma carburizing at 450 °C [37]. This was attributed to the original solid-solution carbon contributing to the formation of the S phase, thus resulting in a thick S phase layer.

3.2. Formation Mechanism of the S Phase during a Continuous Plasma Nitriding after Carburizing Process

Figure 1b shows the XRD patterns of the 40 wt.% tungsten carbide layers obtained by continuous nitriding for 3.5 h after carburizing for 0.5 h, thus indicating the coexistence of the supersaturated carbon S phase (carburizing S phase) and nitrogen S phase (nitriding S phase). Nitrogen in the S phase has a higher solubility concentration than carbon in the S phase, thus resulting in a greater expansion of the austenite lattice [53,54]. Therefore, the diffraction peaks of the nitriding S phase (111) shift to a lower angle than those of the carburizing S phase. In addition, in the carbon depth profiles (Figure 4c), the carbon peaks are located in a deeper region compared with the nitrogen peaks, suggesting that the nitriding S phase developed on the outside, whereas the carburizing S phase formed on the inside. Note that the diffusion rate of carbon in the austenite phase is higher than that of nitrogen [55], and the affinity of the dissolved chromium to trap nitrogen is greater than that for carbon [56]; therefore, carbon was pushed inside by the diffused nitrogen. Consequently, the S phase exhibited a dual structure. The pushed carbon included the carbon diffused during the first carburizing process and the original solid-solution carbon, similar to the single carburizing process.

Figure 3 shows that the thickness of the dual S phase is 6.3 μm at 400 °C and 14.7 μm at 450 °C, which is similar to that of the S phase formed by the single carburizing process. Notably, a continuous process combining carburizing and nitriding for austenitic stainless-steel plates increases the thickness of the S phase [38–45]. However, in this study, the amount of carbon increased during the first carburizing, which was conducted for a short time of 0.5 h in the continuous process and was less than the original solid-solution carbon content. Therefore, the thickness of the S phase increased only negligibly during the first carburizing process.

3.3. Hardness of the S Phase by a Single Carburizing Process and Continuous Nitriding after Carburizing Process

The surface hardness of the S phase in the stainless-steel parts was measured using the Vickers hardness test, and the results are shown in Figure 5. The average hardness values of the samples treated by the single carburizing process at 400 and 450 °C for the 20 wt.% layers are 1342 and 1330 HV, respectively, whereas those for the 40 wt.% composite layers are 1210 and 1365 HV, respectively. With the continuous nitriding process after carburizing, the 40 wt.% composite layers exhibit surface hardness values of 1312 and 1361 HV at 400 and 450 °C, respectively. The average hardness of the as-deposited layers is 504 HV and 647 HV for the 20 wt.% and 40 wt.% composite layers, respectively, indicating that the formation of the S phase increases the hardness values. On the other hand, the surface hardness values corresponding to the different tungsten carbide compositions remained almost the same for the single carburizing and continuous processes.

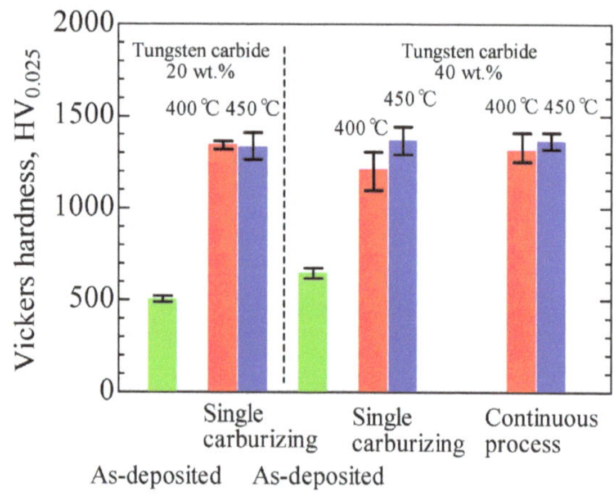

Figure 5. Vickers surface hardness of the AISI 316L steel-based composite layers with various tungsten carbide compositions.

Figure 6 shows the surface hardness profiles of the S phase processed at 450 °C, demonstrating that the S phase resulting from the continuous process was harder than that produced by the single carburizing process up to a depth of approximately 4 μm. The XRD pattern in Figure 1 shows that the peak of (111) associated with the nitrided S phase shifted to a lower angle compared with that of the carburized S phase. Additionally, the GDOES results displayed in Figure 4 indicate that the nitrogen content in the near-surface area up to a depth of 5 μm during the continuous process was 5–16 wt.%, whereas the carbon content after the single carburizing process was comparatively low (3–6 wt.%). This result indicates that the solid-solution nitrogen content obtained from the nitriding in the continuous process is higher than the carbon content from the single carburizing process. Consequently, the near surface of the S phase, produced by the continuous process, exhibited enhanced hardness.

The hardness depth profiles in Figure 6 show that the hardness values of the 40 wt.% layers are 9 GPa in the deep region where the S phase was not formed. The hardness values are higher than the 20 wt.% composite layer values of 6 GPa. The original solid-solution carbon and eutectic carbide concentrations in the 40 wt.% composite layers are higher than those in the 20 wt.% composite layers, which may have resulted in the enhanced hardness of the stainless-steel parts.

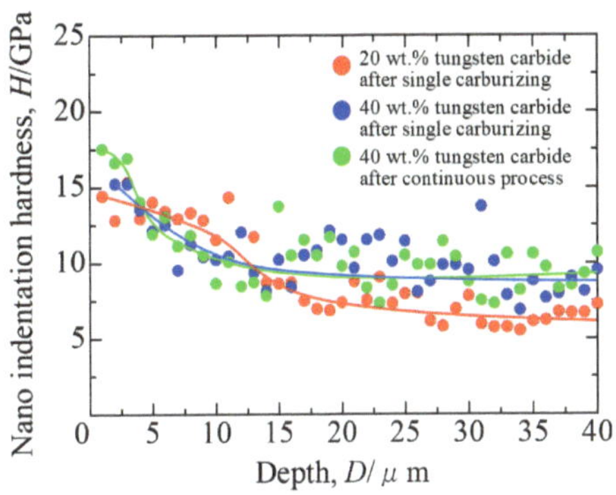

Figure 6. Hardness depth profiles of the AISI 316L steel-based composite layers with various tungsten carbide compositions treated via different plasma processes at 450 °C.

3.4. Anodic Polarization Measurement of the S Phase by a Single Carburizing Process and Continuous Nitriding after Carburizing Process

The anodic polarization measurements were conducted in a 3.5 wt.% NaCl solution. Table 1 shows the corrosion potential and current density from the polarization curves in Figure 7. The corrosion current densities of the as-deposited layers are slightly larger than those of the plasma-processed layers with the same tungsten carbide composition except for the 20 wt.% layer carburized at 450 °C. When the potential is above 0.6 V, the current densities of the as-deposited layers are 100 A m^{-2}, but those of the plasma-processed layers are less than 1 A m^{-2}. Figure 8 shows the surface morphologies after the anodic polarization measurements. The pitting and crevice corrosion is observed around the boundary of the tested area in the as-deposited layers of (a) and (b), especially the 40 wt.% layer was severely corroded. Although some discoloration is observed in the plasma-processed layers of (c)–(h), no sever corrosion is evident, confirming the high corrosion resistance of the S phase on the composite layers. In the continuous process, the nitriding S phase was formed at the surface. Moreover, as nitrogen dissolved in stainless steel strongly increases the PREN [29], the S phase produced by the continuous process suppresses sever pitting corrosion.

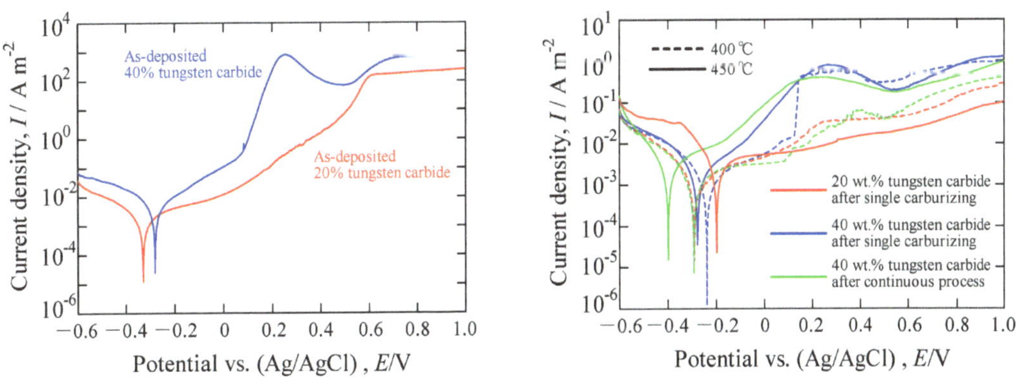

(**a**) (**b**)

Figure 7. Anodic polarization curves of the AISI 316L steel-based composite layers with various tungsten carbide compositions of (**a**) as-deposited and (**b**) treated various plasma processes.

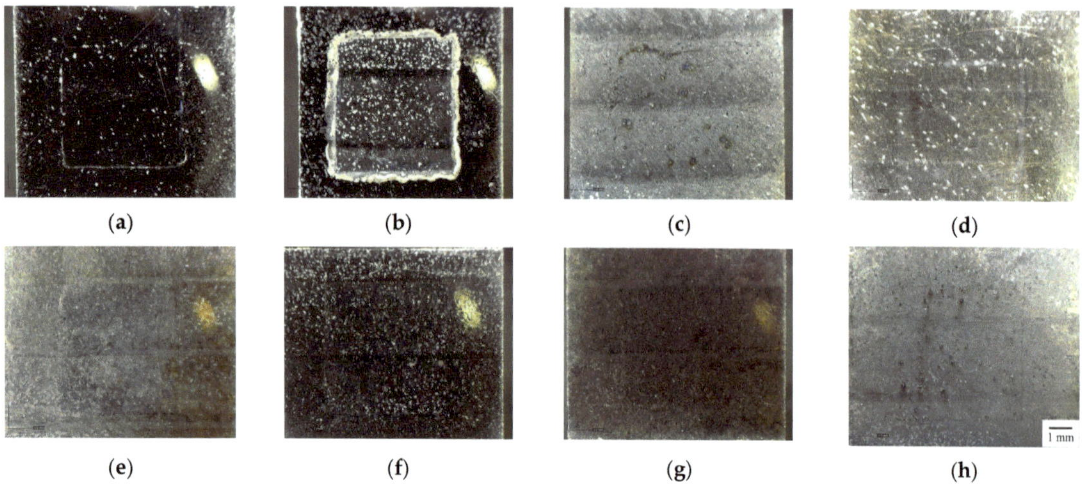

Figure 8. Surface morphologies of the AISI 316L steel-based composite layers with various tungsten carbide compositions after the anodic polarization measurements of (**a**) 20 wt.% as-deposited, (**b**) 40 wt.% as-deposited, (**c**) 20 wt.% carburized at 400 °C, (**d**) 20 wt.% carburized at 450 °C, (**e**) 40 wt.% carburized at 400 °C, (**f**) 40 wt.% carburized at 450 °C, (**g**) 40 wt.% continuously processed at 400 °C, and (**h**) 40 wt.% continuously processed at 450 °C.

Table 1. Corrosion potential and current density in the anodic polarization curves of the AISI 316L steel-based composite layers.

Sample	Corrosion Potential (V)	Corrosion Current Density ($A\ m^{-2}$)
20 wt.% as-deposited	−0.33	2.0×10^{-3}
40 wt.% as-deposited	−0.28	4.4×10^{-3}
20 wt.% carburized at 400 °C	−0.29	1.5×10^{-3}
20 wt.% carburized at 450 °C	−0.20	4.1×10^{-3}
40 wt.% carburized at 400 °C	−0.24	1.7×10^{-3}
40 wt.% carburized at 450 °C	−0.28	1.8×10^{-3}
40 wt.% continuously processed at 400 °C	−0.29	2.2×10^{-3}
40 wt.% continuously processed at 450 °C	−0.40	3.8×10^{-3}

Figure 9 shows the depth profiles of the atomic concentrations obtained by XPS. The spectrum of the as-deposited 40 wt.% layer surface exhibits an oxygen peak with a low-intensity chromium peak, suggesting the presence of a passive film. The spectrum of the S phase in the 40 wt.% composite layer treated with the single carburizing process at 450 °C also exhibits an oxygen peak with a low-intensity chromium peak, indicating that although sever pitting corrosion is suppressed, a strong passive film is not formed. Low-temperature carburizing reportedly improves pitting corrosion resistance. This implies that instead of the passive film, the S phase containing dissolved excess carbon inhibited the pitting corrosion [50,57,58]. Therefore, the suppression of sever pitting corrosion in the composite layers treated by the single carburizing process can also be attributed to the supersaturated carbon S phase.

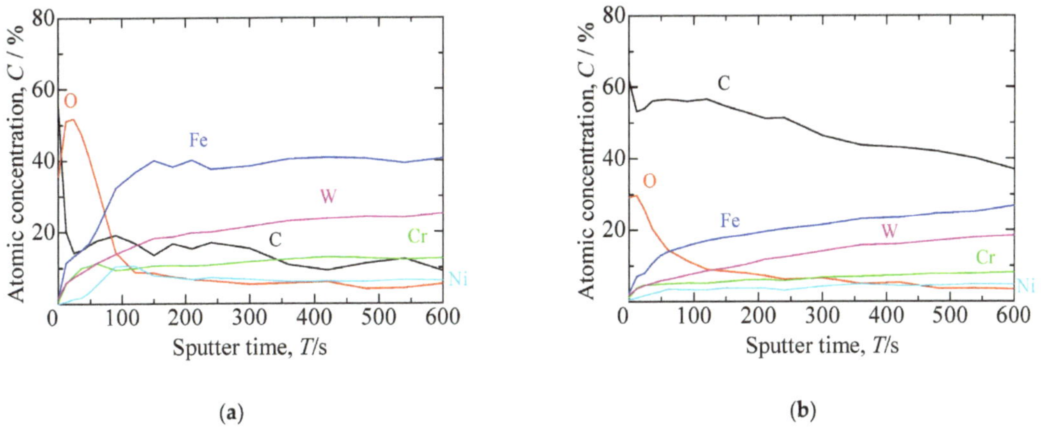

Figure 9. XPS depth profiles of carbon, oxygen, chromium, iron, nickel, and tungsten in the 40 wt.% tungsten carbide composite layers: (**a**) as-deposited and (**b**) carburized at 450 °C.

It should be noted that the corrosion current densities of the plasma-processed layers in Figure 7b do not exhibit any distinct difference or trend with varying tungsten carbide composition or plasma process conditions. Therefore, it is difficult to determine which plasma process condition is most effective for corrosion resistance.

4. Conclusions

In this study, AISI 316L stainless-steel-based tungsten carbide composite layers were fabricated via laser metal deposition and treated with single carburizing for 4 h and continuous nitriding for 3.5 h after carburizing for 0.5 h at 400 and 450 °C to form an expanded austenite phase (S phase). The formation mechanism of the S phase was investigated considering the influence of the original solid-solution carbon and eutectic carbides, which were produced by the thermal decomposition of the tungsten carbide particles during the laser metal deposition process. In addition, the hardness and anodic polarization (in a NaCl solution) of the S phase were evaluated depending on the tungsten carbide composition and plasma treatment conditions.

(1) In the single carburizing process, the S phase was formed with the addition of diffused carbon to the solid-solution carbon originally present in the composite layers. The eutectic carbides in the as-deposited composite layers prevented carbon diffusion, and the S phase in the 40 wt.% composite layers, which included more eutectic carbides, was thinner than that in the 20 wt.% composite layers. However, the original solid-solution carbon was involved in the formation of the S phase. Therefore, the thickness of the S phase layer in the composites was higher than that of the S phase on the AISI 316L stainless-steel plates.

(2) In the continuous process, the first carburizing step induced carbon diffusion, and the diffused carbon and original solid-solution carbon were pushed inside by the diffused nitrogen during the second nitriding process. This phenomenon resulted in the formation of a dual layer S phase, viz., a nitriding S phase on the outside and a carburizing S phase inside. The thickness of the S phase produced by the continuous process was similar to that generated by the single carburizing process.

(3) The average Vickers hardness of the S phase at the surface was 1210–1365 HV, and significant differences were not observed for different tungsten carbide compositions or plasma process conditions. On the other hand, the surface hardness profile of the S phase resulting from the continuous process at 450 °C was harder than that produced by the single carburizing process up to a depth of approximately 4 μm.

(4) The anodic polarization curves, obtained in a NaCl solution, show that the corrosion current densities of the plasma-processed layers are slightly lower than those of the as-deposited layers except for the 20 wt.% layer carburized at 450 °C. The pitting and crevice corrosion occurred in the as-deposited layers. The surface morphologies of the S phase after the anodic polarization measurements exhibited slight discoloration. However, no sever corrosion was observed in the plasma-processed layers implying that the carburizing and nitriding S phase exhibited comparable corrosion resistance enhancements.

(5) The results confirm that the single plasma carburizing process and continuous plasma process nitriding after carburizing at 400 and 450 °C improved the surface hardness and corrosion resistance of the composite layers. However, the metallurgical structure was very different from that of ordinary stainless-steel plates. These plasma treatments can be used as surface modification methods for additive manufacturing stainless-steel-based tungsten carbide composites that can withstand harsh environments.

Author Contributions: Conceptualization, S.A.; methodology, S.A.; investigation, S.A., T.Y., K.T., T.N. and N.U.; resources, S.A.; data curation, S.A.; formal analysis, S.A.; writing—original draft preparation, S.A.; writing—review and editing, S.A.; visualization, S.A.; supervision, S.A.; project administration, S.A.; funding acquisition, S.A. All authors have read and agreed to the published version of the manuscript.

Funding: This research was funded by JSPS KAKENHI (grant number: 18K04792).

Data Availability Statement: The data are available on request from the corresponding author.

Conflicts of Interest: The authors declare no conflict of interest.

References

1. Gnaase, S.; Niggemeyer, D.; Lehnert, D.; Bödgerand, C.; Tröster, T. Comparative Study of the Influence of Heat Treatment and Additive Manufacturing Process (LMD & L-PBF) on the Mechanical Properties of Specimens Manufactured from 1.2709. *Crystals* **2023**, *13*, 157. [CrossRef]
2. Kaligar, A.B.; Kumar, H.A.; Ali, A.; Abuzaid, W.; Egilmez, M.; Alkhader, M.; Abed, F.; Alnaser, A.S. Femtosecond Laser-Based Additive Manufacturing: Current Status and Perspectives. *Quantum Beam Sci.* **2022**, *6*, 5. [CrossRef]
3. Ladani, L.; Sadeghilaridjani, M. Review of Powder Bed Fusion Additive Manufacturing for Metals. *Metals* **2021**, *11*, 1391. [CrossRef]
4. Dass, A.; Moridi, A. State of the Art in Directed Energy Deposition: From Additive Manufacturing to Materials Design. *Coatings* **2019**, *9*, 418. [CrossRef]
5. Singh, A.; Kapil, S.; Das, M. A comprehensive review of the methods and mechanisms for powder feedstock handling in directed energy deposition. *Addit. Manuf.* **2020**, *35*, 101388. [CrossRef]
6. Li, N.; Huang, S.; Zhang, G.; Qin, R.; Liu, W.; Xiong, H.; Shi, G.; Blackburn, J. Progress in additive manufacturing on new materials: A review. *J. Mater. Sci. Technol.* **2019**, *35*, 242–269. [CrossRef]
7. Ligabo, S.A.; Braga, V.; Ferreira, C.C.A.; Siqueira, R.H.M.; Lourenço, J.C.; Abdalla, A.J.; Lima, M.S.F. Microstructure and Corrosion Behavior of AISI 316 Steel Layers Deposited on AISI 347 Steel Substrate by Laser Metal Deposition. *Metals* **2022**, *12*, 2161. [CrossRef]
8. Zinovieva, O.; Romanova, V.; Balokhonov, R. Effects of scanning pattern on the grain structure and elastic properties of additively manufactured 316L austenitic stainless steel. *Mater. Sci. Eng. A* **2022**, *832*, 142447. [CrossRef]
9. Avanzini, A. Fatigue Behavior of Additively Manufactured Stainless Steel 316L. *Materials* **2023**, *16*, 65. [CrossRef]
10. Haghdadi, N.; Laleh, M.; Moyle, M.; Primig, S. Additive manufacturing of steels: A review of achievements and challenges. *J. Mater. Sci.* **2021**, *56*, 64–107. [CrossRef]
11. Bajaj, P.; Hariharan, A.; Kini, A.; Kürnsteiner, P.; Raabe, D.; Jägle, E.A. Steels in additive manufacturing: A review of their microstructure and properties. *Mater. Sci. Eng. A* **2020**, *772*, 138633. [CrossRef]
12. Röttger, A.; Boes, J.; Theisen, W.; Thiele, M.; Esen, C.; Edelmann, A.; Hellmann, R. Microstructure and mechanical properties of 316L austenitic stainless steel processed by different SLM devices. *J. Adv. Manuf. Technol.* **2020**, *108*, 769–783. [CrossRef]
13. Bassis, M.; Ron, T.; Leon, A.; Kotliar, A.; Kotliar, R.; Shirizly, A.; Aghion, E. The Influence of Intralayer Porosity and Phase Transition on Corrosion Fatigue of Additively Manufactured 316L Stainless Steel Obtained by Direct Energy Deposition Process. *Materials* **2022**, *15*, 5481. [CrossRef] [PubMed]
14. Fetni, S.; Enrici, T.M.; Niccolini, T.; Tran, S.H.; Dedry, O.; Jardin, R.; Duchene, L.; Mertens, A.; Habraken, A.M. 2D thermal finite element analysis of laser cladding of 316L + WC composite layer coatings. *Procedia Manuf.* **2020**, *50*, 86–92. [CrossRef]

15. Enrici, T.M.; Dedry, O.; Boschini, F.; Tchuindjang, J.T.; Mertens, A. Microstructural and thermal characterization of 316L + WC composite layer coatings obtained by laser cladding. *Adv. Eng. Mater.* **2020**, *22*, 2000291. [CrossRef]
16. Fetni, S.; Enrici, T.M.; Niccolini, T. Thermal model for the directed energy deposition of composite layer coatings of 316L stainless steel enriched with tungsten carbides. *Mater. Des.* **2021**, *204*, 109661. [CrossRef]
17. Benarji, K.; Kumar, Y.R.; Jinoop, A.N.; Paul, C.P.; Bindra, K.S. Effect of WC Composition on the Microstructure and Surface Properties of Laser Directed Energy Deposited SS 316-WC Composite layers. *J. Mater. Eng. Perform.* **2021**, *30*, 6732–6742. [CrossRef]
18. Li, H.; Hu, Y.; Di, R.; Yuan, R.; Shi, C.; Lei, J. Effects of WC particles on microstructure and mechanical properties of 316L steel obtained by laser melting deposition. *Ceram. Int.* **2022**, *48*, 20388–20399. [CrossRef]
19. Borgioli, F. The "Expanded" Phases in the Low-Temperature Treated Stainless Steels: A Review. *Metals* **2022**, *12*, 331. [CrossRef]
20. Moskvina, V.; Maier, G.; Melnikov, E.; Astafurov, S.; Zagibalova, E.; Panchenko, M.; Reunova, K.; Nikolaev, A.; Ramazanov, K.; Astafurova, E. The effect of thin surface layer of nitrogen-expanded austenite on bulk γ-α' phase transformation in low-temperature deformation of 316L stainless steel. *Mater. Lett.* **2021**, *304*, 130676. [CrossRef]
21. Borgioli, F.; Galvanetto, E.; Bacci, T. Surface Modification of a Nickel-Free Austenitic Stainless Steel by Low-Temperature Nitriding. *Metals* **2021**, *11*, 1845. [CrossRef]
22. Funch, C.V.; Somlo, K.; Christiansen, T.L.; Somers, M.A. Thermochemical post-processing of additively manufactured austenitic stainless steel. *Surf. Coat. Technol.* **2022**, *441*, 128495. [CrossRef]
23. Lindner, T.; Kutschmann, P.; Löbel, M.; Lampke, T. Hardening of HVOF-Sprayed Austenitic Stainless-Steel Coatings by Gas Nitriding. *Coatings* **2018**, *8*, 348. [CrossRef]
24. Adachi, S.; Egawa, M.; Yamaguchi, T.; Ueda, N. Low-temperature plasma nitriding for austenitic stainless steel layers with various nickel contents fabricated via direct laser metal deposition. *Coatings* **2020**, *10*, 365. [CrossRef]
25. Liu, Y.; Liu, D.; Zhang, X.; Li, W.; Ma, A.; Fan, K.; Xing, W. Effect of Alloying Elements and Low Temperature Plasma Nitriding on Corrosion Resistance of Stainless Steel. *Materials* **2022**, *15*, 6575. [CrossRef]
26. Borgioli, F. From Austenitic Stainless Steel to Expanded Austenite-S Phase: Formation, Characteristics and Properties of an Elusive Metastable Phase. *Metals* **2020**, *10*, 187. [CrossRef]
27. Borgioli, F.; Galvanetto, E.; Bacci, T. Effects of Surface Modification by Means of Low-Temperature Plasma Nitriding on Wetting and Corrosion Behavior of Austenitic Stainless Steel. *Coatings* **2020**, *10*, 98. [CrossRef]
28. Yazıcı, M.; Çomaklı, O.; Yetim, T.; Yetim, A.F.; Çelik, A. The effect of plasma nitriding temperature on the electrochemical and semiconducting properties of thin passive films formed on 316 L stainless steel implant material in SBF solution. *Surf. Coat. Technol.* **2015**, *261*, 181–188. [CrossRef]
29. Adachi, S.; Yamaguchi, T.; Ueda, N. Formation and Properties of Nitrocarburizing S-Phase on AISI 316L Stainless Steel-Based WC Composite Layers by Low-Temperature Plasma Nitriding. *Metals* **2021**, *11*, 1538. [CrossRef]
30. Savrai, R.A.; Skorynina, P.A. Structural-phase transformations and changes in the properties of AISI 321 stainless steel induced by liquid carburizing at low temperature. *Surf. Coat. Technol.* **2022**, *443*, 128613. [CrossRef]
31. Liu, H.Y.; Che, H.L.; Gao, J.Y.; Li, G.B.; Lei, M.K. Low-pressure hollow cathode plasma source carburizing of AISI 304L austenitic stainless steel at low temperature. *Surf. Coat. Technol.* **2022**, *442*, 128548. [CrossRef]
32. Montanari, R.; Lanzutti, A.; Richetta, M.; Tursunbaev, J.; Vaglio, E.; Varone, A.; Verona, C. Plasma Carburizing of Laser Powder Bed Fusion Manufactured 316 L Steel for Enhancing the Surface Hardness. *Coatings* **2022**, *12*, 258. [CrossRef]
33. Liu, Z.; Wang, S.; Zhang, S.; Feng, Y.; Peng, Y.; Gong, J. Deformation response of gradient low-temperature gaseous carburized case in austenitic stainless steel during cyclic nanoindentation. *Mater. Today Commun.* **2021**, *28*, 102714. [CrossRef]
34. Maistro, G.; Kante, S.; Nyborg, L.; Cao, Y. Low-temperature carburized high-alloyed austenitic stainless steels in PEMFC cathodic environment. *Surf. Interfaces* **2021**, *24*, 101093. [CrossRef]
35. Cheon, H.; Kim, K.S.; Kim, S.; Heo, S.B.; Lim, J.H.; Kim, J.H.; Yoon, S.Y. Effect of Deformation Structure of AISI 316L in Low-Temperature Vacuum Carburizing. *Metals* **2021**, *11*, 1762. [CrossRef]
36. Molleja, J.G.; Nosei, L.; Ferrón, J.; Bemporad, E.; Lesage, J.; Chicot, D.; Feugeas, J. Characterization of expanded austenite developed on AISI 316L stainless steel by plasma carburization. *Surf. Coat. Technol.* **2010**, *204*, 3750–3759. [CrossRef]
37. Adachi, S.; Ueda, N. Surface hardness improvement of plasma-sprayed AISI 316L stainless steel coating by low-temperature plasma carburizing. *Adv. Powder Technol.* **2013**, *24*, 818–823. [CrossRef]
38. Adachi, S.; Ueda, N. Combined plasma carburizing and nitriding of sprayed AISI 316L steel coating for improved wear resistance. *Surf. Coat. Technol.* **2014**, *259*, 44–49. [CrossRef]
39. Adachi, S.; Ueda, N. Wear and corrosion properties of cold-sprayed AISI 316L coatings treated by combined plasma carburizing and nitriding at low temperature. *Coatings* **2018**, *8*, 456. [CrossRef]
40. Azmi, F.; Basak, A.K.; Adenan, M.S.; Haruman, E.; Saedon, J.B. Responses of hybrid S phase layer to nanoscratching. *Surf. Coat. Technol.* **2022**, *441*, 128509. [CrossRef]
41. Ullah, N.; Naeem, M.; Shafiq, M.; Mujahid, Z.; Díaz-Guillén, J.C.; Lopez-Badillo, C.M.; Zakaullah, M. Effect of methane concentration on surface properties of cathodic cage plasma nitrocarburized AISI-304. *Appl. Phys.* **2021**, *127*, 529. [CrossRef]
42. Böcker, J.; Puth, A.; Dalke, A.; Röpcke, J.; Helden, J.P.H.; Biermann, H. Influence of the Active Screen Plasma Power during Afterglow Nitrocarburizing on the Surface Modification of AISI 316L. *Coatings* **2020**, *10*, 1112. [CrossRef]

43. Wu, D.; Ge, Y.; Kahn, H.; Ernst, F.; Heuer, A.H. Diffusion profiles after nitrocarburizing austenitic stainless steel. *Surf. Coat. Technol.* **2015**, *279*, 180–185. [CrossRef]
44. Scheuer, J.C.; Rodrigo, J.D.; Cardoso, P.; Sílvio, D.; Brunatto, F. Sequential low-temperature plasma-assisted thermochemical treatments of the AISI 420 martensitic stainless steel. *Surf. Coat. Technol.* **2021**, *421*, 127459. [CrossRef]
45. Lindner, T.; Mehner, T.; Lampke, T. Surface modification of austenitic thermal-spray coatings by low-temperature nitrocarburizing. *IOP Conf. Ser. Mater. Sci. Eng.* **2016**, *118*, 012008. [CrossRef]
46. Godec, M.; Donik, Č.; Kocijan, A.; Podgornik, B.; Skobir Balantič, D.A. Effect of post-treated low-temperature plasma nitriding on the wear and corrosion resistance of 316L stainless steel manufactured by laser powder-bed fusion. *Addit. Manuf.* **2020**, *32*, 101000–101008. [CrossRef]
47. Kovács1, D.; Kemény, D.M. Effect of plasma nitriding of austenitic stainless steel produced by direct metal laser sintering. *Acta Metall. Slovaca* **2021**, *27*, 190–194. [CrossRef]
48. Czerwiec, T.; He, H.; Weber, S.; Dong, C.; Michel, H. On the occurrence of dual diffusion layers during plasma-assisted nitriding of austenitic stainless steel. *Surf. Coat. Technol.* **2006**, *200*, 5289–5295. [CrossRef]
49. Micheler, T. Influence of plasma nitriding on hydrogen environment embrittlement of 1.4301 austenitic stainless steel. *Surf. Coat. Technol.* **2008**, *202*, 1688–1695. [CrossRef]
50. Ceschini, L.; Chiavari, C.; Lanzoni, E.; Martini, C. Low-temperature carburised AISI 316L austenitic stainless steel: Wear and corrosion behaviour. *Mater. Des.* **2012**, *38*, 154–160. [CrossRef]
51. Peng, Y.; Gong, J.; Chen, C.; Liu, Z.; Jiang, Y. Numerical Analysis of Stress Gradient and Traps Effects on Carbon Diffusion in AISI 316L during Low Temperature Gas Phase Carburization. *Metals* **2018**, *8*, 214. [CrossRef]
52. Ernst, F.; Li, D.; Kahn, H.; Michal, G.M.; Heuer, A.H. The carbide M_7C_3 in low-temperature-carburized austenitic stainless steel. *Acta Mater.* **2011**, *5*, 2268–2276. [CrossRef]
53. Christiansen, T.; Somers, M.A.J. Low temperature gaseous nitriding and carburising of stainless steel. *Surf. Eng.* **2005**, *21*, 445–455. [CrossRef]
54. Thaiwatthana, S.; Li, X.; Dong, H.B.; Bell, T. Comparison Studies on Properties of Nitrogen and Carbon S Phase on Low Temperature Plasma Alloyed AISI 316 Stainless Steel. *Surf. Eng.* **2002**, *18*, 433–437. [CrossRef]
55. Sun, Y. Hybrid Plasma Surface Alloying of Austenitic Stainless Steels with Nitrogen and Carbon. *Mater. Sci. Eng. A* **2005**, *404*, 124–129. [CrossRef]
56. Gu, X.; Michal, M.G.; Frank, E.; Kahn, H. Numerical Simulations of Carbon and Nitrogen Composition Depth Profiles in Nitrocarburized Austenitic Stainless Steels. *Metall. Mater. Trans. A* **2014**, *45A*, 4268–4279. [CrossRef]
57. Buhagiar, J.; Spiteri, A.; Sacco, M.; Sinagra, E.; Dong, H. Augmentation of crevice corrosion resistance of medical grade 316LV stainless steel by plasma carburizing. *Corros. Sci.* **2012**, *59*, 169–178. [CrossRef]
58. Borgioli, F. The Corrosion Behavior in Different Environments of Austenitic Stainless Steels Subjected to Thermochemical Surface Treatments at Low Temperatures: An Overview. *Metals* **2023**, *13*, 776. [CrossRef]

metals

Article

Synergic Effects of Nanosecond Laser Ablation and PVD-Coating on Cemented Carbides: Assessment on Surface and Mechanical Integrity

Shiqi Fang [1,*], Luis Llanes [2], Y. B. Guo [1] and Dirk Bähre [3]

[1] Department of Mechanical and Aerospace Engineering, Rutgers University-New Brunswick, Piscataway, NJ 08854, USA
[2] CIEFMA—Department of Materials Science and Metallurgical Engineering, EEBE—Campus Diagonal Besòs, Universitat Politècnica de Catalunya, 08019 Barcelona, Spain
[3] Institute of Production Engineering, Saarland University, 66123 Saarbrücken, Germany
* Correspondence: shiqi.fang@rutgers.edu

Abstract: Emerging laser precision machining, particularly using pulsed lasers, enlightens the innovation and functionalization of cemented carbides. These backbone materials of the tooling industry are usually considered difficult to machine or shape using conventional mechanical approaches. The coating of cemented carbide tools, deemed to improve their mechanical and thermal properties, is a common supplementary surface treatment prior to their application. This work aims to study the synergic effects of nanosecond laser ablation and coating deposition on the surface, as well as the mechanical integrity of cemented carbides. In this regard, two plain WC–Co grades with different metallic binder content ($10\%_{wt}$Co and $15\%_{wt}$Co) were first processed by a short-pulsed nanosecond laser. Subsequently, an AlTiN film was physically vapor-deposited on the laser-processed surfaces. The resulting surface integrity was assessed in terms of topographical, morphological, and microstructural changes. Mechanical integrity was evaluated in terms of indentation and sliding contact responses using Vickers hardness and scratch tests, respectively, the latter including frictional, penetrating, and sliding performances under selected surface processing conditions. In general, the nanosecond laser ablation proved to be beneficial for the mechanical integrity of coated cemented carbides in most studied cases, as it increased surface hardness, reduced penetration depth, and hindered damage during sliding. This was the case despite a slight increase in surface roughness, as well as minor morphological and microstructural changes at the coating–substrate interface, discerned.

Keywords: cemented carbide; laser ablation; coating; surface integrity; hardness; scratch; friction; wear

Citation: Fang, S.; Llanes, L.; Guo, Y.B.; Bähre, D. Synergic Effects of Nanosecond Laser Ablation and PVD-Coating on Cemented Carbides: Assessment on Surface and Mechanical Integrity. *Metals* **2024**, *14*, 34. https://doi.org/10.3390/met14010034

Academic Editors: Francesca Borgioli, Umberto Prisco, Tomasz Tański and Denis Benasciutti

Received: 25 November 2023
Revised: 25 December 2023
Accepted: 26 December 2023
Published: 28 December 2023

1. Introduction

Cemented carbides (also referred to as hardmetals) are a group of ceramic–metal composite materials consisting of hard carbides of transition metals (e.g., WC, TiC, TaC, with a grain size usually <5 μm) embedded by a metallic binder (e.g., Co, Ni) [1–4]. This ceramic–metal nature provides the opportunity to achieve desired combinations of hardness, wear resistance, and toughness through microstructural arrangements of grain size and binder content. In general, as grain size or binder content increases, toughness increases, while hardness and wear resistance decrease.

As a newly developing technique for precision machining, ablation using short/ultra-short pulse lasers is receiving more and more attention, and its application is emerging in various fields, including manufacturing, biology, and medicine, among others [5–9]. It is a process of removing atoms from a solid by irradiating it with a pulsed laser beam. When processing hard materials, especially when shaping tool materials or processing their surfaces, the laser stands out from other non-conventional processing methods because of its precision, speed, and cleanliness attributes [10–13]. During laser processing, the energy beam is usually focused on the target materials. As the diameter of the laser spot can be as small as several microns, extremely high energy intensity can then be attained at the focus point. Such a tiny 'cutting point' effectively ensures excellent precision and efficiency. Benefiting from the extremely short reaction time of the nanosecond laser pulse with the material ($<10^{-6}$ s) [14,15], the ablation is only slight. Macroscopically, the material is then removed rapidly by a large multitude of laser pulses applied in a very short time. Compared to some common non-abrasive processing methods for hard materials such as electrical discharging machining (e.g., Ref. [16]), the thermal reaction can be effectively reduced by the pulse laser. As a result, possible side effects associated with the thermal reaction, such as melting or material redeposition, can be reduced on the machined surfaces, e.g., Ref. [17].

Meanwhile, the deposition of ceramic thin films is a common surface treatment implemented on hard substrates to improve the mechanical properties of cutting tools, especially hardness and wear resistance [18–22]. Hence, it is of practical interest to investigate the synergic effects of laser ablation and coating deposition with respect to the surface and mechanical integrity. This is indeed the main goal of the investigation presented here. It was conducted on two different cemented carbides, which were first processed by a nanosecond laser at two different energy levels and subsequently coated with a ceramic film. The surface and mechanical integrity were assessed by combining advanced characterization techniques, including focused ion beam milling (FIB) and scanning electron microscopy (SEM) inspection, as well as Vickers hardness and micro-scratch testing. The impact of laser ablation for selected coated processing conditions was evaluated, documented, and critically analyzed.

2. Experimental Aspects

2.1. Materials Studied

Two experimental cemented carbide grades, supplied by Hyperion Materials & Technologies, were studied: A (WC-10%wt Co) and B (WC-15%wt Co). Samples were produced by pressing and liquid phase sintering (sinter-HIP process) following conventional industrial-like processing stages. They did not show any undesirable phases (graphite or eta phase) after being consolidated. The nominal values of the microstructural and mechanical characteristics of the studied materials are listed in Table 1. They both exhibited carbide grains of similar length scale, as inferred from the mean carbide size values measured on SEM micrographs by using the linear intercept method [23]. Hardness was determined using a Vickers diamond pyramidal indenter and applying a load of 294 N. Fracture toughness was determined by testing to failure single-edge pre-cracked notch beam specimens [24].

Table 1. Nominal values of microstructural characteristics and basic mechanical properties of the studied cemented carbide grades (date from [23,24]).

Grade	Mean Carbide Grain Size (μm)	Co (wt%)	HV (GPa)	K_{Ic} (MPa$\sqrt{m}$)
A	2.33 ± 1.38	10	11.4 ± 0.2	15.8 ± 0.3
B	1.70 ± 1.08	15	10.2 ± 0.1	17.0 ± 0.2

The working surfaces of the cemented carbide samples had a dimension of 15 mm × 15 mm, and they were polished to a level, R_a, of around 0.03–0.04 μm before the laser ablation. The polishing program included four steps: rough-grinding, fine-grinding, polishing, and fine-polishing. The machine ATM Saphir 520 (ATM Qness GmbH, Mammelzen, Germany) and four different grinding and polishing disks, i.e., Cameo Disk (Lam Plan, Gaillard, France), MD-Allegro (Struers, Willich, Germany), MD-Dac (Struers, Willich, Germany), and MD-Nap (Struers, Willich, Germany), were used. The abrasive sizes of the polishing suspension used were 6 μm, 3 μm, and 1 μm, respectively. The normal force used in each step was 35 N, 30 N, 30 N, and 25 N, respectively, and the working time of each step was 7 min, 5 min, 15 min, and 3 min, respectively.

2.2. Surface Treatment: Laser Ablation and Coating

A Nd: YLF nanosecond laser was implemented to induce ablation on the target cemented carbide surfaces. The source emitted a laser beam with a pulse duration of 5 ns, wavelength of 349 nm, and frequency of 1000 Hz. Due to their different thermal properties, the two constituents of WC–Co cemented carbides may be ablated in different ways. On the one hand, binders can be selectively removed by the laser beam with a fluence of around 2.5 J/cm^2. On the other hand, both binder and carbide grains may be simultaneously removed by the laser beam at a fluence level higher than 10 J/cm^2 [13,25–27]. Accordingly, two different pulse energies, i.e., 10.3 μJ and 58 μJ, were chosen in this study, corresponding to 2.9 J/cm^2 and 16.6 J/cm^2, respectively.

During ablation, the laser beam scanned the target surface in a line-by-line mode. Multiple scan repetitions, 18 and 5—corresponding to the two laser energy levels, were applied to remove a material volume with a thickness of 10 μm. Subsequently, these surfaces were coated using an industrial physical vapor deposition (PVD) installation (CC800/9 HiPIMS, CemeCon AG, Würselen, Germany). The coating layer is an AlTiN composite with a thickness of 4 μm, and it has a nominal hardness of 37 GPa (HV). It can be used for machining steel and difficult-to-machine materials such as Inconel and titanium, and it can also be used in dry machining due to its high Al content [28,29]. The laser ablation conditions and coating operations, as well as the nomenclature used in the work, are given in Table 2. The surfaces without laser processing were also studied as reference conditions.

Table 2. Investigated surface processing conditions.

Condition	Machining Operation
C	PVD-Coating (reference condition)
LL + C	Nanosecond Laser ablation with low energy + PVD-Coating
LH + C	Nanosecond Laser ablation with high energy + PVD-Coating

2.3. Surface and Mechanical Integrity Assessment: Surface Roughness Measurement, SEM-FIB Inspection, Vickers Hardness, and Scratch Tests

Surface integrity was assessed to evaluate topographical, morphological, and microstructural changes. Two surface roughness parameters (R_a and R_z) were determined using a surface texture instrument, the Mahrsurf XR20 (Mahr GmbH, Göttingen, Germany). Morphological and microstructural changes, including induced damage, were inspected using a Zeiss Neon40 scanning electron microscope (SEM, Carl Zeiss AG, Jena, Germany), combined with focused ion beam (FIB) milling. Meanwhile, mechanical integrity was evaluated in terms of indentation and sliding contact responses, using Vickers hardness and scratch tests, respectively. The former was assessed using a micro indentation testing unit (EMCO TEST DuraScan 20, Emco-Test, Kuchl, Austria), a Vickers diamond tip, and applying two different normal loads, i.e., 0.3 kgf (2.94 N) and 3 kgf (29.4 N). Average values were obtained from three tests conducted on each surface. Scratch tests were conducted using a Rockwell diamond indenter of 200 μm radius (Revetest, CSM Instruments, Needham Heights, MA, USA). The tests were carried out by incremental loading, from 0 N to 100 N, using a constant loading rate of 10 N/min and over a scratch length of 5 mm. The resulting scratch tracks were finally examined using the SEM-FIB to determine critical loads, document wear scenarios, and/or discern morphological changes. Additionally, reference comparative tests over a scratch length of 2 mm, corresponding to a more pronounced increase in load per unit distance, were also conducted in the A (WC-10%$_{wt}$Co) grade.

3. Results and Discussion

3.1. Surface Integrity Assessment

3.1.1. Topographical Changes: Surface Roughness Analysis

The obtained arithmetic average and mean roughness values, R_a and R_z, are listed in Table 3. The laser beams yielded surface roughness values about one order of magnitude higher than for the reference condition. Considering these values are similar to those commonly achieved on ground cemented carbides (e.g., Ref. [30]), it can be stated that surface roughness induced by laser ablation in cemented carbides is acceptable. Such a roughening effect was slightly more pronounced when both cemented carbides were ablated with high laser energy.

Table 3. Surface roughness resulting from the different processing conditions for the two cemented carbides under consideration.

Sample	A (WC-10%$_{wt}$Co)			B (WC-15%$_{wt}$Co)		
	C	LL + C	LH + C	C	LL + C	LH + C
R_a (μm)	0.03	0.41	0.68	0.04	0.48	0.56
R_z (μm)	0.28	2.70	3.85	0.42	3.36	3.83

3.1.2. Morphological and Microstructural Changes: SEM-FIB Inspection

Morphological and microstructural changes were characterized using SEM-FIB [31]. In the non-lasered cases (Figure 1a,d), microstructural assemblages of the cemented carbide substrates were discerned, indicating a neat growth of the coating over them. Film morphology on the lasered substrates exhibited bubble-like features of different sizes, as shown in Figure 1b,c,e,f. These features were found on all the laser-ablated surfaces, and they were more pronounced on those surfaces processed with high laser energy. Meanwhile, such features appeared to be less representative in grade B than in grade A, indicating hindered formation with increasing binder content.

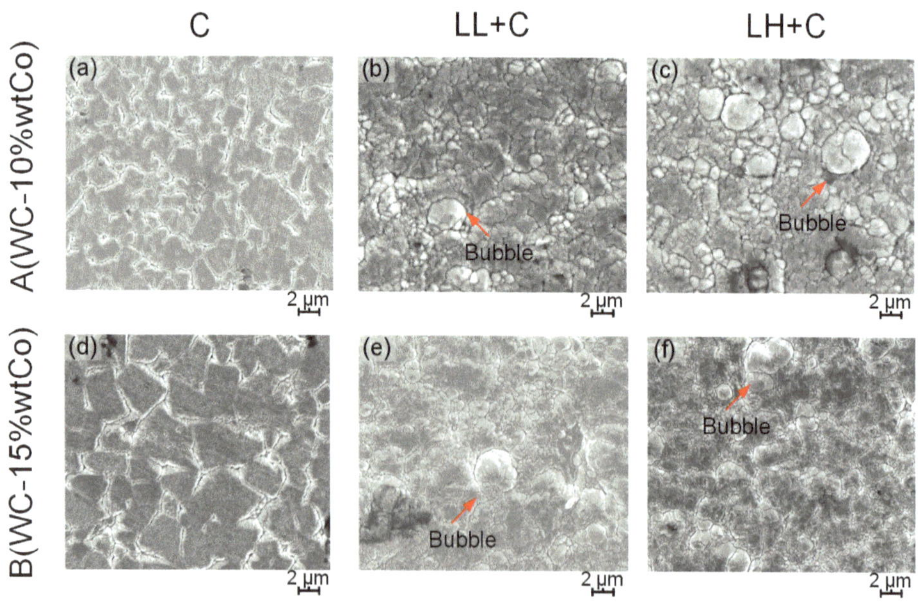

Figure 1. SEM micrographs showing surface morphology of the two cemented carbide grades processed by: (**a**,**d**) coating deposition only, (**b**,**e**) laser ablation with low energy followed by coating, and (**c**,**f**) laser ablation with high energy followed by coating.

Cross-sectional inspection was carried out by means of FIB milling to characterize microstructural and coating–substrate interface features (Figure 2). It is evidenced that the ceramic layers grew with high parallelism on the polished (reference) surfaces (Figure 2a,d). Meanwhile, on the laser-ablated surfaces, the coatings developed on the substrate profile, and thus inherited the corresponding surface topographical features.

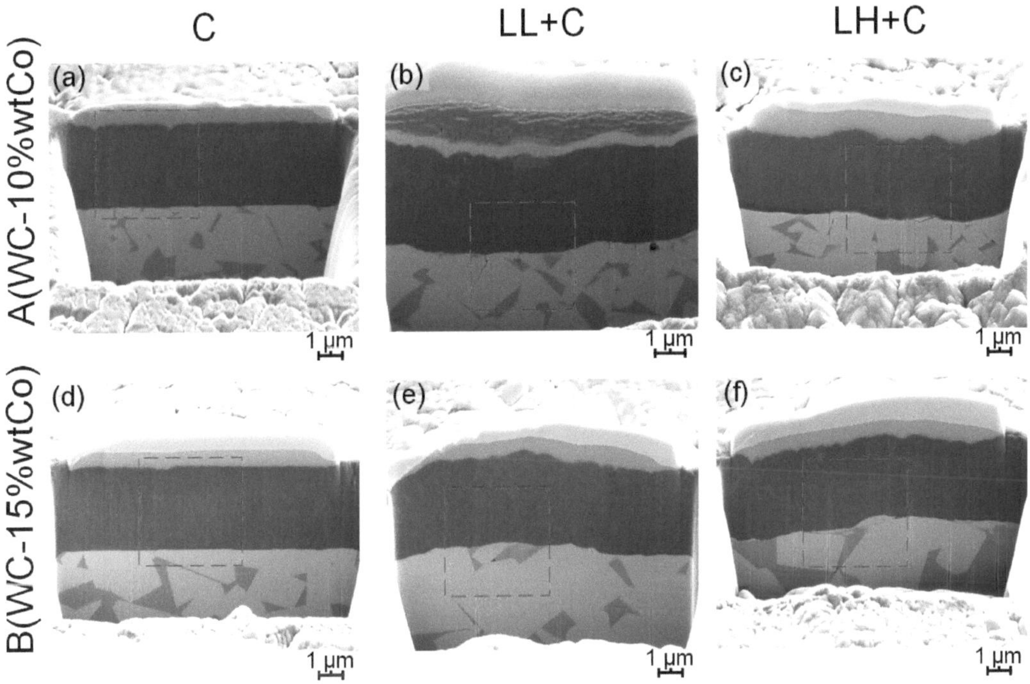

Figure 2. SEM-FIB micrographs showing coating–substrate interface for the two cemented carbide grades processed by: (**a**,**d**) coating deposition only, (**b**,**e**) laser ablation with low energy followed by coating, and (**c**,**f**) laser ablation with high energy followed by coating.

Detailed information about microstructural changes induced by laser near the coating–substrate interface is given in Figure 3. Different from the scenario found in the reference conditions (Figure 3a,d), it may be discerned that some damage was induced by laser ablation, including transgranular cracks running both perpendicular and parallel to the interface, as well as resolidification-like heterogeneities (e.g., Figure 3b,c,e,f). However, these changes were mostly localized within the substrate subsurface, at a very shallow depth (less than 1 µm). In this regard, the metallic binder was found to be more susceptible to laser ablation than the carbide grains, and thus excessive ablation of the binder left some pocket-like structures on the surface (Figure 3e). These areas have minuscule depths, and they could be partly responsible for the increased surface roughness obtained in these cases. The affected layers were thicker on the surface processed with high laser energy, promoting more resolidification of molten material (Figure 3e,f).

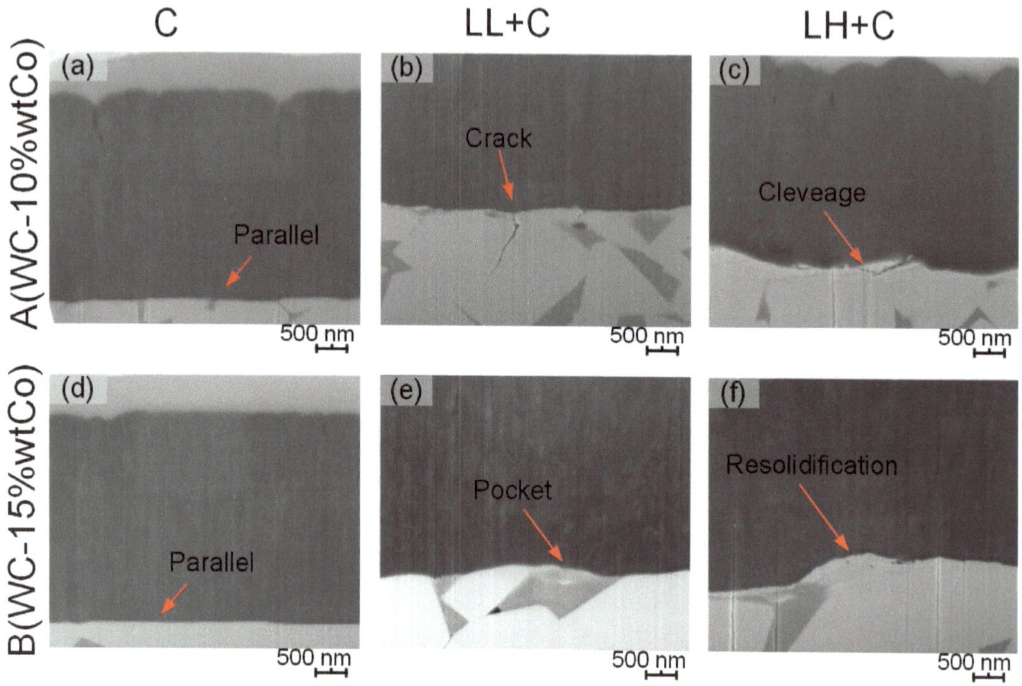

Figure 3. Details of local microstructural characterization at the marked positions in Figure 2. An explanation of specific features highlighted in (**a**–**f**) is given in the text.

3.2. Mechanical Integrity Assessment

3.2.1. Indentation Contact Response: Vickers Hardness

The results of the Vickers hardness tests are compared in Figure 4. Coating deposition improved surface hardness for both cemented carbide grades, as compared to the bulk ones. Moreover, laser-processed and coated surfaces were hardened more than those exclusively coated. However, measured hardness values for both grades varied, depending on the applied load. In general, and as expected, the measured microhardness (HV0.3) values were greater than the macrohardness (HV3) ones.

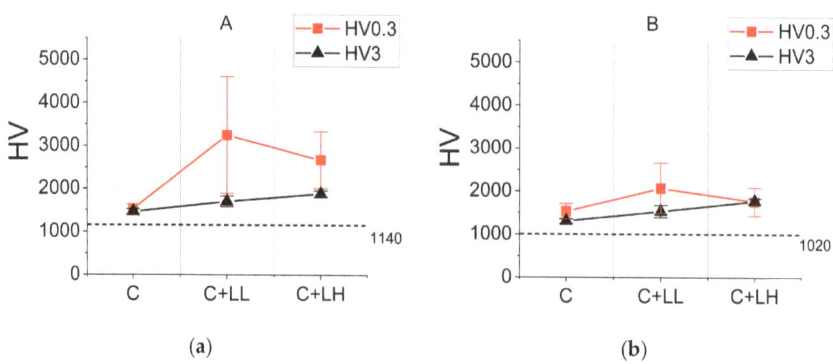

Figure 4. Comparison of the Vickers hardness for the two cemented carbide grades studied, as a function of surface modification condition: (**a**) A (WC-10%$_{wt}$Co) and (**b**) B (WC-15%$_{wt}$Co), with nominal values indicated by the dotted lines.

Although the microhardness (HV0.3) values were expected to involve the intrinsic and higher load-bearing capability of the deposited coatings, the interface conditions could have an impact on the obtained results. In this regard, it is observed that the discrepancy between HV0.3 and HV3 values was more pronounced on the surfaces processed with low laser energy, as compared to those processed with high energy. This could be taken as an indicator that better surface finishing was achieved by the low-energy laser beam. Under the same surface finishing conditions, the difference in the measured hardness using the two applied loads, HV0.3 and HV3, should be the same on the same cemented carbide grade. On the contrary, the change in the measured hardness differences induced by the applied loads could indicate different surface finishing conditions. A 'bad' surface finishing may result in rough and fragile surface conditions weakening the HV0.3 measurement more than the HV3 measurement under the normal load, HV0.3. Based on the idea, the more pronounced discrepancy could indicate a better surface finish using low laser energy. On the one hand, this led to moderate ablation, as it was able to remove the binder exclusively, but not the surrounding carbide grains. Such slightly unbalanced ablation between binder and carbides resulted in a localized agglomeration of ceramic particles on the surface, and accordingly an increase in surface hardness [26,32]. On the other hand, both binder and grains were simultaneously ablated by the high-energy beam; thus, the high-energy beam could induce more pronounced material removal that may have resulted in a more 'damaged' interface. A more 'damaged' interface turned out to be less effective for carrying the imposed load during indentation. As a result, selective laser ablation of the binder with moderate energy may be pointed out, regarding surface integrity and enhanced hardness in the laser-coating system, as the most suitable option of the ones studied. From this perspective, hardening was found to be enhanced in grade A, as compared to grade B, as a lower binder content was more beneficial for the formation of contiguous carbide agglomerates. The macroscale (HV3) hardness for both grades increased as the laser energy was increased. In these cases, the hardness values obtained were closer to the ones exhibited by the substrates, indicating that the indenter was now able to penetrate the coating layers; thus, the interface condition was less relevant for defining the resistance of the material against permanent deformation. Residual stresses, which can influence mechanical properties, might have been induced by thermal diffusion during laser ablation, and such influence may become more important as the laser energy increases [33]. However, the assessment of residual stress effects was outside the scope of this work.

3.2.2. Sliding Contact Response: Scratch Testing

Taking into account the results attained, reported, and analyzed in the previous section, further investigation was limited to surface conditions involving ablation using low-laser energy exclusively. Accordingly, scratch tests were only conducted on surfaces processed by low-energy laser ablation and coating, i.e., LL + C, as well as the corresponding reference conditions, i.e., C (Table 2).

Frictional Performance: Coefficient of Friction

Friction is the force resisting the relative motion of two surfaces sliding against each other. Surface conditions may be effectively compared by evaluating frictional performance: the ratio between the normal force, F_n, and frictional (tangential) force, F_t, is defined as the coefficient of friction [34–36]. A low value of the coefficient of friction indicates a smooth sliding between the two contact surfaces. In this study, coefficient of friction (COF) values were obtained within the range of stable sliding, i.e., between 1 mm and 4 mm of the sliding distance. There, all the studied samples exhibited very similar average COF values, ranging from 0.056 to 0.063, and laser ablation did not induce any important impact on the frictional performance of the coatings during the scratch tests. The average COF values obtained during the scratch tests are given in Table 4.

Table 4. Average values of COF were determined during scratch testing for the two cemented carbides and surface modification conditions studied.

Sample		COF	σ	Sample		COF	σ
A (WC-10%$_{wt}$Co)	C	0.057	0.003	B (WC-15%$_{wt}$Co)	C	0.059	0.003
	LL + C	0.056	0.005		LL + C	0.063	0.005

Indentation and Sliding Contact Performance: Penetration Depths (P_d) and Critical Loads (L_c)

Penetration depth was measured at three specific positions within the scratch tracks, corresponding to sliding distances of 1.25 mm, 2.50 mm, and 3.75 mm, and normal load values of 25 N, 50 N, and 75 N, respectively (Figure 5). The average values were calculated from experimental findings gathered in duplicated tests conducted under the same conditions. Independent of the microstructural assemblage of the two studied grades, less penetration occurred on the laser-ablated and coated surfaces, as compared to the reference ones. It allows us to point out that the indentation performance of the coated cemented carbides is effectively reinforced by the laser ablation of the substrate. Nevertheless, it should be highlighted that the penetration difference between the lasered and non-lasered conditions varied on the two grades, an experimental fact that could be affected by the synergistic effect of different factors such as hardness, carbide size, binder content, and binder mean free path, among others.

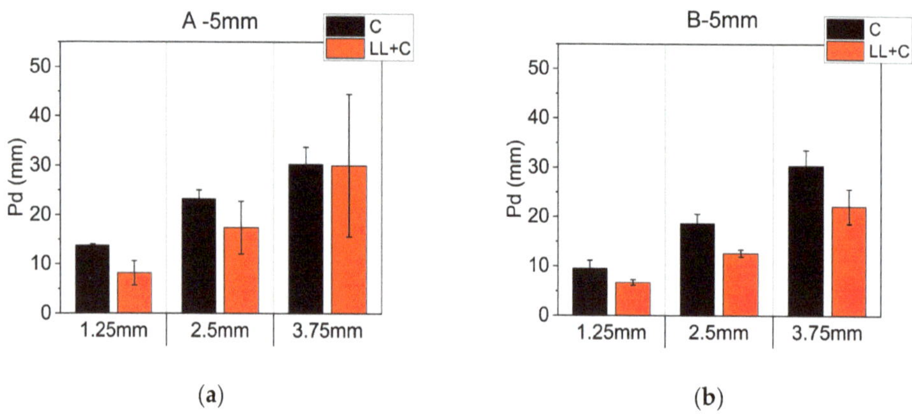

Figure 5. Average penetration depths (P_d) at specific positions within the scratch tracks for (**a**) A (WC-10%$_{wt}$Co) and (**b**) B (WC-15%$_{wt}$Co), processed under different conditions.

The sliding contact resistance of the laser-coated cemented carbide grades was assessed by determining, through SEM inspection, the critical loads for the emergence of specific damage events. In this regard, three standardized critical loads are defined in a scratch test, namely, L_{c1}, L_{c2}, and L_{c3}, as follows [37]:

- First critical load, L_{c1}: linked to the initiation of forward chevron cracks within the scratch track;
- Second critical load, L_{c2}: associated with the first failure event involving local or gross interfacial spallation; and
- Third critical load, L_{c3}: defined by the first point where the substrate is visible along the center of the scratch track in a crescent that goes completely through the track, with gross interfacial spallation.

The positions of the critical loads on the scratch tracks (5 mm) were inspected by SEM and are accordingly marked in Figures 6 and 7. Their values are also summarized in Table 5. Several observations may be highlighted. First, none of the conditions studied exhibited any forward chevron cracks within the scratch track; thus, the L_{c1} could not be determined in any case. Second, local and gross interfacial spallation, both at the border and within a scratch track, were specific failure events identified for the non-laser-processed surface (C) for both grades, as well as for the laser-ablated one (LL + C) involving grade B. In these cases, the critical loads were higher for the reference A grade condition, as compared to that for the B grade condition. However, the laser-ablated condition (LL + C) for the latter exhibited higher values of L_{c2} and L_{c3} than the just coated one. Finally, the laser ablated condition (LL + C) involving grade A showed the best sliding contact response, as no damage or failure event was discerned for this condition within the scratch track. In general, these experimental findings allow us to state that laser ablation led to reinforced sliding contact resistance, this being more pronounced for the grade with lower binder content.

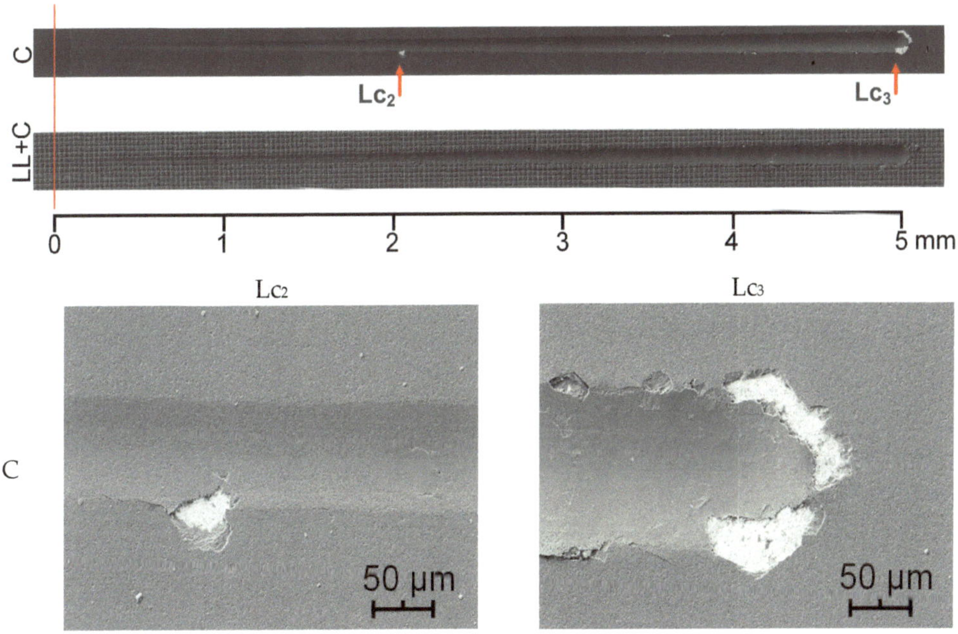

Figure 6. SEM images of the 5 mm scratch tracks for grade A (WC-10%$_{wt}$Co) were processed with two different surface modification conditions, indicating positions associated with the emergence of specific failure events, and thus critical loads. Details of the discerned damage features are shown in high-magnification images. Scratch direction from left to right.

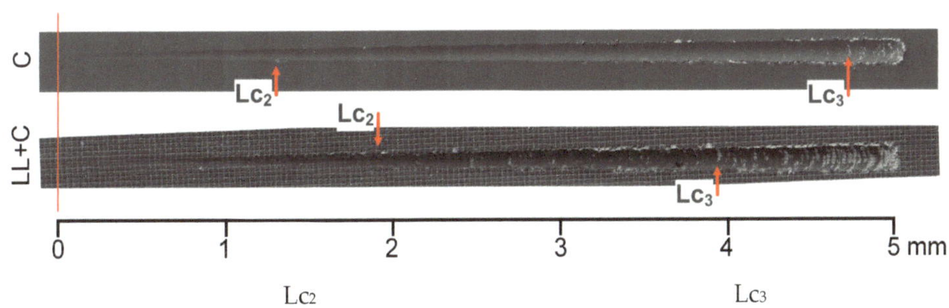

Figure 7. *Cont.*

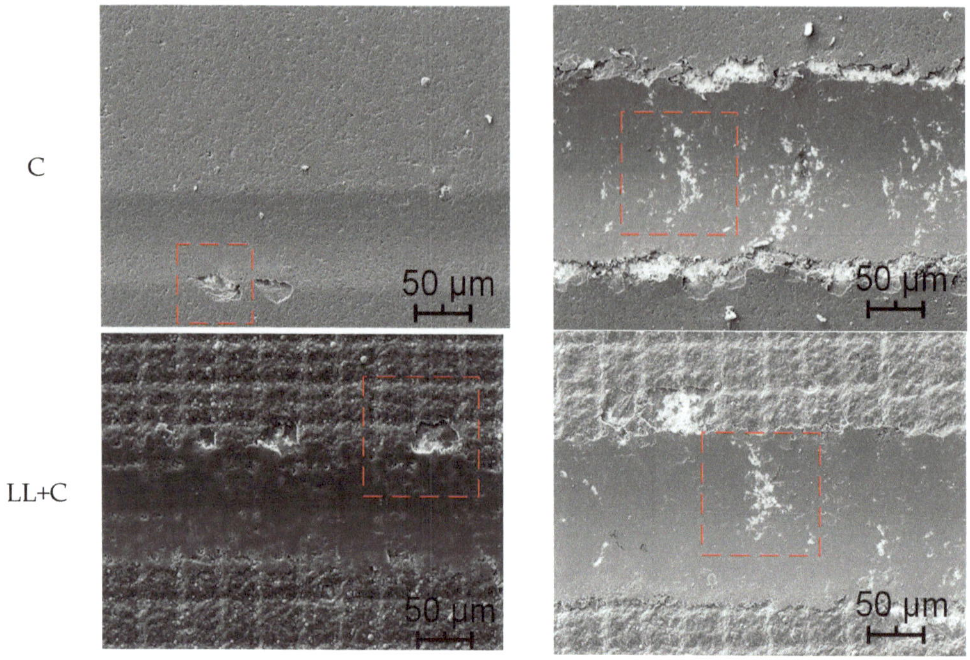

Figure 7. SEM images of scratch tracks for grade B (WC-15%$_{wt}$Co) were processed with two different surface modification conditions, indicating positions associated with the emergence of specific failure events, and thus critical loads. Details of the discerned damage features are shown in high-magnification images. Scratch direction from left to right.

Table 5. Values of critical loads determined on the selected scratch tracks (5 mm).

Sample		Lc_1 (mm)	F (N) at Lc_1	Lc_2 (mm)	F (N) at Lc_2	Lc_3 (mm)	F (N) at Lc_3
A	C	-	-	2.0	40.5	4.9	99.1
	LL + C	-	-	-	-	-	-
B	C	-	-	1.3	25.8	4.7	94.5
	LL + C	-	-	1.9	38.5	3.9	78.6

Reference Case Study (Large Increase in Load per Unit Distance)

A reference case study was conducted on the lasered and coated grade A (WC-10%$_{wt}$Co) to document and analyze its indentation and sliding performances with a large increase in load per unit distance, i.e., over a sliding distance of 2 mm. Similar to the analysis done for the scratch test run over a length of 5 mm, penetration depths were measured, and critical loads on the selected scratch tracks (2 mm for this case study) were determined using SEM.

Average penetration depths were obtained at three specific sliding distances on the scratch tracks, i.e., 0.5 mm, 1.0 mm, and 1.5 mm, corresponding to normal loads of 25 N, 50 N, and 75 N, respectively (Figure 8). As for the 5 mm tests, it was found that penetrations were deeper in the scratch tracks for the reference conditions than for the laser-ablated ones. The positions of critical loads on the selected scratch tracks (2 mm) are marked in Figure 9. The positions and values of the critical loads are also summarized in Table 6. The scenarios observed were very similar to those found in previous scratch tests. The laser-ablated grade A did not show any obvious damage on the surface, and thus, no critical loads could be determined. Both scratch tests (5 mm and 2 mm) attest that the indentation and sliding performances of the coatings on grade A were reinforced by laser ablation.

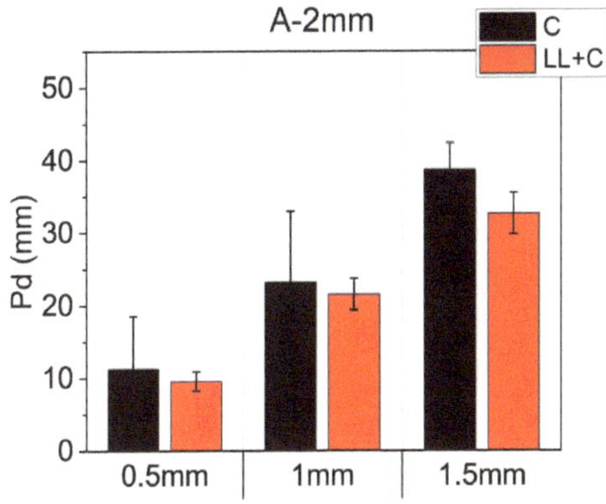

Figure 8. Average penetration depths (P_d) at specific positions within the scratch tracks for A (WC-10%$_{wt}$Co), processed under different conditions.

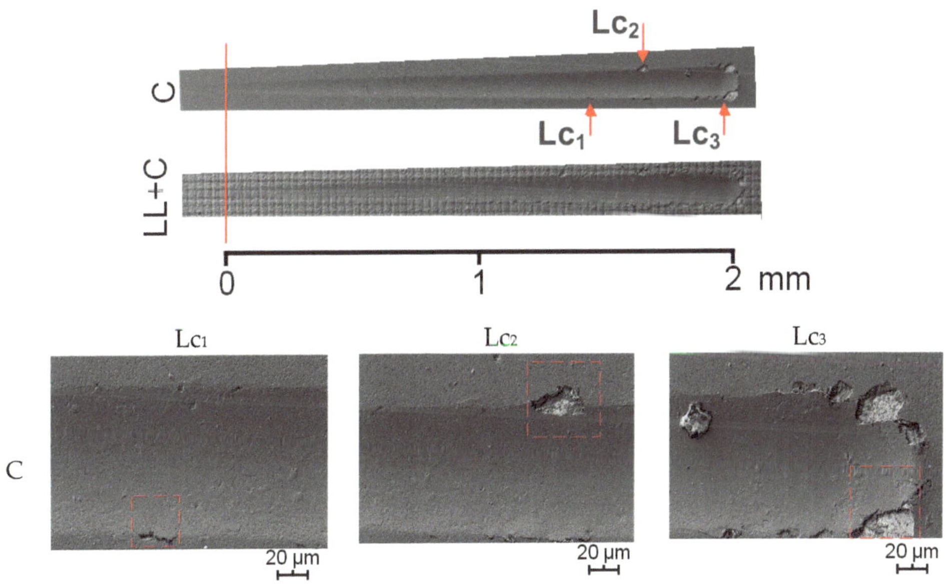

Figure 9. SEM images of the 2mm scratch tracks for grade A (WC-10%$_{wt}$Co) were processed with two different surface modification conditions, indicating positions associated with the emergence of specific failure events, and thus critical loads. Details of the discerned damage features are shown in high-magnification images. Scratch direction from left to right.

Table 6. Values of critical loads determined on the selected scratch tracks (2 mm).

Sample		Lc_1 (mm)	F (N) at Lc_1	Lc_2 (mm)	F (N) at Lc_2	Lc_3 (mm)	F (N) at Lc_3
A	C	1.4	71.8	1.6	81.1	2.0	98.2
	LL + C	-	-	-	-	-	-

Characteristic Wear Scenarios

Based on the inspection of the scratch tracks, characteristic wear scenarios were studied in the following cases, namely:

Case 0: Wear on the coated grades A (WC-10%$_{wt}$Co) and B (WC-15%$_{wt}$Co), without laser processing: A + C and B + C;
Case 1: Wear on the coated grades A (WC-10%$_{wt}$Co), exclusively processed with low laser energy: A + LL + C;
Case 2: Wear on the coated grades B (WC-15%$_{wt}$Co), exclusively processed with low laser energy: B + LL + C;

Case 0: A + C and B + C.

For both coated grades A and B, only minor damage, such as spallation along the track edge and subsequent substrate exposure, was found. The first damage appeared at the positions of 2.0 mm and 1.3 mm on grades A and B, respectively. Meanwhile, the first substrate exposure was discerned at a very late stage, close to the end of the track, i.e., at the positions of 4.9 mm and 4.7 μm in grades A and B, respectively (Figures 6 and 7). The locations where the first substrate exposure was identified, in grades A and B, were examined in detail (Figure 10):

(1) At the center of the scratch track of the coated grade A, cracking of the coating layer was observed (Figure 10a).
(2) At a similar position on the coated grade B, exposed WC grains and adhered part of the broken coating layer were evidenced (Figure 10b).

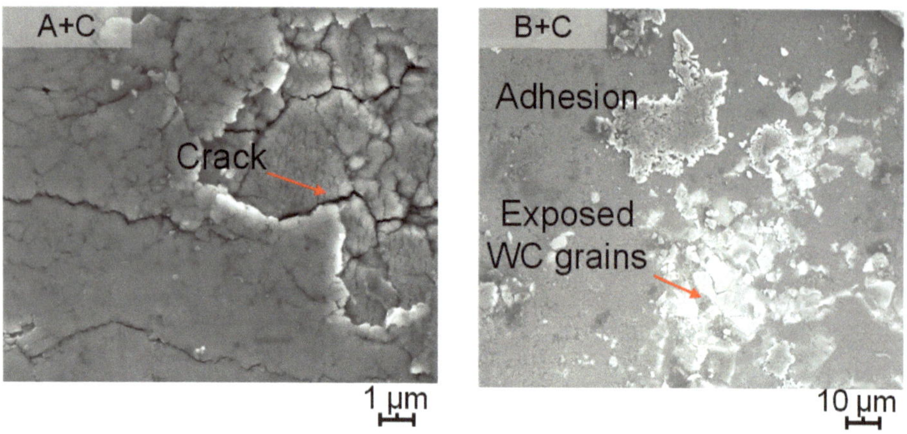

(**a**) (**b**)

Figure 10. Damage and failure events were identified: (**a**) cracking due to the crushing on the coated grade A, and (**b**) substrate exposure and adhesion on the coated grade B.

Case 1: A + LL + C

Laser ablation not only delayed damage emergence, but also reduced its frequency on the coated grade A during the scratch tests. The coatings were, in general, able to resist the sliding movement without having any noticeable damage, like spallation or substrate exposure. Tiny changes were found in the scratch tracks due to pressing and sliding of the indenter. For instance, at the track end for the (A + LL + C) sample, the coating material was stacked in front of the indenter (Figure 11a).

Case 2: B + LL + C

Regarding grade B, some mild damage, such as stacking and cleavage of the layers either on the border or in the center of scratch tracks, together with more severe substrate damage, was found. Near the location where the first substrate exposure was discerned in the track (B + LL + C), the coating was completely stripped. Furthermore, the substrate was not only exposed but also damaged. As a result, WC grains were loosened and cracks nucleated (Figure 11b).

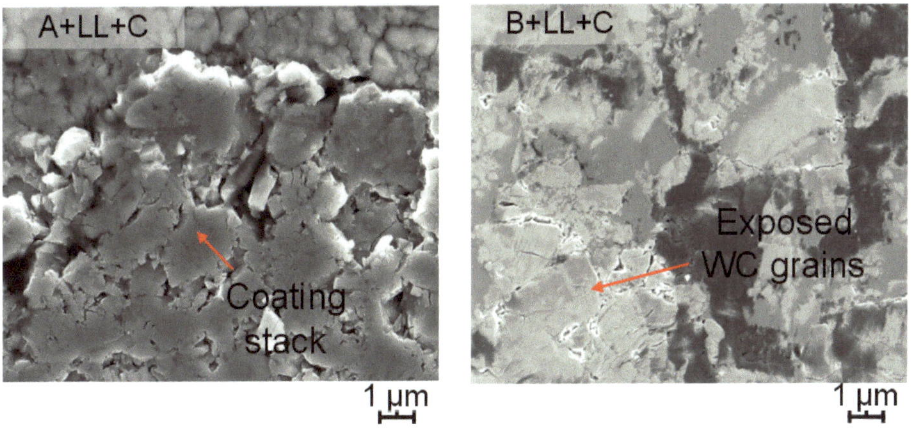

(**a**) (**b**)

Figure 11. Damage and failure events identified: (**a**) stacking of the coating on grade A, and (**b**) complete removal of coating and substrate exposure on the laser-ablated and coated grade B.

3.2.3. Comprehensive Evaluation of the Mechanical Integrity

Based on the results of the aforementioned measurements and characterization, a comprehensive evaluation of the mechanical integrity of the laser-ablated and coated cemented carbides was attempted. It was performed by comparing several key technical indicators, i.e., Vickers hardness (HV0.3, HV3), coefficient of friction (COF), penetration depth (P_d), and critical load (L_{c2} and L_{c3}). Three scales, namely excellent (E), good (G), and moderate (M), were defined, and the results are given in Table 7. Based on the assessment conducted, it is found that:

(1) Cemented carbide with low binder content, i.e., A (WC-10%$_{wt}$Co) grade, generally performed better than the one with a high amount of metallic phase, i.e., B (WC-15%$_{wt}$Co) grade.
(2) Laser ablation in most cases improved the performance of the coated cemented carbides in terms of their indentation and sliding contact responses.
(3) Laser ablation is commonly more beneficial to reinforce the mechanical properties of the low binder content grade, i.e., A (WC-10%$_{wt}$Co), as compared to the other one studied, i.e., B (WC-15%$_{wt}$Co).

Table 7. Comprehensive evaluation within each coated cemented carbide grade *.

Sample		Indenting Response		Sliding Response (5 mm)				Results
		HV0.3	HV3	COF	P_d at 50N (mm)	L_{c2} (mm)	L_{c3} (mm)	
A	C	M (1532.0)	M (1463.0)	E (0.057)	M (23.25)	G (2.03)	G (4.95)	1E2G3M
	LL + C	E (3249.4)	E (1702.2)	E (0.056)	G (17.4)	E (-)	E (-)	5E1G0M
B	C	M (1533.6)	M (1308.0)	G (0.059)	G (18.65)	M (1.29)	G (4.73)	0E3G3M
	LL + C	G (2073.6)	G (1533.8)	M (0.063)	E (12.60)	G (1.93)	M (3.93)	1E3G2M

* Explanation of the evaluating indicators: E: excellent, G: good, M: moderate. Vickers hardness (HV0.3, HV3): the higher the value is, the better the performance is. Coefficient of friction (COF): the smaller the value is, the better the performance is. Penetration depth (P_d): evaluation was made according to the measured values at the normal load of 50 N, corresponding to the middle point of the scratch track, 2.5 mm, respectively. The smaller the value is, the better the performance is. Critical load (L_{c2} and L_{c3}): the higher the value is, the better the performance is.

4. Conclusions

In this work, the surface and mechanical integrity of two laser-processed and subsequently coated cemented carbides were evaluated using different characterization methods and approaches. Surface processing conditions varied with two different laser energy levels. From the results attained, the following conclusions may be drawn:

- Laser ablation affects the surface integrity of cemented carbides, as it induces changes in the roughness, morphology, and microstructure of the substrate. However, these changes were rather minor, although they became slightly more important with increased laser energy.
- Laser ablation promoted surface hardening on both coated grades, and the relative increase tended to be proportional to the applied laser energy. Although the micro- and macrohardness data followed similar trends, relative differences between their values were discerned, these being more pronounced on the surfaces machined by low laser energy.
- The investigation just conducted on the grades processed with low laser energy allowed us to conclude laser ablation enhances both indentation and sliding contact responses. It was discerned from the findings of reduced penetration depth and higher critical load values for the emergence of specific damage/failure events, as compared to the behavior determined for the reference surface conditions. Within this context, the improvement in sliding resistance was more pronounced on the grade with lower binder content studied, i.e., A (WC-10$\%_{wt}$Co). Meanwhile, laser ablation had no significant influence on the frictional performance, as all the coated grades exhibited very similar COF values.

Author Contributions: Conceptualization, S.F.; methodology, S.F. and L.L.; investigation, S.F.; resources, L.L. and D.B.; writing—original draft, S.F.; writing—review and editing, L.L., Y.B.G. and D.B. All authors have read and agreed to the published version of the manuscript.

Funding: This research was funded by the Feodor Lynen Research Fellowship of the Alexander von Humboldt Foundation, by the Spanish Ministerio de Ciencia, Innovación y Universidades MICINN—FEDER (Spain) grant number PID2022-137274NB-C32, and by the German Research Foundation grant number INST 256/510-1 FUGG.

Data Availability Statement: The original contributions presented in the study are included in the article, further inquiries can be directed to the corresponding author.

Acknowledgments: The work leading to this publication was supported by the Feodor Lynen Research Fellowship of the Alexander von Humboldt Foundation. It was also partly funded by the Spanish Ministerio de Ciencia, Innovación y Universidades MICINN—FEDER (Spain) through grant number PID2022-137274NB-C32. The authors acknowledge funding for the FIB/SEM instrument by the German Research Foundation (INST 256/510-1 FUGG). The authors thank Hyperion Materials & Technologies for providing the samples and Flubetech S.L. for providing the PVD coating facilities.

Conflicts of Interest: The authors declare no conflict of interest.

References

1. García, J.; Collado Ciprés, V.; Blomqvist, A.; Kaplan, B. Cemented carbide microstructures: A review. *Int. J. Refract. Met. Hard Mater.* **2019**, *80*, 40–68. [CrossRef]
2. Wolfe, T.A.; Jewett, T.J.; Gaur, R.P.S. Powder synthesis. In *Comprehensive Hard Materials*; Elsevier: Amsterdam, The Netherlands, 2014; pp. 185–212. [CrossRef]
3. Exner, H.E. Physical and chemical nature of cemented carbides. *Int. Met. Rev.* **1979**, *24*, 149–170. [CrossRef]
4. Prakash, L. Fundamentals and General Applications of Hardmetals. In *Comprehensive Hard Materials*; Elsevier: Amsterdam, The Netherlands, 2014; pp. 29–90. [CrossRef]
5. Yang, Q.; Zhang, P.; Lu, Q.; Yan, H.; Shi, H.; Yu, Z.; Sun, T.; Li, R.; Wang, Q.; Wu, Y.; et al. Application and development of blue and green laser in industrial manufacturing: A review. *Opt. Laser Technol.* **2023**, *170*, 110202. [CrossRef]
6. Orazi, L.; Romoli, L.; Schmidt, M.; Li, L. Ultrafast laser manufacturing: From physics to industrial applications. *CIRP Ann.* **2021**, *70*, 543–566. [CrossRef]

7. Shukla, P.; Waugh, D.G.; Lawrence, J.; Vilar, R. Laser surface structuring of ceramics, metals and polymers for biomedical applications. In *Laser Surface Modification of Biomaterials*; Elsevier: Amsterdam, The Netherlands, 2016; pp. 281–299.
8. Bäuerle, D. (Ed.) *Laser Processing and Chemistry*; Springer: Berlin/Heidelberg, Germany, 2011; pp. 279–313.
9. Etsion, I. State of the Art in Laser Surface Texturing. *J. Tribol.* **2005**, *127*, 248–253. [CrossRef]
10. Hazzan, K.E.; Pacella, M.; See, T.L. Laser Processing of Hard and Ultra-Hard Materials for Micro-Machining and Surface Engineering Applications. *Micromachines* **2021**, *12*, 895. [CrossRef]
11. Breidenstein, B.; Denkena, B.; Bergmann, B.; Krödel, A. Laser material removal on cutting tools from different materials and its effect on wear behavior. *Met. Powder Rep.* **2018**, *73*, 26–31. [CrossRef]
12. Fang, S.; Llanes, L.; Bähre, D. Laser surface texturing of a WC-CoNi cemented carbide grade: Surface topography design for honing application. *Tribol. Int.* **2018**, *122*, 236–245. [CrossRef]
13. Dumitru, G.; Lüscher, B.; Krack, M.; Bruneau, S.; Hermann, J.; Gerbig, Y. Laser processing of hardmetals: Physical basics and applications. *Int. J. Refract. Met. Hard Mater.* **2005**, *23*, 278–286. [CrossRef]
14. Geiger, M.; Roth, S.; Becker, W. Influence of laser-produced microstructures on the tribological behaviour of ceramics. *Surf. Coat. Technol.* **1998**, *100–101*, 17–22. [CrossRef]
15. Momma, C.; Chichkov, B.N.; Nolte, S.; von Alvensleben, F.; Tünnermann, A.; Welling, H.; Wellegehausen, B. Short-pulse laser ablation of solid targets. *Opt. Commun.* **1996**, *129*, 134–142. [CrossRef]
16. Jahan, M.P.; Rahman, M.; Wong, Y.S. A review on the conventional and micro-electrodischarge machining of tungsten carbide. *Int. J. Mach. Tools Manuf.* **2011**, *51*, 837–858. [CrossRef]
17. Bergs, T.; Chrubasik, L.; Petersen, T.; Klink, A.; Klocke, F. Experimental analysis of surface integrity of cemented carbides resulting from contemporary sinking EDM technology. *Prod. Eng.* **2019**, *13*, 511–517. [CrossRef]
18. Schalk, N.; Tkadletz, M.; Mitterer, C. Hard coatings for cutting applications: Physical vs. chemical vapor deposition and future challenges for the coatings community. *Surf. Coat. Technol.* **2022**, *429*, 127949. [CrossRef]
19. Bobzin, K. High-performance coatings for cutting tools. *CIRP J. Manuf. Sci. Technol.* **2017**, *18*, 1–9. [CrossRef]
20. Gao, P.; Guo, Q.; Xing, Y.; Guo, Y. Structural, Mechanical, and Tribological Properties of Hard Coatings. *Coatings* **2023**, *13*, 325. [CrossRef]
21. Bouzakis, K.D.; Michailidis, N.; Skordaris, G.; Bouzakis, E.; Biermann, D.; M'Saoubi, R. Cutting with coated tools: Coating technologies, characterization methods and performance optimization. *CIRP Ann.—Manuf. Technol.* **2012**, *61*, 703–723. [CrossRef]
22. Byrne, G.; Dornfeld, D.; Denkena, B. Advancing cutting technology. *CIRP Ann.—Manuf. Technol.* **2003**, *52*, 483–507. [CrossRef]
23. Fang, S.; Salán, N.; Pauly, C.; Llanes, L.; Mücklich, F. Critical Assessment of Two-Dimensional Methods for the Microstructural Characterization of Cemented Carbides. *Metals* **2022**, *12*, 1882. [CrossRef]
24. Tarragó, J.M.; Roa, J.J.; Valle, V.; Marshall, J.M.; Llanes, L. Fracture and fatigue behavior of WC-Co and WC-CoNi cemented carbides. *Int. J. Refract. Met. Hard Mater.* **2015**, *49*, 184–191. [CrossRef]
25. Fang, S.; Lima, R.; Sandoval, D.; Bähre, D.; Llanes, L. Ablation Investigation of Cemented Carbides Using Short-Pulse Laser Beams. *Procedia CIRP* **2018**, *68*, 172–177. [CrossRef]
26. Li, T.; Lou, Q.; Dong, J.; Wei, Y.; Liu, J. Selective removal of cobalt binder in surface ablation of tungsten carbide hardmetal with pulsed UV laser. *Surf. Coat. Technol.* **2001**, *145*, 16–23. [CrossRef]
27. Fang, S.; Hsu, C.J.; Klein, S.; Llanes, L.; Bähre, D.; Mücklich, F. Influence of laser pulse number on the ablation of cemented Tungsten Carbides (WC-CoNi) with different grain size. *Lubricants* **2018**, *6*, 11. [CrossRef]
28. Claver, A.; Randulfe, J.J.; Palacio, J.F.; Fernández de Ara, J.; Almandoz, E.; Montalá, F.; Colominas, C.; Cot, V.; García, J.A. Improved adhesion and tribological properties of altin-tisin coatings deposited by dcms and hipims on nitrided tool steels. *Coatings* **2021**, *11*, 1175. [CrossRef]
29. Dalibon, E.L.; Prieto, G.; Tuckart, W.R.; Brühl, S.P. Tribological behaviour of a hyperlox coating deposited over nitrided martensitic stainless steel. *Surf. Topogr. Metrol. Prop.* **2022**, *10*, 034003. [CrossRef]
30. Yang, J.; Roa, J.J.; Odén, M.; Johansson-Jõesaar, M.P.; Esteve, J.; Llanes, L. Substrate surface finish effects on scratch resistance and failure mechanisms of TiN-coated hardmetals. *Surf. Coat. Technol.* **2015**, *265*, 174–184. [CrossRef]
31. Fang, S.; Llanes, L.; Bähre, D. Wear characterization of cemented carbides (WC-CoNi) processed by laser surface texturing under abrasive machining conditions. *Lubricants* **2017**, *5*, 20. [CrossRef]
32. Uhlmann, E.; Bergmann, A.; Gridin, W. Investigation on Additive Manufacturing of Tungsten Carbide-cobalt by Selective Laser Melting. *Procedia CIRP* **2015**, *35*, 8–15. [CrossRef]
33. Denkena, B.; Breidenstein, B.; Wagner, L.; Wollmann, M.; Mhaede, M. Influence of shot peening and laser ablation on residual stress state and phase composition of cemented carbide cutting inserts. *Int. J. Refract. Met. Hard Mater.* **2013**, *36*, 85–89. [CrossRef]
34. Trzepiecinski, T. A study of the coefficient of friction in steel sheets forming. *Metals* **2019**, *9*, 988. [CrossRef]
35. Popov, V.L. *Contact Mechanics and Friction: Physical Principles and Applications*; Springer: Berlin/Heidelberg, Germany, 2010. [CrossRef]
36. Blau, P.J. The significance and use of the friction coefficient. *Tribol. Int.* **2001**, *34*, 585–591. [CrossRef]
37. *ASTM C 1624-05*; Standard Test Method for Adhesion Strength and Mechanical Failure Modes of Ceramic Coatings by Quantitative Single Point Scratch Testing. ASTM International Standards: West Conshohocken, PA, USA, 2005.

MDPI AG
Grosspeteranlage 5
4052 Basel
Switzerland
Tel.: +41 61 683 77 34

Metals Editorial Office
E-mail: metals@mdpi.com
www.mdpi.com/journal/metals

www.ingramcontent.com/pod-product-compliance
Lightning Source LLC
Chambersburg PA
CBHW041407210926
51850CB00013B/3199
9783725829064